Environmental Science and Engineering
Subseries: Environmental Science

Series Editors: R. Allan • U. Förstner • W. Salomons

Christoph S. Garbe, Robert A. Handler,
Bernd Jähne
(Eds.)

Transport at the Air-Sea Interface

Measurements, Models and Parametrizations

With 127 Figures and a CD-ROM

Springer

EDITORS:

PRIV.-DOZ. DR. C.S. GARBE
PROF. DR. B. JÄHNE
INTERDISCIPLINARY CENTER FOR
SCIENTIFIC COMPUTING
UNIVERSITY OF HEIDELBERG
IM NEUENHEIMER FELD 368
69120 HEIDELBERG
GERMANY

AND

INSTITUTE OF ENVIRONMENTAL PHYSICS
UNIVERSITY OF HEIDELBERG
IM NEUENHEIMER FELD 229
69120 HEIDELBERG
GERMANY

E-mail:
Christoph.Garbe@
iwr.uni-heidelberg.de
Bernd.Jaehne@iwr.uni-heidelberg.de

DR. R.A. HANDLER
Naval Research Laboratory
4555 Overlook Avenue SW
Washington DC 20375
USA

E-mail: robert.handler@nrl.navy.mil

Additional material to this book can be downloaded from http://extras.springer.com

ISSN 1863-5520

ISBN 13 978-3-642-43928-5 **Springer Berlin Heidelberg New York**

Springer is a part of Springer Science+Business Media
springeronline.com

Softcover re-print of the Hardcover 1st edition 2011

Cover design: deblik, Berlin
Production: A. Oelschläger
Typesetting: Camera-ready by the Editors
Printed on acid-free paper 30/2132/AO 543210

Preface

The sea surface represents the interface between the ocean and the atmosphere. As the interface is approached from either the atmospheric or ocean side, the transport mechanisms are shifted from turbulent to diffusive diffusion. Hence the viscous boundary layers at both sides of the water surface represents the major resistance to the transport of energy, mass and impulse between atmosphere and ocean. This has implications for the composition of the atmosphere and has gained importance especially for radiatively (climate)-active gases such as CO_2, CH_4, N_2O and DMS. But also, the transport of gases across the interface is significant for the fate and the atmospheric/marine budgets of many man-made pollutants, in particular of volatile organic compounds and mercury.

The two key variables which are required for the determination of the gas exchange fluxes are the partial pressure difference of the considered gas at the sea surface, and the gas exchange transfer velocity k. Partial pressure differences are either obtained from measurements in the surface water and in the atmosphere or from biogeochemical models. Whereas our knowledge about the partial pressure difference distributions, in particular for CO_2, has increased considerably during the past years, the choice of an appropriate transfer velocity is still a matter of controversy. This is because the transport mechanisms across the free, wind-driven water surface are still only known superficially. This is not surprising because both the experimental investigation as well as the modeling is very challenging.

Experimental techniques and modeling efforts have evolved separately with little quantitative comparisons. Recently, refined measurement techniques have advanced which allows researchers to gain novel insights into the boundary layer processes. Likewise, computer simulations have improved significantly both in terms of resolution and model complexity. This has made it feasible to compare model output of realistic boundary conditions to actual measurements. Through these comparisons, models can be verified, leading to a deepening of our knowledge of the transport of energy and mass between ocean and atmosphere. Only by linking experimental measurements with computer models, can our understanding of air-sea interactions be enhanced. In turn, through insights into the underlying transport processes, physically sound parameterizations can be found. Better parameterizations are needed in order to improve global models of our climate and predict climatic change.

In order to bridge the gap between current models and measurements, as well as spark new ideas for novel simulation and experimental efforts, an “International Workshop on Transport at the Air Sea Interface” was organized by the editors of this volume. The focus of the workshop was on small scale processes directly at the interface. The workshop took place at the University of Heidelberg from September 6–8, 2006. Leading scientist from around the world came together and focused on different aspects of the transport across the air-water interface. The current state of the art of research was presented and current and future research interests and problems were discussed.

The program of the workshop is listed after this preface. This volume contains peer reviewed, extended and updated versions of selected talks that also reflect the discussions during the workshop. The editors cordially thank all reviewers for their detailed responses and their efforts to improve the quality of the papers.

Christoph Garbe

Heidelberg, *Robert Handler*

April 2007 *Bernd Jähne*

Contributions to "International Workshop on Transport at the Air-Sea Interface"

Session 1.: Shear Free Surface

Herlina
Turbulent gas flux measurements near the air-water interface in a grid-stirred tank

Evan Variano, Edwin A. Cowen
Quantitative imaging of CO2 transfer at an unsheared free surface

Aldo Tamburrino, Claudius Aravena, John S. Gulliver
Visualization of 2-D divergence on the free surface and its relation to gas transfer

Session 2.: Small Scale Processes

Patrick Rosendahl
Modelling the influence of small-scale processes in the upper water layer on air-sea CO2 Exchange

Martin Gade
Microwave remote sensing of small-scale features at the water surface that influence air-sea CO2 exchange

Alastair D. Jenkins
The interaction of ocean surface processes, waves, and turbulence in the adjacent boundary layers

Session 3.: Gas Exchange

Gerhard Peters
Estimation of "small" surface fluxes by eddy covariance

Achim Falkenroth, Alexandra Herzog, Bernd Jähne
Visualization of concentration fields by oxygen quenching and pH indicators

Kai Degreif, Bernd Jähne
The Schmidt number dependency of air sea gas transfer: new results and models

Session 4.: Heat Transfer

Chris Fairall
Measurement and Parameterization of Latent heat transfer over the Open Ocean

Hannah Linag, Kapil Phadnis, Mohamed Atmane, Christopher Zappa, Mark Loewen, William Asher, Andrew Jessup
A laboratory study of passive and active IR techniques to measure heat flux

Bernd Jähne, Christopher Popp, Uwe Schimpf, Christoph Garbe
Analysis of the heat transfer process across the aqueous heat boundary layer by active thermography: mean transfer velocities and intermittence

Christoph S. Garbe, Bernd Jähne
Measuring and modeling parameters of heat transfer from surface flow fields by IR image sequence analysis

Session 5.: Temperature Structure of the Interface

Nicholas Scott, Geoffrey Smith, Robert Handler
The structure of the surface temperature field at an air-water interface at low to moderate wind speeds

Brian Ward
Thermometric measurements of the molecular sublayer at the air-water interface

Session 6.: Wave Breaking

Johannes Gemmrich
Momentum flux and energy dissipation associated with breaking waves.

W.L. Peirson, C. Welch, J.W. Walker, M.L. Banner
Understanding the enhancement of air-water interfacial oxygen exchange rate by microscale breaking waves

Christopher J. Zappa, Felix A. Tubiana, Wade R. McGillis, J. Bent, Gerrit de Leeuw, Marcel M. Moerman
Investigating wave processes important to air-sea fluxes using infrared techniques

Gerrit de Leeuw et al.
Eddy correlation measurements of sea spray aerosol fluxes

Session 7.: Turbulence

Guillemette Caulliez, Richard Dupont, Victor I. Shrira
Turbulence generation processes in the wind-driven subsurface water flow

Haitao Xu, Nicholas T. Ouellette, Mickaël Bourgoin, Ewe-Wei Saw, Evan Variano, Raymond Shaw and Eberhard Bodenschatz
Experimental investigations of turbulent relative dispersion and the spray characteristics of a waterfall

Tetsu Hara, John Wendelbo, E. Vaninwegen, Christoph Garbe, Uwe Schimpf, Nelson Frew
Estimation of air-sea gas and heat fluxes from infrared imagery based on near surface turbulence models

Wu-ting Tsai
Numerical simulation of turbulent boundary layer beneath a wavy water surface

Session 8.: Rain

David Ho
Rain and air-water gas exchange

S. Komori, N. Takagaki, R. Saiki, N. Suzuki
The effects of raindrops on interfacial turbulence and air-water gas transfer

Anne-Kristin Anweiler
Lab experiments on the influence of rain on air-sea CO2 exchange

Session 9.: Parameterizations

Sanjoy Banerjee
The air water interface: turbulence and scalar exchange

Wade McGillis
Using meteorological techniques to parameterize processes controlling air-water gas fluxes

Rik Wanninkhof
The impact of different gas exchange parameterizations on global air-sea CO2 fluxes

Contents

List of Contributors

Claudio Aravena
Public Work Ministry
Santiago, Chile
claudio.aravena.r@
moptt.gov.CL

William E. Asher
University of Washington
Applied Physics Lab
61 Route 9W
Seattle, WA 98105-6698
USA
asher@apl.washington.edu

Sanjoy Banerjee
Department of Chemical Engineering
Department of Mechanical Engineering
Bren School of Environmental
Science and Management
University of California,
Santa Barbara, CA 93106-5080 USA
banerjee@
engineering.ucsb.edu

Michael L. Banner
The University of New South Wales
School of Mathematics
Sydney, NSW 2052
Australia
m.banner@unsw.edu.au

Jonathan T. Bent
Lamont-Doherty Earth Observatory
University of Columbia
Palisades, NY 10964
USA
bent@ldeo.colmunbia.edu

Guillemette Caulliez
Institut de Recherche sur les
Phénomènes Hors Equilibre
163, av. de Luminy - case 903
13288 Marseille Cedex 9
France
caulliez@irphe.univ-mrs.fr

Edwin A. Cowen
Department of Civil
and Environmental Engineering
Hollister Hall
Cornell University
Ithaca, NY 14853
USA
eac20@cornell.edu

John W.H. Dacey
Department of Biology
Woods Hole Oceanographic
Institution
Woods Hole, Massachusetts 02543
USA
jdacey@whoi.edu

Gerrit de Leeuw
Finnish Meteorological Institute
Research and Development
Climate and Global Change Unit
Erik Palmenin Aukio 1
P.O. Box 503
FI-00101 Helsinki
Finland
Gerrit.Leeuw@fmi.fi
and
Dept. of Physical Sciences
University of Helsinki
P.O. Box 64
FI- 00014, Finland
gerrit.deleeuw@helsinki.fi

Kai Degreif
Institute for Environmental Physics
University of Heidelberg
Im Neuenheimer Feld 229
D-69120 Heidelberg
Germany
Kai.Degreif@
iwr.uni-heidelberg.de

Richard Dupont
Institut de Recherche sur les
Phénomènes Hors Equilibre
163, av. de Luminy - case 903
13288 Marseille Cedex 9
France

James B. Edson
University of Connecticut
Avery Point
Department of Marine Sciences
1080 Shennecossett Road
Groton, CT 06340
USA
james.edson@uconn.edu

Achim Falkenroth,
Institute for Environmental Physics
University of Heidelberg
Im Neuenheimer Feld 229
D-69120 Heidelberg
Germany
Achim.Falkenroth@
iwr.uni-heidelberg.de

Nelson M. Frew
Department of Marine Chemistry
and Geochemistry
Woods Hole Oceanographic
Institution
Woods Hole, MA 02543
USA
nfrew@whoi.edu

Christoph S. Garbe
Interdisciplinary Center for
Scientific Computing (IWR)
University of Heidelberg
Heidelberg
Germany
Christoph.Garbe@
iwr.uni-heidelberg.de

Johannes Gemmrich
University of Victoria
Department of Physics and
Astronomy
P.O.Box 3055
Victoria, Bristish Columbia, V8W
3P6
Canada
gemmrich@uvic.ca

John S. Gulliver
Department of Civil Engineering
University of Minnesota
5000 Pillsbury Drive S. E.
Minneapolis, MN 55455-0116
USA
gulli003@umn.edu

Robert A. Handler
Naval Reserach Laboratory
4555 Overlook Avenue SW
Washington, DC 20375
USA
robert.handler@nrl.navy.mil

Tetsu Hara
University of Rhode Island
Graduate School of Oceanography
Narragansett, RI 02882-1197
USA
thara@uri.edu

Herlina
Institut für Hydromechanik
Universität Karlsruhe
Kaiserstr. 12
76128 Karlsruhe
Germany
herlina@ifh.uni-karlsruhe.de

David T. Ho
Lamont-Doherty Earth Observatory
of Columbia University
61 Route 9W - PO Box 1000
Palisades, NY 10964-8000
USA
david@ldeo.columbia.edu

Li-ping Hung
Department of Civil Engineering
National Chiao Tung University
Hsinchu, 300, Taiwan

Bernd Jähne
Interdisciplinary Center for
Scientific Computing (IWR)
University of Heidelberg
Heidelberg
Germany
Bernd.Jaehne@
iwr.uni-heidelberg.de

Alastair D. Jenkins
Bjerknes Center for Climate
Reserach
Geophysical Institute
Allégaten 70
5007 Bergen
Norway
alastair.jenkins@
bjerknes.uib.no

Gerhard H. Jirka
Institute for Hydromechanics
University of Karlsruhe
Kaiserstr.12
D-76128 Karlsruhe
Germany
jirka@uka.de

Satoru Komori
Department of Mechanical
Engineering and Science
and
Advanced Institute of
Fluid Science and Engineering
Kyoto University, Kyoto 606-8501
Japan
komori@mech.kyoto-u.ac.jp

Wade R. McGillis
Lamont-Doherty Earth Observatory
University of Columbia
Palisades, NY 10964
USA
wrm2102@columbia.edu

Marcel M. Moerman
TNO Physics and Electronics
Laboratory
P.O. Box 96864
2509 JG, The Hague
The Netherlands
moerman@fel.tno.nl

Sarah Norris
School of Earth and Environment
Environment Building
The University of Leeds
Leeds. LS2 9JT
United Kingdom
s.norris@see.leeds.ac.uk

William L. Peirson
Water Research Laboratory
School of Civil and
Environmental Engineering

The University of New South Wales
King St. Manly Vale, NSW 2093
Australia
w.peirson@unsw.edu.au

Gerhard Peters
Universität Hamburg
Meteorologisches Institut
Bundesstr. 55
20146 Hamburg
Germany
gerhard@miraculix.dkrz.de

Christopher Popp
Institute for Environmental Physics
University of Heidelberg
Im Neuenheimer Feld 229
D-69120 Heidelberg
Germany
Christopher.Popp@
iup.uni-heidelberg.de

Peter A. Raymond
Yale School of Forestry
and Environmental Studies
Environmental Science Center
21 Sachem Street
New Haven, CT 06511
USA
peter.raymond@yale.edu

Rina Saiki
Department of Mechanical
Engineering and Science
Kyoto University, Kyoto 606-8501
Japan
saiki@mech.kyoto-u.ac.jp

Uwe Schimpf
Institute for Environmental Physics
University of Heidelberg
Im Neuenheimer Feld 229
D-69120 Heidelberg
Germany
Uwe.Schimpf@
iwr.uni-heidelberg.de

Nicholas Scott
Naval Reserach Laboratory
4555 Overlook Avenue SW
Washington, DC 20375
USA
Nicholas.Scott@nrl.navy.mil

Victor I. Shrira
Deptartment of Mathematics
Keele University
Keele
Staffordshire, ST5 5BG
UK
v.i.shrira@keele.ac.uk

Geoffrey Smith
Naval Reserach Laboratory
4555 Overlook Avenue SW
Washington, DC 20375
USA
geoffrey.smith@nrl.navy.mil

Michael Smith
School of Earth and Environment
Environment Building
The University of Leeds
Leeds. LS2 9JT
United Kingdom
m.smith@see.leeds.ac.uk

Naoya Susuki
Department of Mechanical
Engineering and Science
Kyoto University, Kyoto 606-8501
Japan
susuki@mech.kyoto-u.ac.jp

Naohisa Takagaki
Department of Mechanical
Engineering and Science
Kyoto University, Kyoto 606-8501
Japan
takagaki@mech.kyoto-u.ac.jp

Aldo Tamburrino
Department of Civil Engineering
University of Chile
Chile
atamburr@ing.uchile.cl

Kenji Tanno
Department of Mechanical
Engineering and Science
Kyoto University, Kyoto 606-8501
Japan
kenji@mech.kyoto-u.ac.jp

Wu-ting Tsai
Institute of Hydrological Sciences
National Central University
Jungli, Taoyuan, 32001
Taiwan
wttsai@ncu.edu.tw

Eric VanInwegen
University of Rhode Island
Graduate School of Oceanography
Narragansett, RI 02882-1197
USA
egv@gso.uri.edu

Evan A. Variano
Department of Civil and Environmental Engineering
Hollister Hall
Cornell University
Ithaca, NY 14853
USA
ev42@cornell.edu

James W. Walker
Water Research Laboratory
School of Civil and Environmental
Engineering
The University of New South Wales
King St. Manly Vale, NSW 2093
Australia
James.Walker@unsw.edu.au

Rik Wanninkhof
NOAA/AOML
4301 Rickenbacker Causeway
Miami, Florida 33149
USA
Rik.Wanninkhof@noaa.gov

Brian Ward
Center for Coastal Physical
Oceanography
Department of Ocean, Earth and
Atmospheric Sciences
Old Dominion University
Norfolk, Virginia 23529
USA
bward@ccpo.odu.edu

Jonathan D. Ware
Department Applied Ocean Physics
and Engineering
The Woods Hole Oceanographic
Institution
Woods Hole, MA 02543-1847 USA
jware@whoi.edu

Chani Welch
SMEC International Pty Ltd.
Australia
chani.welch@smecbd.com

John Wendelbo
University of Rhode Island
Graduate School of Oceanography
Narragansett, RI 02882-1197
USA

Christopher Zappa
Lamont-Doherty Earth Observatory
University of Columbia
Palisades, NY 10964
USA
zappa@ldeo.colmunbia.edu

List of Reviewers

Anne-Kristin Anweiler
University of Hamburg
Center for Marine and Atmospheric Research (ZMAW)
Institute of Oceanography (IfM)
Bundesstr. 53
D-20146 Hamburg, Germany
anweiler@ifm.uni-hamburg.de

William E. Asher
University of Washington
Applied Physics Lab
61 Route 9W
Seattle, WA 98105-6698
USA
asher@apl.washington.edu

Günther Balschbach
Institute for Environmental Physics
University of Heidelberg
Im Neuenheimer Feld 229
D-69120 Heidelberg
Germany
Guenther.Balschbach@iwr.uni-heidelberg.de

Michael L. Banner
The University of New South Wales
School of Mathematics
Sydney, NSW 2052
Australia
m.banner@unsw.edu.au

Guillemette Caulliez
Institut de Recherche sur les Phénomènes Hors Equilibre
163, av. de Luminy - case 903
13009 Marseille Cedex 9
caulliez@irphe.univ-mrs.fr

Edwin A. Cowen
Department of Civil and Environmental Engineering
Hollister Hall
Cornell University
Ithaca, NY 14853
USA
eac20@cornell.edu

Kai Degreif
Institute for Environmental Physics
University of Heidelberg
Im Neuenheimer Feld 229
D-69120 Heidelberg
Germany
Kai.Degreif@iwr.uni-heidelberg.de

Achim Falkenroth,
Institute for Environmental Physics
University of Heidelberg
Im Neuenheimer Feld 229
D-69120 Heidelberg
Germany
Achim.Falkenroth@iwr.uni-heidelberg.de

Christopher W. Fairall
NOAA Environmental Technology Laroratory
325 Broadway, R/ETL7
Boulder, CO 80303
USA
chris.fairall@noaa.gov

Christoph S. Garbe
Interdisciplinary Center for Scientific Computing (IWR)
University of Heidelberg
Heidelberg
Germany
Christoph.Garbe@iwr.uni-heidelberg.de

Johannes Gemmrich
University of Victoria
Department of Physics and Astronomy
P.O.Box 3055
Victoria, Bristish Columbia, V8W 3P6
Canada
gemmrich@uvic.ca

John Gulliver
Department of Civil Engineering
University of Minnesota
5000 Pillsbury Drive S. E.
Minneapolis, MN 55455-0116
USA
gulli003@umn.edu

Robert Handler
Naval Reserach Laboratory
4555 Overlook Avenue SW
Washington, DC 20375
USA
robert.handler@nrl.navy.mil

Tetsu Hara
University of Rhode Island
Graduate School of Oceanography
Narragansett, RI 02882-1197
USA
thara@uri.edu

Herlina
Institut für Hydromechanik
Universität Karlsruhe
Kaiserstr. 12
76128 Karlsruhe
Germany
herlinas@hotmail.com

Alexandra Herzog
Institute for Environmental Physics
University of Heidelberg
Im Neuenheimer Feld 229
D-69120 Heidelberg
Germany
Alexandra.Herzog@iwr.uni-heidelberg.de

Bernd Jähne
Interdisciplinary Center for Scientific Computing (IWR)
University of Heidelberg
Heidelberg
Germany
Bernd.Jaehne@iwr.uni-heidelberg.de

Alastair Jenkins
Bjerknes Center for Climate Reserach
Geophysical Institute
Allégaten 70
5007 Bergen
Norway
alastair.jenkins@bjerknes.uib.no

Gerhard H. Jirka
Institute for Hydromechanics
University of Karlsruhe
Kaiserstr.12
D-76128 Karlsruhe
Germany
jirka@uka.de

Satoru Komori
Department of Mechanical Engineering and Science
and
Advanced Institute of Fluid Science and Engineering
Kyoto University, Kyoto 606-8501
Japan
komori@mech.kyoto-u.ac.jp

Wade R. McGillis
Lamont-Doherty Earth Observatory
University of Columbia
Palisades, NY 10964
USA
wrm2102@columbia.edu

George Marmorino
Naval Reserach Laboratory
4555 Overlook Avenue SW
Washington, DC 20375
USA
George.Marmorino@nrl.navy.mil

Tobias Naegler
Institut für Umweltphysik
University of Heidelberg
Im Neuenheimer Feld 229
D-69120 Heidelberg
Germany tobias.naegler@iup.uni-heidelberg.de

Ryuichi Nagaosa
Research Planning Headquarters
AIST Tsukuba Central 2
1-1-1 Umezono, Tsukuba 305-8568
Japan
ryuichi.nagaosa@aist.go.jp

William L. Peirson
Water Research Laboratory
School of Civil and Environmental Engineering
The University of New South Wales
King St. Manly Vale, NSW 2093
Australia
w.peirson@unsw.edu.au

Gerhard Peters
Universität Hamburg
Meteorologisches Institut
Bundesstr. 55
20146 Hamburg
Germany
gerhard@miraculix.dkrz.de

Roland Rocholz
Institute for Environmental Physics
University of Heidelberg
Im Neuenheimer Feld 229
D-69120 Heidelberg
Germany
Roland.Rocholz@iwr.uni-heidelberg.de

Uwe Schimpf
Institute for Environmental Physics
University of Heidelberg
Im Neuenheimer Feld 229
D-69120 Heidelberg
Germany
Uwe.Schimpf@iwr.uni-heidelberg.de

Bernd Schneider
Institut für Ostseeforschung Warnemünde
Seestrasse 15
D-18119 Rostock
Germany
bernd.schneider@io-warnemuende.de

Geoffrey Smith
Naval Reserach Laboratory
4555 Overlook Avenue SW
Washington, DC 20375
USA
geoffrey.smith@nrl.navy.mil

Taro Takahashi
Lamont-Doherty Earth Observatory
University of Columbia
Palisades, NY 10964
USA

Wu-ting Tsai
Institute of Hydrological Sciences
National Central University
Jungli, Taoyuan, 32001
Taiwan
wttsai@ncu.edu.tw

Evan Variano
Department of Civil
and Environmental Engineering
Hollister Hall
Cornell University
Ithaca, NY 14853
USA
ev42@cornell.edu

Rik Wanninkhof
NOAA/AOML
4301 Rickenbacker Causeway
Miami, FL 33149
USA
Rik.Wanninkhof@noaa.gov

Gary A. Wick
NOAA ETL
325 Broadway
R/ET6
Boulder, CO 80305, USA
Gary.A.Wick@noaa.gov

Xin Zhang
Scripps Institution of Oceanography
University of California, San Diego
9500 Gilman Drive
La Jolla CA, 92093-0213
USA
xzhang@ucsd.edu

Christoph Zülicke
Leibniz-Institut für
Ostseeforschung Warnemünde
(IOW)
Marine Chemistry Department
Seestraße 15,
18119 Rostock-Warnemünde,
Germany
christoph.zuelicke@io-warnemuende.

1

The Impact of Different Gas Exchange Formulations and Wind Speed Products on Global Air-Sea CO_2 Fluxes

Rik Wanninkhof

NOAA/Atlantic Oceanographic and Meteorological Laboratory,
4301 Rickenbacker Causeway, Miami, Florida

Abstract Significant advances have been made over the last decade in estimating air-sea CO_2 fluxes over the ocean by the bulk formulation that expresses the flux as the product of the gas transfer velocity and the concentration difference of aqueous CO_2 over the liquid boundary layer. This has resulted in a believable global monthly climatology of air-sea CO_2 fluxes over the ocean on a 4° by 5° grid [38]. It is shown here that the global air-sea CO_2 fluxes are very sensitive to estimates of gas transfer velocity and the parameterization of gas transfer with wind. Wind speeds can now be resolved at sufficient temporal and spatial resolution that they should not limit the estimates, but the absolute magnitudes of winds for different wind products differ significantly. It is recommended to use satellite-derived wind products that have the appropriate resolution instead of assimilated products that often do not appropriately resolve variability on sub-daily and sub-25-km space scales. Parameterizations of gas exchange with wind differ in functional form and magnitude but the difference between the most-used quadratic relationships is about 15%. Based on current estimates of uncertainty of the air-water CO_2 concentration differences, the winds, and the gas exchange-wind speed parameterization, each parameter contributes similarly to the overall uncertainty in the flux that is estimated at 25%.

1.1 Introduction

In order to determine the role of the ocean in the global cycles of climate-relevant gases such as carbon dioxide (CO_2), the flux of these gases across the air-sea interface must be quantified. The ocean sequesters 20-30% of the excess CO_2 produced by fossil-fuel burning, thereby mitigating the greenhouse effect [19]. The projected future amount of CO_2 in the atmosphere is thus critically dependent on the amount of exchange between the ocean and atmosphere. Quantifying the air-sea gas CO_2 is, therefore, a major research objective of various international global change research programs.

C.S. Garbe, R.A. Handler, B. Jähne (eds.): Transport at the Air Sea Interface
pp. 1-23, 2007, © Springer-Verlag Berlin, Heidelberg 2007

Global air-sea gas flux estimates of slightly soluble gases are routinely determined from the product of the concentration gradient of the gas in question across the liquid boundary layer and the gas transfer velocity,

$$F = k\Delta C = kK_o\Delta pC \tag{1.1}$$

Equation (1.1) is often referred to as the bulk formulation where F is the flux [mol m^{-2} day^{-1}]; k is the gas transfer velocity [m d^{-1}]; ΔC is the concentration gradient [mol m^{-3}]; K_o is the solubility [mol m^{-3} atm^{-1}]; and ΔpC is the partial pressure (or fugacity) difference across the air-water interface [atm^{-1}]. The ΔC and ΔpC are often approximated from measurements in the surface ocean mixed layer at 1-5 m depth and in air well above the interface. While this bulk formulation is frequently used in this form, there are known issues with these approximations that are discussed elsewhere [26, 29]. One of the most referenced global applications utilizing this approach is the global air-sea CO_2 flux estimate of Takahashi et al. [38] based on a monthly global climatology of the partial pressure difference of CO_2 and ΔpCO_2.

Here I will focus on how uncertainty in the gas transfer velocity, in particular, its relationship with wind speed, affects the global CO_2 flux. First, a brief background is provided on determination of the gas transfer velocity in wind-wave tanks and in the field, and the functional relationship of gas transfer to wind speed. The sensitivity of the global CO_2 flux estimates to changes in wind, ΔpCO_2, and functional dependence on wind is shown. The issue of applying gas transfer velocities derived from other trace gases to CO_2 exchange is presented. The impact of the recent reassessment of the inventory of excess-^{14}C in the ocean is assessed. Excess-^{14}C is the ^{14}C produced by nuclear bomb tests corrected for dilution by ^{14}C-free fossil fuel emissions. Henceforth, the excess-^{14}C is referred to as bomb-^{14}C. The effect of high-resolution satellite wind speeds on the gas exchange wind speed relationship is discussed. The paper concludes with a brief summary of current estimates of interannual variability in CO_2 flux.

1.2 Discussion

1.2.1 A Summary of Gas Exchange Wind Speed Relationships

Gas transfer velocities have been determined in many field and wind-wave tank experiments. The laboratory studies benefit from full experimental control, but scaling considerations and possible artifacts due to the limited size and configuration of the experimental setups have raised questions about the applicability of the wind-wave tank results to the open ocean with respect to the absolute magnitude of the derived relationship of gas exchange with wind speed [16]. The work in wind-wave tanks has shown a strong dependence of gas exchange with wind. A notable finding is that there are distinct regimes in gas exchange and wind speed that are delineated by wave state [5]. Over smooth surfaces there is a weak dependence of gas exchange with wind that closely follows theoretical considerations of transfer across a smooth wall [9]. Once capillary and capillary-gravity waves form, the linear dependency strengthens appreciably. The onset of breaking waves enhances the gas transfer and gas transfer shows a solubility dependence with gases of lower solubility, showing a stronger enhancement. The transitions from smooth to rough surfaces and to breaking waves occur at wind speeds of about 3 and 13 m s^{-1}, respectively, depending on cleanliness and configuration of the tank. In field studies this clear delineation is not seen because of wind speed variability on short time scales, and variable thresholds for onset of capillary waves and breaking waves in the natural environment. Liss et al. [25] provide a comprehensive review of the status of air-sea gas exchange research in the 1980s.

Because of limitations of wind-wave tank studies, most empirical gas exchange-wind speed relationships are either derived from observations over the ocean or scaled to such studies. Initial studies over the ocean were performed using the ^{222}Rn disequilibrium method. The results showed no discernable trend with wind [36]. Factors that cause the absence of a clear correlation include experimental shortcomings of insufficient sampling at a particular location over the averaging time for the ^{222}Rn deficit method (four days) and inability to quantify losses and gains of ^{222}Rn in the mixed layer [23]. Use of deliberate tracers, in particular, the dual tracer technique with sulfur hexafluoride (SF_6) and the light isotope of helium (^3He), has proven to be a powerful approach to assess gas transfer in the coastal and open ocean [18, 32, 43]. Several, but not all, of the limitations of the ^{222}Rn are circumvented using injected tracers into the surface mixed layer. Major advantages of the dual tracer technique over the ^{222}Rn deficit method include the ability to do the studies in shallow coastal seas, ease of quantifying losses other than gas exchange, and the Lagrangian nature of the approach.

Using other gases as proxies for air-sea CO_2 transfer velocity works well for transfer over the smooth or turbulent interface in the absence of wave breaking since the gas transfer velocities can be scaled to their Schmidt number, which is defined as the kinematic viscosity of water divided by the molecular diffusion coefficient of the gas in question in water, according to

$$\frac{k_{CO_2}}{k_x} = \left(\frac{Sc_{CO_2}}{Sc_x}\right)^{-2/3} \quad \text{for smooth surfaces } \left(U_{10} \lesssim 3 \text{ m s}^{-1}\right) \tag{1.2}$$

and

$$\frac{k_{CO_2}}{k_x} = \left(\frac{Sc_{CO_2}}{Sc_x}\right)^{-1/2} \quad \text{for wavy surfaces} \tag{1.3}$$

While these dependencies are well established based on theoretical and experimental considerations [10, 20], it is important to consider that the interrelationships will break down under conditions of bubble entrainment. This is of particular concern when the results of the dual tracer technique using the gases ^3He and SF_6 that have very low solubilities are used to estimate the exchange of CO_2 which has a higher solubility. Comparisons in the field at low to intermediate winds have confirmed that results can be scaled using a $Sc^{-1/2}$ dependence [32], but laboratory and theoretical considerations suggest that Schmidt number normalized gas transfer velocities of SF_6 and ^3He are appreciably higher than CO_2 transfer at high winds due to bubble exchange [1, 44].

The effect of solubility for a particular pair of gases has been included in the Schmidt number parameterization through an apparent Schmidt number dependence [2] but this is seldom applied when converting the tracer results to CO_2 exchange. For example, Ho et al. [18] suggest a parameterization of

$$k_{600} = 0.266 \cdot U_{10}^2 \tag{1.4}$$

based on a dual deliberate tracer study in the open ocean near New Zealand. The k_{600} is the gas transfer velocity, k, normalized to a Schmidt number of 600 according to Eq. (1.3). As shown in Fig. 8 of [1], the apparent Schmidt number for the combination of ^3He and CO_2 decreases monotonically from -0.5 to -0.65 over a wind speed range from 5 to 25 m s^{-1}. Accounting for this change would lead to a dependence for CO_2 that can be well approximated by:

$$k_{CO_2,600} = 0.230 \cdot U_{10}^2 \tag{1.5}$$

The 15% difference in coefficients is relatively small considering the differences in the relationships discussed below. However, a change in the coefficient from 0.266 to 0.23 will decrease the global uptake of CO_2 by 15%. The adjustment procedures are strictly only applicable for situations where the gases are far from equilibrium. More work needs to be done in these comparisons [2], but it is clear that comparison of exchange rates of gases with differing solubilities must be done with some caution.

To estimate global air-sea CO_2 flux, constraints on the global gas transfer velocities are critical. While these constraints can be obtained from atmospheric measurements of CO_2 along stable carbon isotopes, or N_2/O_2 ratios [4], they are commonly obtained from the inventory of bomb-^{14}C in the ocean [6, 7]. This method takes advantage of the rapid increase of $^{14}CO_2$ in the atmosphere in the 1960s due to testing of thermo-nuclear devices. The atmospheric ^{14}C anomaly is followed as it penetrates into the ocean.

One of the first invasion rate estimates, I [mol m^{-2} yr^{-1}], was derived from optimizing for inventory and surface concentration of bomb-^{14}C in a multi-box ocean model for time dependent inventories and surface concentrations [6]. Wanninkhof [40] used this estimate, along with an inferred quadratic functional dependence with wind, to obtain a global parameterization of gas exchange with wind speed. The gas transfer velocity, k, was determined from I through

$$k = \frac{I}{K_o \cdot pCO_{2,a}} \tag{1.6}$$

where $pCO_{2,a}$ is the partial pressure of CO_2 in air. In this case, the invasion rate of CO_2 was assumed equivalent to that of $^{14}CO_2$, and the average mixing ratio of CO_2 in the atmosphere in 1964, at the peak of nuclear bomb testing, was used. A global average wind speed normalized to 10-m height (U_{10}) of 7.4 m s^{-1} from ship-based observations [13] yields the relationship

$$k_{av} = 0.39 \cdot U_{10,av}^2 \left(\frac{660}{Sc}\right)^{1/2} \tag{1.7}$$

where k_{av} is the global average gas transfer velocity and $U_{10,av}$ is the global average wind speed. This parameterization, when used in models to estimate air-sea gas fluxes, leads to consistent estimates of changing ocean bomb-^{14}C inventories. This is, in part, due to the fact that many of the older general circulation models are tuned to or validated with the same bomb-^{14}C inventories in the ocean.

The original global bomb-^{14}C inventory estimate did not lend itself to determine regional gas transfer rates because of difficulties accounting for transport of ^{14}C once it entered the ocean. The basin-wide invasion rates [6] are quite similar, and the wind speeds for each basin are similar enough to prevent obtaining meaningful discrete points for different oceans except for the Red Sea [8]. Therefore, while the global gas transfer velocity could be estimated from the invasion rate [6], the functional form of the relationship between gas exchange and wind had to be obtained by other means.

Three functional forms have been commonly used in combination with the bomb-^{14}C constraint:

- linear with a non-zero intercept [6, 39];
- quadratic [40]; and
- cubic [27, 41].

The linear relationship was proposed, in part, because the evidence of any other reasonable functional dependence was lacking from field observations. A quadratic dependence was suggested since this was the approximate dependence observed in wind-wave tanks [40]. Moreover, wind stress scales with U_{10}^2, and some theories suggest that gas exchange scales with stress. Monahan was one of the original proposers of a cubic dependence of gas exchange and wind speed [30]. In this formulation, it is implicitly assumed that bubbles have a controlling role on air-sea gas transfer. Several improvements of these global empirical parameterizations have been developed that include boundary layer stability criteria [12, 15], and both bubble-mediated exchange and exchange over the air-water interface [3].

An important advance over the last decade has been the improved wind speed measurements over the ocean from active and passive microwave sensors on earth-orbiting satellites. These measurements provide coverage of much of the ocean surface, once or twice a day, at a resolution of 25 km. Besides offering, for the first time, comprehensive measurements in the remote ocean, the measurements also provide a good estimate of the variability in wind speed. The variability of the wind affects the calculated k for non-linear dependencies of gas exchange with wind [40, 42]. Wanninkhof [40] proposed different dependencies for steady or "short-term" wind and for "long-term" averaged winds assuming that long-term averaged winds followed a Rayleigh wind speed frequency distribution. While both long-term and short-term dependencies were assumed quadratic with wind, the coefficients of proportionality differed by 26%. It was known that wind speed distributions vary by location and by averaging time, but lack of winds at high resolution prevented an exact solution.

With the remotely sensed winds it is now possible to determine gas transfer velocities without needing to assume a particular wind speed distribution curve. Average gas transfer velocities can be expressed as

$$k_{\text{av},660} = a \sum \frac{u^n}{s} = a \cdot {}^nM \tag{1.8}$$

where $k_{\text{av},660}$ is the average transfer velocity for a Schmidt number of 660; a is a coefficient of proportionality; s is the number of wind speed measurements, $n = 2$ for the quadratic dependence and $n = 3$ for the cubic dependence; and ${}^nM$ is the nth moment that is sometimes expressed as $\langle u^n \rangle$. Changing sea surface temperature (SST) over the period of determination will affect the gas transfer as well through the temperature dependency of the Schmidt number correction $({}^{660}/_{Sc})^{1/2}$ (see Eq. (1.7)), which is non-linear as well. However, using the average SST over the time period of investigation will cause a bias of less than 5%.

1.2.2 The Sensitivity of Global Air-Sea CO_2 Flux

The uncertainty in the global air-sea CO_2 flux determined from the bulk flux method is estimated at +22, -19% [38], but this error estimate is primarily associated with the estimated uncertainty in the ΔpCO_2 field and likely an underestimate of the true error. An illustration of the sensitivity of the global CO_2 flux can be obtained from varying the wind, ΔpCO_2, and the functional dependence by an amount that approximates its uncertainty and determining the resulting change in flux. For this exercise we obtained the winds, sea surface temperatures, and ΔpCO_2 from the monthly global CO_2 climatology [38] and used as default the gas exchange wind speed formulations $k_{660} = 0.31 \cdot {}^2M$ or $k_{660} = 0.0283 \cdot {}^3M$ to take into account the monthly variability of the wind in each pixel. For this exercise the ^2M and ^3M were determined from the six-hour NCEP winds for 1995 re-gridded from the original 2° by 2° grid to the 4° by 5° grid of the Takahahsi monthly global CO_2 climatology as used in [42]. The results of these changes are shown in Table 1.1. Changes in wind speed have a pronounced effect, especially for a cubic dependency. The functional dependency itself can change the flux two-fold. It is also of note that while many of the proposed relationships have a zero intercept, there is little evidence to support this premise. Turbulence and instabilities near the water surface induced by (diurnal) heating and shear [28] are believed to cause a finite gas transfer at low or no wind. This has lead to a reformulation of gas transfer to $k = b + a \cdot U_{10}^n$ where b is referred to as a "background" gas transfer velocity. McGillis et al. [27] suggest a value of $b = 3.2$ cm hr^{-1}. Including this term and adjusting the coefficient a to meet the bomb-^{14}C constraint leads to a decrease in the ocean uptake of 11-15%. This is because, on average, the ocean releases CO_2 at lower winds when the "background" transfer plays a more important role and because the relationships with a non-zero intercept yield lower k at higher winds in order to meet the bomb-^{14}C constraint.

1.2.3 The Impact of Updated Oceanic Bomb-^{14}C Inventories

Using the bomb-^{14}C invasion into the ocean to determine the gas transfer velocity requires knowledge of the time evolution of the atmospheric ^{14}C and the oceanic ^{14}C inventories on a regional basis, and the ^{14}C levels in the surface ocean. Several approximations have been made to estimate the global gas transfer velocity in this manner with poorly quantified effect on the final results. In particular, the means of extrapolation of sparse field measurements has led to uncertainties in the estimate of the ocean ^{14}C inventory. Considerable effort has been put into improving the global bomb-^{14}C inventory that has yielded revised global oceanic bomb-^{14}C based gas transfer estimates. The largest current shortcoming is the uncertainty in the partial pressure of $^{14}CO_2$ in seawater, $p^{14}CO_2sw$, which controls the "back flux" of $^{14}CO_2$. This term is increasingly significant because the atmosphere and ocean are reaching equilibrium with respect to $^{14}CO_2$. In the estimates below, most of the differences in calculated gas transfer rates can be associated with differences in inventory estimates and calculation methods.

The first estimates of the global inventory of bomb-^{14}C in the ocean in the 1980s were based on interpolating relatively few measurements in each ocean basin [6]. Separation of the bomb-^{14}C contribution from the natural background was problematic [35]. The estimates were also subject to interpolation errors and differences in interpolation schemes.

A simple box model used in the original analysis [6] could roughly reproduce the observed surface values and basin inventories obtained dur-

Table 1.1. Sensitivity of global air-sea CO_2 fluxes to changes in wind speed, ΔpCO_2, and wind speed formulation (in Pg C yr^{-1}).

Variable	Adjustment	$k = 0.31 \cdot {}^2M$	$k = 0.0283 \cdot {}^3M$
Wind[a]	+1 m/s (U_{av} = 8.1 m/s)	-1.86 (17%)[b]	-2.58 (34%)
	0 m/s (U_{av} = 7.1 m/s)	-1.59	-1.93
	-1 m/s (U_{av} = 6.1 m/s)	-1.31 (-18%)	-1.4 (-27%)
ΔpCO_2[c]	+1 μatm	-1.38 (-13%)	-1.75 (-9%)
	0 μatm	-1.59	-1.93
	-1 μatm	-1.79 (13%)	-2.12 (10%)
k[d]	+20% (a = 0.37, 0.0339)	-1.90 (19%)	-2.32 (20%)
	0% (a = 0.31, 0.0283)	-1.59	-1.93
	-20% (a = 0.25, 0.0226)	-1.26 (-21%)	-1.55 (-20%)
k[e]	linear = $2.88 \cdot U_{10}$	-1.02 (–36%)	
	quadratic = $0.31 \cdot {}^2M$	-1.59	
	cubic = $0.0238 \cdot {}^3M$	-1.93 (20%)	
k[f]	linear = $3.2 + 2.46 \cdot U_{10}$[f]	-0.92 (–11%)	
	quadratic = $3.2 + 0.26 \cdot {}^2M$	-1.39 (–14%)	
	cubic = $3.2 + 0.0238 \cdot {}^3M$	-1.67 (–15%)	

[a] Change wind speed for each monthly 4° by 5° pixel by 1 or -1 m/s. The winds are six-hour NCEP winds for 1995 re-gridded from the original 2° by 2° grid to a 4° by 5° grid (Doney, pers. com.). The resulting global average winds U_{av} are listed in parentheses.
[b] Percent difference from the standard case.
[c] Change ΔpCO_2 for each monthly 4° by 5° pixel by 1 or -1 μatm.
[d] Change coefficient a in $k = a \cdot {}^2M$ or $k = a \cdot {}^3M$ for each monthly 4° by 5° pixel by the listed amount. The change in a for a quadratic or cubic dependence, respectively, is listed in parentheses.
[e] Change functional dependence of k as listed.
[f] Include a finite "background" gas transfer at low winds.

ing the GEOSECS cruises and offered a means to project future ^{14}C concentrations in the ocean (Figure 1.1). The controversy about the bomb-^{14}C inventory in the ocean and resulting global ^{14}C constraint started when the inventory values [6] were put in question by an independent stratospheric ^{14}C constraint and a global mass balance [17]. In this analysis, the ocean inventory was approximately 25% less for 1974 than the original estimate [6].

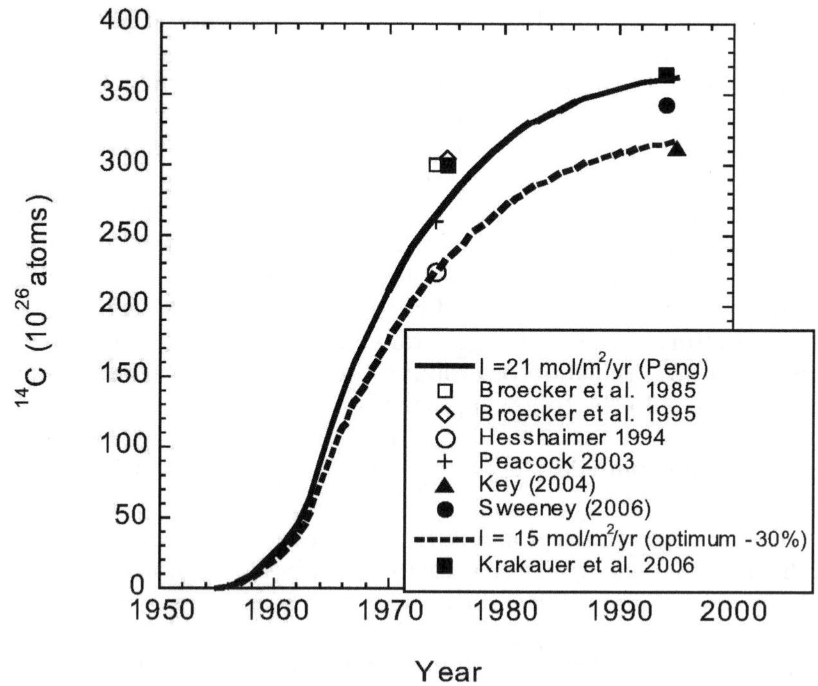

Figure 1.1. Change in bomb-^{14}C inventory over time. The solid line is the result of the box model optimized for basin-wide ^{14}C inventories and surface concentrations [6] as recently rerun by Peng (pers. com.). The dashed line is the model run where the invasion rate, I, was decreased by 30%. The points are the model and data-based estimates listed in Table 1.2.

This was followed by an analysis which suggested the results could be reconciled if a more sophisticated ocean model was used [11]. A more rigorous observation based ^{14}C inventory was performed for the mid-1970s using a model and chlorofluorocarbon (CFC) inventories to estimate the distribution of bomb-^{14}C in the ocean [34]. This estimate was 15% lower than the original estimates [6, 7]. The comparison of estimates based on data from the 1970s is complicated by the rapid rise of ^{14}C in the ocean during this time (Fig. 1.1), the multi-year expeditions that were used to determine the inventories, and the inconsistent estimates of the inventories cited in various publications. Of note is that the original optimized ocean model results [6] fall below the estimates of global inventory (see Fig. 1.1) but that this run is in good agreement with the more recent ocean bomb-^{14}C estimate [34]. For comparison, a model run using the same model as in [6] with an evasion rate 30% below the optimum is also provided.

Currently, inventories are estimated for two time periods from large hydrographic surveys that were conducted in the 1970s (the Geochemical

Ocean Section Study [GEOSECS]) and 1990s (the World Ocean Circulation Experiment/ World Hydrographic Survey [WOCE/WHP]). A summary of the estimates is listed in Table 1.2. Aside from a significantly greater sampling density, the WOCE/WHP survey provided a second point centered at about 1995 to challenge the ocean circulation models used to estimate the inventories and fluxes. The observational bomb-^{14}C inventory is $3.13 \cdot 10^{26}$ for 1995 [21]. As indicated in Table 1.2, this data-based inventory for 1995 is about 10% lower than the models. This is attributed to the fact that the observation based inventory will be biased low because the Arctic Ocean and coastal areas were not sampled in the WOCE/WHP surveys.

A recent effort to determine gas transfer velocities from bomb-^{14}C by Sweeney et al. [37] relies on optimizing the ^{14}C inventories in a general circulation model in the following manner

$$\Delta Inv \left(\text{Bomb } \Delta I\, ^{14}C\right) = \int_{1954}^{1994} \int_{\substack{\text{Ocean}\\ \text{Surface}}} f_{\text{Bomb}^{14}CO_2} dA dt \qquad (1.9)$$

where f represents the regional surface fluxes from which k can be derived. In principle, gas transfer velocities can be determined for each of the 29 regions but, in practice, the average winds do not vary significantly over many of the areas. Also, general uncertainties in ocean transport preclude determining a robust functional dependence. This method relies on ocean circulation models, and the results are directly affected by biases and shortcomings in the circulation schemes.

This approach yielded a gas transfer velocity of 14.6 ± 4.7 cm hr^{-1} [37] that is substantially less than previous estimates of 21.9 ± 3.3 cm hr^{-1} [40]. However, several other works reach different conclusions. The estimate of Krakauer et al. [22] provides a global gas transfer velocity indistinguishable from the original while that of Naegler et al. [31] falls in between (Table 1.3).

In Krakauer et al. [22] regional specific gas transfer rates were obtained. While only small regional differences in wind and gas transfer were observed, the functional dependence of gas transfer was nearly linear, contrary to other recent work that, a priori, assumed a quadratic dependence [31, 37].

An important point that has been largely ignored in many estimates of global CO_2 uptake by the ocean is that the relationships of gas exchange and wind speed should be scaled to the global average squared wind of the wind speed product used [31]. There are differences of up to 1.3 m s^{-1} in global average winds for the global products (see Table 1, in Naegler et al. [31]). If a different wind product is used, the coefficient should be scaled accordingly. For instance, if a global wind estimate of 7.4 m s^{-1} is used with a global gas transfer velocity of 21.9 cm hr^{-1}, it yields a coefficient a in the quadratic relationship $k = a \cdot u^2$ of 0.39 [40]. If an average global wind of 6.6 m s^{-1} of the NCEP wind assimilation product is used instead, a coefficient of $^{21.9}/_{6.6^2} = 0.50$ should be applied.

Table 1.2. Estimates of global bomb-^{14}C inventories.

Source	Estimate [10^{28} atoms]	Comment
Broecker et al. [6]	3.14 ± 11%	Atlantic and part of Pacific GEOSECS[a]
Broecker et al. [7]	2.89 ± (5 – 20)%	GEOSECS, NORPAX, Part TTO[b]
Broecker et al. [7]	3	January 1, 1974 (data)[c]
Broecker et al. [7]	3.05 ± 10%	January 1, 1975 (data)[d]
Hesshaimer et al. [17]	2.25	January 1, 1974 (atm. mass balance)
Peacock [34] [e]	2.60 ± 10%	January 1, 1974 (data and model)
Krakauer et al. [22]	2.93-3.09	1975 (model)
Peng (pers. comm.)	3.62	1995 (model)
Key et al. [21]	3.12 ± 15%	1995 (WOCE data)
Sweeney et al. [37]	3.43 ± 12%	1994 (model)
Krakauer et al. [22]	3.48-3.82	1994 (model)
Krakauer et al. [22]	3.64	1994 (data and model)[f]
Naegler et al. [31]	3.55	1995 (data and model)[f]

[a] The inventory is based on an incomplete global data set comprised of Atlantic and Pacific GEOSECS data.
[b] Inventory based on a complete GEOSECS dataset and additional data in the North Atlantic (TTO) and equatorial Pacific (NORPAX).
[c] Inventory from data listed in (b) but normalized to 1974 using a box model.
[d] Inventory from data listed in (b) but normalized to 1975 using a box model.
[e] Peacock [34] calculates inventories by several approaches but this is the value considered most consistent with available data.
[f] Using model output to fill in the ocean regions not sampled in the WOCE data.

Comparison of inventories (Table 1.2) and calculated gas transfer velocities (Table 1.3) shows that there is no straight correspondence between the inventories in 1995 and the resulting gas transfer velocities. This is caused by the different approaches to determine the global transfer velocities and because the ocean currently is close to atmospheric equilibrium with respect to ^{14}C, making the solutions very susceptible to variations in transport of bomb-^{14}C in the ocean in different circulation models. While important improvements have been made in ^{14}C inventory estimates, separation of bomb ^{14}C from natural ^{14}C, and modeling of ^{14}C in the ocean, the recent analyses bear more scrutiny before considering them as conclusive. Based on this work, it appears that the popular estimate developed in the 1990s [40] is on the high side of the envelope but well within the uncertainty of believable estimates.

Table 1.3. Global gas transfer rates estimated from bomb-^{14}C.

Reference	k [cm hr^{-1}]	Comment
Wanninkhof [40]	21.9 ± 3.3	Sc = 660, quadratic[a]
Naegler et al. [31]	16.7 ± 2.9	Sc = 660, quadratic
Sweeney et al. [37]	14.6 ± 4.7	Sc = 660, quadratic
Krakauer et al. [22]	21.0 ± 2	Sc = 660, linear[b]

[a] Estimate normalized to a Schmidt number of 660 and using a quadratic relationship of gas exchange and wind.
[b] Estimate normalized to a Schmidt number of 660 and using a linear relationship of gas exchange and wind.

1.2.4 The Effect of Wind Speed Variability

A major advance in the application of the gas transfer wind speed relationship has been the high-resolution wind speed measurements obtained over the global ocean using observations from satellites. High-resolution data are critical for non-linear relationships between wind speed and gas exchange in which the distribution of winds is important, in addition to the mean wind. This can be gleaned from

$$k_{\mathrm{av},600} = \frac{a}{s}\sum\left(U_{10,s}\right)^n \neq \frac{a}{s}\left(U_{10,av}\right)^n \tag{1.10}$$

where $k_{\mathrm{av},660}$ is the average gas transfer at Schmidt number of 660, a is the coefficient of proportionality, s is the number of sampling intervals, $U_{10,s}$ is the "instantaneous" or "short term" wind speed, and $U_{10,av}$ is the average wind over the time in question. Note that $\sum\left(U_{10,s}\right)^n/s$ is equivalent to the nth moment, ^nM. The current QuikSCAT scatterometer coverage provides near-global observations at 25-km resolution once or twice per day[1] that, in itself, is insufficient to fully capture the variability. The SSM/I product provides a higher temporal coverage but lower resolution and more data gaps due to clouds, as it uses a passive instrument[2]. These products are combined to provide a dataset henceforth called the QuikSCAT-SSM/I product. Using the combined dataset and by invoking the time/space similarity of the Taylor theorem provides a sufficient number of data points to accurately determine the moments on the 4° by 5° grid that are used in estimating global air-sea CO_2 fluxes. The average number of observations per pixel per month is 8500 but at high latitudes this value can be as low as 50. This hybrid dataset has shortcomings with regard to possible biases, particularly as this dataset encompasses four years before QuikSCAT data were available. However, inspection of the temporal and

[1] http://podaac-www.jpl.nasa.gov/products/product108.html
[2] http://podaac-www.jpl.nasa.gov/products/product079.html

regional trends and moments do not show any apparent changes greater than plausible compared with assimilation products. The ten year's worth of data are considered important to avoid possible artifacts due to interannual variability.

Previous work showed that the moments from the six-hour NCEP wind speed assimilation product for 1995 differed regionally [42]. This is reassessed here using the QuikSCAT-SSM/I winds. The comparison of the 1995 NCEP data and the 2005 QuikSCAT-SSM/I data versus latitude are shown in Fig. 1.2, which indicates that while both products show the same zonal trends in the average winds, ^2M, and ^3M, the offset is appreciable. It is of note that assimilation schemes such as NCEP and ECWMF periodically get updated when new wind products, pressure fields, and modeling approaches become available. Also, algorithms for satellite winds get improved as well, and the data undergo periodic reprocessing. This is the reason that the NCEP winds for 2004 and those of QuikSCAT-SSM/I for 2004 show better agreement, while the older NCEP product gave lower winds at mid-latitudes.

1.2.5 Interannual Variability of Global Air-Sea CO_2 Fluxes

Interannual variability in wind speed, ^2M, and ^3M can provide an estimate of the effect of changes in winds on year-to-year changes in flux. The second parameter that affects the flux is the ΔpCO_2. Both factors have been taken into consideration in previous estimates of interannual variability for 1981-2001 [33]. Here we look at variability from 1995-2005 and utilize more comprehensive wind statistics that are available for the more recent data. Annual averaged wind speeds, ^2M, and ^3M and enhancement factors for the QuikSCAT-SSM/I merged product are shown in Fig. 1.3. The QuikSCAT data came online in mid-1999 such that the datasets before and after 2000 are disparate. Any changes should be interpreted with caution. The results in Fig. 1.3 show relatively little change in these parameters other than the increase from 1998-2000 that was probably caused by improved coverage and data quality once the QuikSCAT data became available.

Similar year-to-year global average winds do not exclude regional variability. To investigate the extent of the interannual variability, the standard deviation of the winds for every 4° by 5° pixel from 60°N to 60°S is determined. The resulting zonal averages are shown in Fig. 1.4 as a percent standard deviation of the mean, ^2M, and ^3M. The average standard deviation for all pixels over the last decade is 5.2% for the mean, 9.5% for the ^2M, and 13.0% for the ^3M of the wind with the greatest variability at the high northerly and low latitudes. With a non-linear dependence on gas exchange, it will be the high latitudes with higher winds that will experience the greatest year-to-year absolute variability in gas transfer velocity.

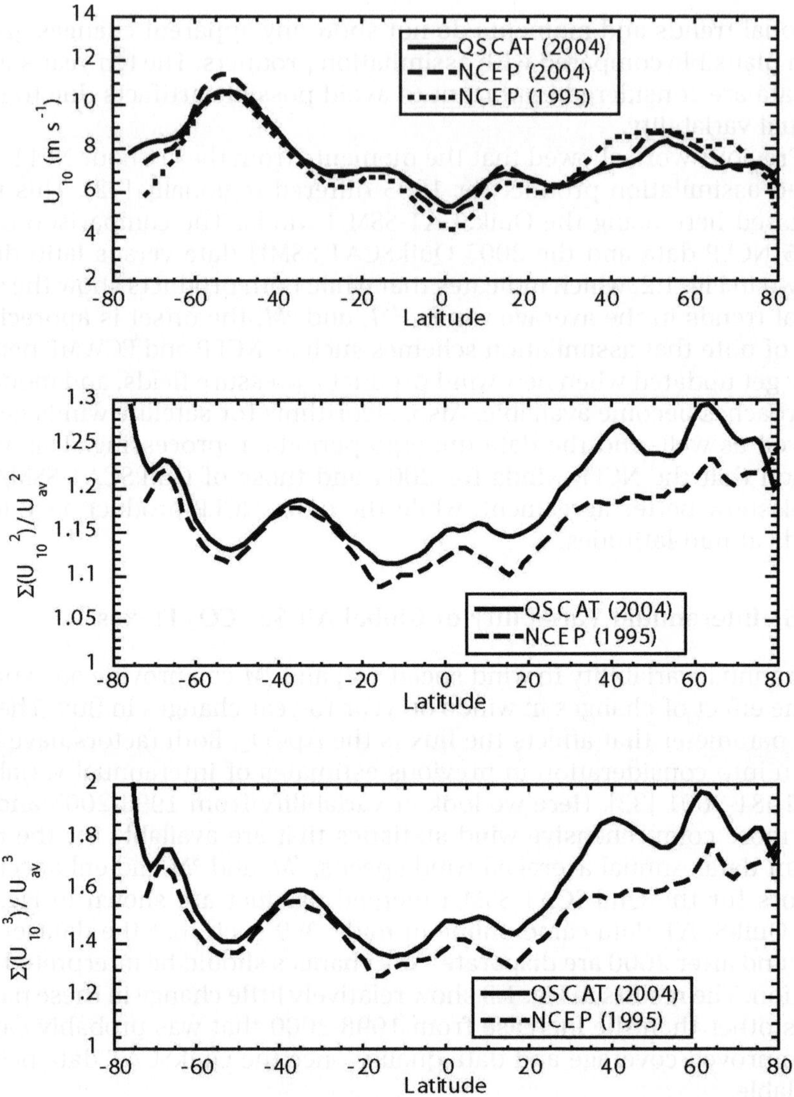

Figure 1.2. Zonal wind statistics based on satellite observations (QuikSCAT-SSM/I) and assimilation models (NCEP): comparison of NCEP winds in 1995 and 2004 and a melded product of SSM/I and QuikSCAT for 2004 (Top). The difference between the 1995 and 2004 NCEP data is believed to be due to changes in the assimilation routines for the 1995 product and that of 2004. The second moment, ${}^2M$, divided by the average wind squared, which is sometimes referred to as an enhancement factor for a quadratic wind speed dependence (Middle). The third moment, ${}^3M$, divided by the average wind cubed, which is sometimes referred to as an enhancement factor for a cubic wind speed dependence (Bottom).

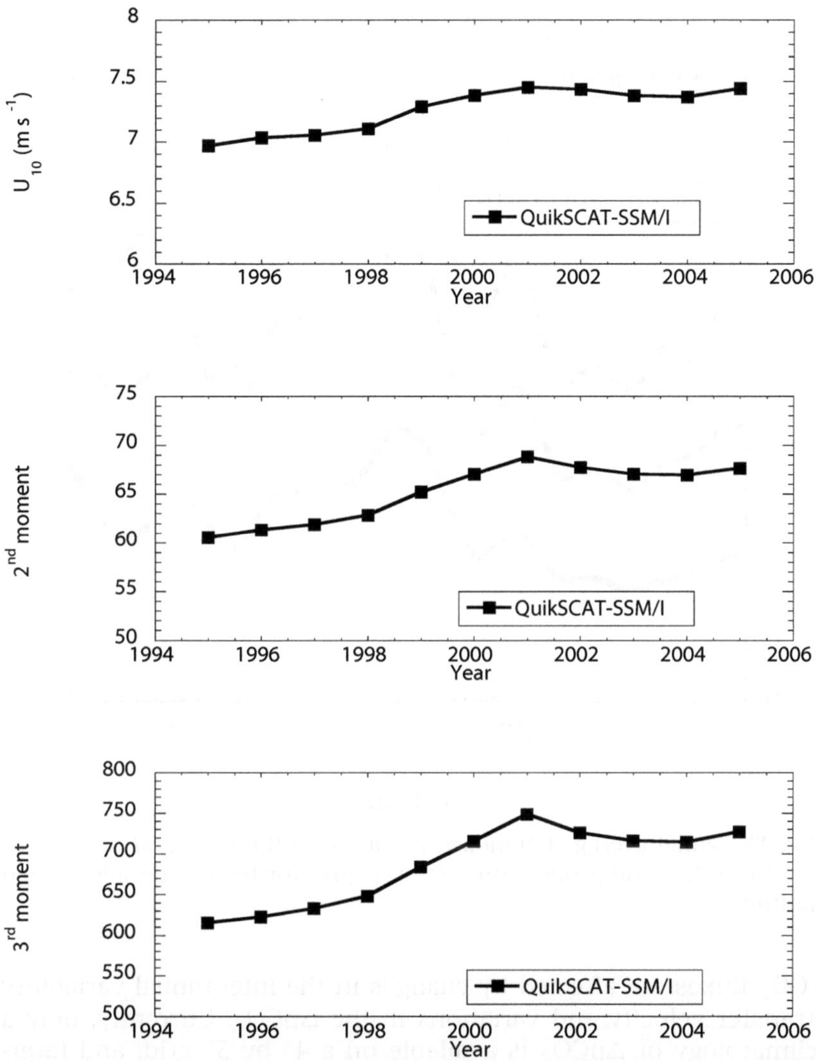

Figure 1.3. Temporal trend in global average winds (top), the second moment, ^2M (middle), and the third moment, ^3M (bottom) over the last decade using a melded product of SSM/I and QuikSCAT data. The change from 1998-2000 is attributed to the QuikSCAT data coming online.

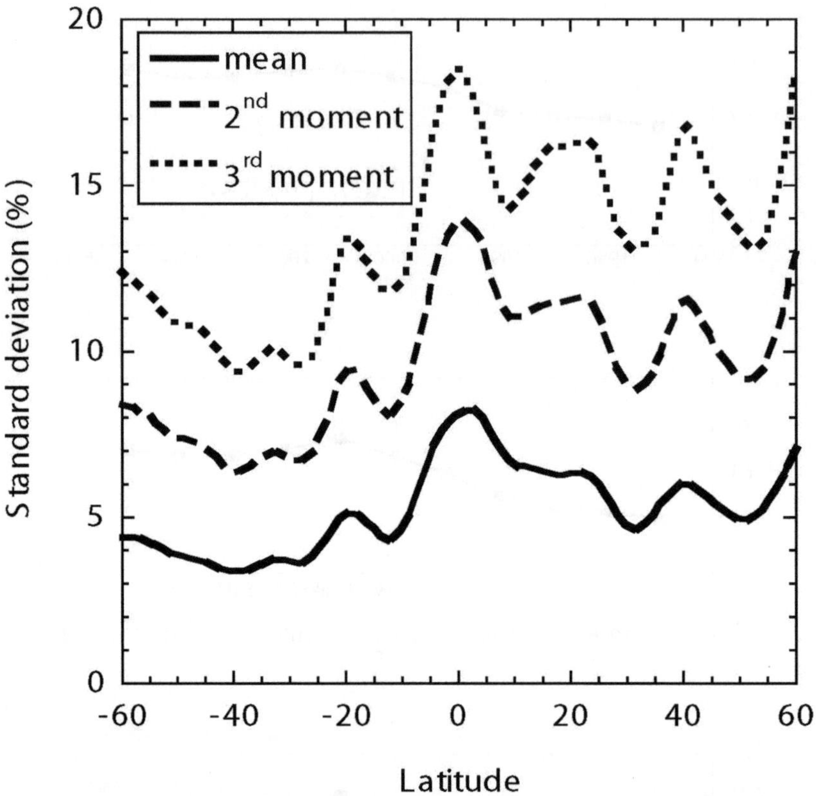

Figure 1.4. The zonal averaged standard deviation in the mean, ${}^2M$, and ${}^3M$ of the QuikSCAT-SSM/I wind product on a 4° by 5° grid for the last decade plotted against latitude.

The CO_2 fluxes are affected by changes in the interannual variability of gas transfer velocity and variations in the ΔpCO_2. Currently, only a global climatology of ΔpCO_2 is available on a 4° by 5° grid, and interannual changes in ΔpCO_2 are estimated through models or by empirical means. To assess the interannual variability in air-sea CO_2 flux, the seasonal changes in ΔpCO_2 are correlated with SST, and these relationships are used along with interannual SST changes [24, 33] to determine monthly ΔpCO_2 maps based on the ΔpCO_2 climatology. For k, the ${}^2M$ of the QuikSCAT-SSM/I product is used, along with the parameterization, $k_{660} = 0.31 \cdot {}^2M$. Annual fluxes from 1995-2004 are shown in Fig. 1.5.

There is no significant correlation between the global fluxes and the global average winds (Fig. 1.5), indicating that the regional wind patterns are of importance for estimating global CO_2 fluxes and that the changes in the ΔpCO_2 field have a significant effect on the fluxes. The interannual variability shown in a previous work (Park et al. [33]) of 0.18 Pg C yr^{-1} for

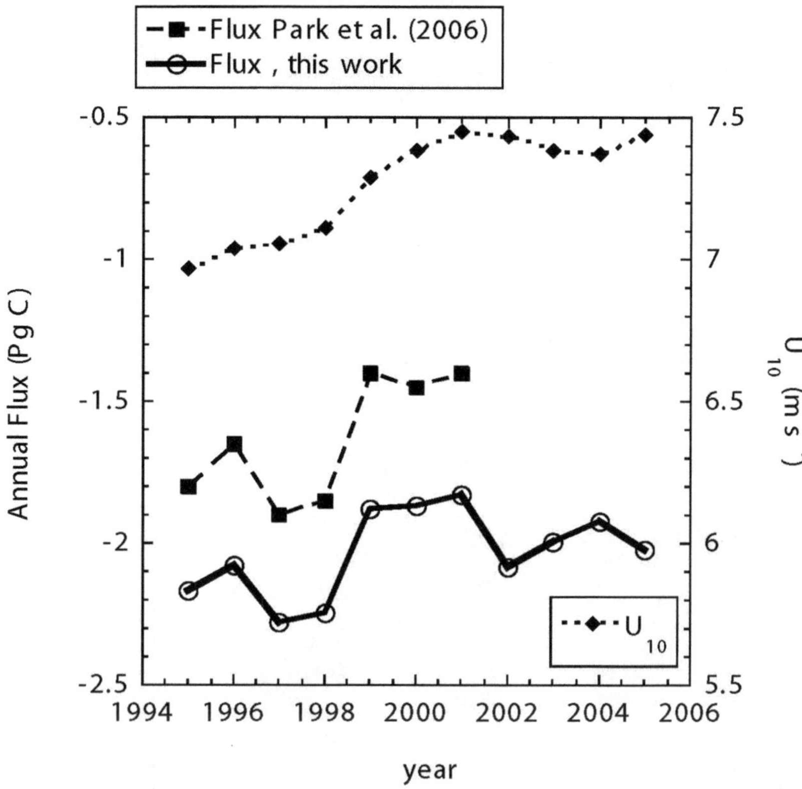

Figure 1.5. Interannual variability of global air-sea CO_2 fluxes estimated by Park et al. (2006) (dashed line) [33] and in this work (solid line). The difference is attributed to use of different wind products. The annual QuikSCAT-SSM/I winds are shown as diamonds with the scale on the right axis.

the time period of 1981-2001 is indistinguishable from the 0.16 Pg C yr^{-1} determined here. The average annual flux in the previous work is 1.7 Pg C yr^{-1} (Park et al. [33]) and 2.0 Pg C yr^{-1} in this exercise. The same pCO_2 climatology was used, and the same approach of estimating changes of pCO_2 through a relationship with SST was applied. The difference is attributed to the different wind speed product used and failure to normalize the coefficient for the averaged global winds as outlined in Naegler et al. [31] and above. Moreover, in the previous work the variability in wind speed was not fully accounted for.

Most of the interannual variability in CO_2 fluxes is in the equatorial Pacific and is attributed to El Niño-Southern Oscillation (ENSO) induced changes [14]. Reduced upwelling and lower winds during El Niño events greatly decrease the efflux of CO_2 in the region. This change is well captured by the technique. Figure 1.6 shows the global efflux during January-

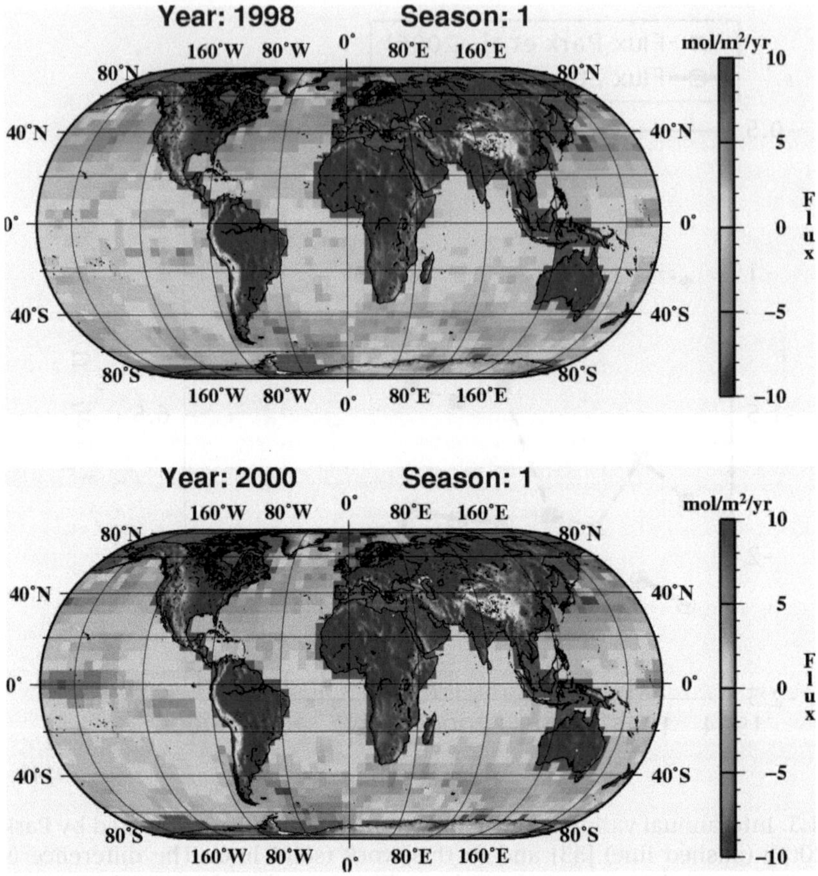

Figure 1.6. Differences in air-sea CO_2 fluxes for season 1 (January-March) of an El Niño year (1998) (top) and season 1 (January-March) of a normal year (2000) (bottom) illustrating the large differences in the equatorial Pacific.

March of 2000, considered a normal year, and during January-March of 1998, an El Niño year, illustrating the large anomaly in the equatorial Pacific area. The decrease in efflux is largely caused by changes in ΔpCO_2, but wind speed changes play a secondary reinforcing role in causing the lower fluxes from the ocean [14].

1.3 Conclusions

Winds play a controlling role in modulating global air-sea fluxes. While it is clear that the processes at the air-sea interface that dictate the air-sea gas transfer of CO_2 are affected by other factors as well, there have been no other compelling parameters that have proven to be superior to utilizing winds for global gas exchange parameterizations. There are several issues that have to be considered when using relationships of gas exchange and wind speed.

The current estimates using global constraints based on bomb-^{14}C are in flux (no pun intended). Consensus should be reached on the bomb-^{14}C inventories and how to extrapolate in undersampled areas such as the Arctic region and coastal oceans. Recent papers [22, 37] have suggested that the often-used parameterization [40] needs revision, but the changes differ. Based on a re-evaluation of bomb-^{14}C inventories, Krakauer et al. [22] suggest that the magnitude of the global gas transfer velocity is in accord with previous estimates but that the functional dependence with wind speed is incorrect; in contrast, Sweeney et al. [37] advocate a quadratic functional dependence but with a global average that is about 30% lower. These recent works point to a common issue of incorrect use of the gas transfer-wind speed relationships with different wind speed products than were used to create the relationship [31, 37]. A straightforward approach to deal with this mismatch is provided [31].

A second recommendation in these papers is that the appropriate moment of the wind should be used (^1M, which equals the mean, for linear dependencies through the origin; ^2M for quadratic relationships; and ^3M for cubic dependencies) rather than the average wind speed. With the advent of high-resolution wind products, accurate moments are now available.

Global wind speed products have biases that can be accounted for in calculating the gas transfer velocity. Biases in products that change over time are a problematic issue that hinders assessment of long-term trends in winds and fluxes. A high-resolution melded SSM/I and QuikSCAT product bears out the issue of a change in global winds of 0.3 m s^{-1} when the QuikSCAT data came online. For the determination of global CO_2 fluxes, regional assessments of wind and ΔpCO_2 are critical. Initial estimates of interannual variability in CO_2 flux show significant regional variability but global interannual anomalies that are relatively small and in line with models and atmospheric inferences.

Acknowledgement. Joaquin Trinanes provided the merged SSM/I and QuikSCAT wind products. Dr. T.-H. Peng dug up the old model code to recreate the model runs in [6]. Dr. T. Takahashi of LDEO, Columbia University, kindly provided the ΔpCO_2 climatology used in this work. Gail Derr of AOML edited and formatted the manuscript. Their assistance is greatly appreciated. Discussions with participants of the International Workshop on Transport at the Air Sea Interface, in Heidelberg (September 6-8, 2006), in particular, David Ho, Tobias Naegler, Wade McGillis, and Colm Sweeney, were very helpful in producing this work. Reviews by T. Takahashi and two anonymous reviewers are gratefully acknowledged.

References

[1] Asher W.E., Wanninkhof R. (1998a) Transient tracers and air-sea gas transfer. *J Geophys Res* 103:15939-15958
[2] Asher W.E., Wanninkhof R. (1998b) The effect of bubble-mediated gas transfer on purposeful dual gaseous-tracer experiments. *J Geophys Res* 103:10555-10560
[3] Asher W.E., Edson J., McGillis W.R., Wanninkhof R., Ho D.T., Litchendorf T. (2002) Fractional area whitecap coverage and air-sea gas transfer velocities measured during Gas Ex-98. In: Donelan M., Drennan W., Saltzman E., Wanninkhof R. (eds), *Gas transfer at water surfaces*. AGU, Geophysical Monograph 127, Washington, DC, pp 199-205
[4] Battle M., Bender M.L., Tans P.P., White J.W.C., Ellis J.T., Conway T., Francey R.J. (2000) Global carbon sinks and their variability inferred from atmospheric O_2 and $\delta^{13}C$. *Science* 287:2467-2470
[5] Broecker H.C., Peterman J., Siems W. (1978) The influence of wind on CO_2 exchange in a wind-wave tunnel, including the effects of mono layers. *J Mar Res* 36:595-610
[6] Broecker W.S., Peng T.-H., Östlund G., Stuiver M. (1985) The distribution of bomb radiocarbon in the ocean. *J Geophys Res* 99:6953-6970
[7] Broecker W.S., Sutherland S., Smethie W., Peng T.-H., Östlund G. (1995) Oceanic radiocarbon: Separation of the natural and bomb components. *Global Biogeochem Cycles* 9:263-288
[8] Cember R. (1989) Bomb radiocarbon in the Red Sea: A medium-scale gas exchange experiment. *J Geophys Res* 94:2111-2123
[9] Deacon E.L. (1977) Gas transfer to and across an air-water interface. *Tellus* 29:363-374
[10] Degreif K. (2006) "Untersuchungen zum Gasaustausch - Entwicklung und Applikation eines zeitlich aufgelösten Massenbilanzverfahrens," PhD thesis, Universität Heidelberg, Heidelberg
[11] Duffy D.B., Caldeira K. (1995) Three-dimensional model calculation of ocean uptake of bomb ^{14}C and implications for the global budget of bomb ^{14}C. *Global Biogeochem Cycles* 9:373-375
[12] Erickson III D.J. (1993) A stability-dependent theory for air-sea gas exchange. *J Geophys Res* 98:8471-8488
[13] Esbensen S.K., Kushnir Y. (1981) The heat budget of the global ocean: An atlas based on estimates from surface marine observations. *Climatic Res Inst Rept* 29, Oregon State Univ, Corvalis OR
[14] Feely R.A., Takahashi T., Wanninkhof R., McPhaden M.J., Costa C.E., Sutherland S.C., Carr M.-E. (2006) Decadal variability of the air-sea CO_2 fluxes in the equatorial Pacific Ocean. *J Geophys Res* 111: doi: 10.1029/2005JC003129
[15] Hare J.E., Fairall C.W., McGillis W.R., Edson J.B., Ward B., Wanninkhof R. (2004) Evaluation of National Oceanic and Atmospheric Administration / Coupled-Ocean Atmospheric Response Experiment (NOAA

/ COARE) air-sea gas transfer parameterization using GasEx data. *J Geophys Res* 109: doi: 10.1029/2003JC001831

[16] Hasse L. (1990) On the mechanism of gas exchange at the air-sea interface. Tellus 42B:250-253

[17] Hesshaimer V., Heimann M., Levin I. (1994) Radiocarbon evidence for a smaller oceanic carbon dioxide sink than previously believed. *Nature* 370:201-203

[18] Ho D.T., Law C.S., Smith M.J., Schlosser P., Harvey M., Hill P. (2006) Measurements of air-sea gas exchange at high wind speeds in the Southern Ocean: Implications for global parameterizations. *Geophys Res Let* 33: doi: 10.1029/2006GL026817

[19] Houghton J.T., Meira Filho L.G., Bruce J.P., Lee H., Callander B.A., Haites E.F. (eds) (1995) Climate change 1994: Radiative forcing of climate change and an evaluation of the IPCC IS92 emission scenarios. Cambridge University Press, Cambridge

[20] Jähne B., Münnich K.O., Bösinger R., Dutzi A., Huber W., Libner P. (1987) On parameters influencing air-water gas exchange. *J Geophys Res* 92:1937-1949

[21] Key R.M., Kozyr A., Sabine C.L., Lee K., Wanninkhof R., Bullister J.L., Feely R.A., Millero F.J., Mordy C., Peng T.-H. (2004) A global ocean carbon climatology: Results from the Global Data Analysis Project (GLODAP). Global Biogeochem Cycles 18: doi: 10.1029/2004GB002247

[22] Krakauer N.Y., Randerson J.T., Primau F.W., Gruber N., Denemenlis D. (2006) Carbon isotope evidence for the latitudinal distribution and wind speed dependence of the air-sea gas transfer velocity. *Tellus*, in press

[23] Kromer B., Roether W. (1983) Field measurements of air-sea gas exchange by the radon deficit method during JASIN (1978) and FGGE (1979). Meteor Forsch Ergebnisse A/B 24:55-75

[24] Lee K., Wanninkhof R., Takahashi T., Doney S., Feely R.A. (1998) Low interannual variability in recent oceanic uptake of atmospheric carbon dioxide. *Nature* 396:155-159

[25] Liss P.S., Merlivat L. (1986) Air-sea gas exchange rates: Introduction and synthesis. In: Buat-Menard P. (ed) *The role of air-sea exchange in geochemical cycling*. Reidel, Boston, pp 113-129

[26] McGillis W.R., Wanninkhof R. (2006) Aqueous CO_2 gradients for air-sea flux estimates. *Mar Chem* 98:100-108

[27] McGillis W.R., Edson J.B., Hare J.E., Fairall C.W. (2001) Direct covariance air-sea CO_2 fluxes. *J Geophys Res* 106:16729-16745

[28] McGillis W.R., Edson J.B., Zappa C.J., Ware J.D., McKenna S.P., Terray E.A., Hare J.E., Fairall C.W., Drennan W., Donelan M., DeGrandpre M.D., Wanninkhof R., Feely R.A. (2004) Air-sea CO_2 exchange in the equatorial Pacific. *J Geophys Res* 109: doi: 10.1029/2003JC002256

[29] McNeil C.L., Merlivat L. (1996) The warm oceanic surface layer: Implications for CO_2 fluxes and surface gas measurements. *Geophys Res Let* 23:3575-3578
[30] Monahan E.C., Spillane M.C. (1984) The role of oceanic whitecaps in air-sea gas exchange. In: Brutsaert W., Jirka G.H. (eds), *Gas transfer at water surfaces.* Reidel, Boston, pp 495-503
[31] Naegler T., Ciais P., Rodgers K., Levin I. (2006) Excess radiocarbon constraints on air-sea gas exchange and the uptake of CO_2 by the oceans. *Geophys Res Let* 33:doi:10.1929/2005GL025408
[32] Nightingale P.D., Malin G., Law C.S., Watson A.J., Liss P.S., Liddicoat M.I., Boutin J., Upstill-Goddard R.C. (2000) In situ evaluation of air-sea gas exchange parameterizations using novel conservative and volatile tracers. *Global Biogeochem Cycles* 14:373-387
[33] Park G.-H., Lee K., Wanninkhof R., Feely R.A. (2006) Empirical temperature-based estimates of variability in the oceanic uptake of CO_2 over the past two decades. *J Geophys Res* 111:doi: 10.1029/2005JC003090
[34] Peacock S. (2004) Debate over the ocean bomb radiocarbon sink: Closing the gap. *Global Biogeochem Cycles* 18: doi: 10.1029/2003GB002211
[35] Rubin S.I., Key R.M. (2002) Separating natural and bomb-produced radiocarbon in the ocean: The potential alkalinity method. *Global Biogeochem Cycles* 16: doi: 10.1029/2001GB001432
[36] Smethie W.M., Takahashi T., Chipman D.W., Ledwell J.R. (1985) Gas exchange and CO_2 flux in the tropical Atlantic Ocean determined from ^{222}Rn and pCO_2 measurements. *J Geophys Res* 90:7005-7022
[37] Sweeney C., Gloor M., Jacobson A.J., Key R.M., McKinley G., Sarmiento J.L., Wanninkhof R. (2006) Constraining global air-sea gas exchange for CO_2 with recent bomb ^{14}C measurements. *Global Biogeochem Cycles,* accepted
[38] Takahashi T., Sutherland S.C., Sweeney C., Poisson A., Metzl N., Tilbrook B., Wanninkhof R., Feely R.A., Sabine C., Olafsson J., Nojiri Y. (2002) Global sea-air CO_2 flux based on climatological surface ocean pCO_2 and seasonal biological and temperature effects. *Deep-Sea Res II* 49:1601-1622
[39] Tans P.P., Fung I.Y., Takahashi T. (1990) Observational constraints on the global atmospheric CO_2 budget. *Science* 247:1431-1438
[40] Wanninkhof R. (1992) Relationship between gas exchange and wind speed over the ocean. *J Geophys Res* 97:7373-7381
[41] Wanninkhof R., McGillis W.R. (1999) A cubic relationship between gas transfer and wind speed. *Geophys Res Let* 26:1889-1893
[42] Wanninkhof R., Doney S.C., Takahashi T., McGillis W.R. (2002) The effect of using time-averaged winds on regional air-sea CO_2 fluxes. In: Donelan M., Drennan W., Saltzman E., Wanninkhof R. (eds) *Gas trans-*

fer at water surfaces. AGU, Geophysical Monograph 127, Washington DC, pp 351-357

[43] Wanninkhof R., Sullivan K.F., Top Z. (2004) Air-sea gas transfer in the Southern Ocean. *J Geophys Res* 109: doi: 10.1029/2003JC001767

[44] Woolf D.K. (1997) Bubbles and their role in gas exchange. In: Liss P.S., Duce R.A. (eds) *The sea surface and global change.* Cambridge University Press, Cambridge, pp 173-206

[45] Woolf D.K. (2006) Recent developments in parameterization of air-sea gas exchange. *Flux News (WCRP)*, pp 9-10

2

Turbulent Gas Flux Measurements near the Air-Water Interface in a Grid-Stirred Tank

Herlina and Gerhard H. Jirka

Institute for Hydromechanics
University of Karlsruhe
Kaiserstraße 12
76131 Karlsruhe, Germany

Abstract The gas transfer process across the air-water interface induced by far-field homogeneous turbulence generated in the water phase was investigated experimentally. The measurements were performed in a grid-stirred tank using a combined Particle Image Velocimetry (PIV) - Laser Induced Fluorescence (LIF) (PIV-LIF) technique, which enables simultaneous and spatially synoptic measurements of velocity and gas concentration fields. The techniques allowed visualization of the velocity and concentration fields with good spatial and temporal resolution and thus provided good insight into the gas transfer mechanisms. Detailed quantification of the gas concentration distribution within the thin aqueous boundary layer as well as the near surface hydrodynamics were obtained. With the combined PIV-LIF technique, the turbulent mass flux covariance term $c'w'$ was quantified directly. Comparing the turbulent mass flux with the total mean mass flux determined from the reaeration (bulk) measurements, it could be shown that the contribution of $c'w'$ is indeed significant.

2.1 Introduction

Oxygen is a fundamental parameter for natural water bodies to sustain aquatic life and to take up organic pollutant loadings. The absorption of oxygen into the water body is thus a very important process in order to recover the deficit of dissolved oxygen, especially in polluted rivers. In rivers with weak wind speeds, the dominant driving mechanism for gas transfer is the turbulence generated by bottom-shear. Our study focuses on this type of gas transfer process.

The gas transfer rate is commonly parameterized by the transfer velocity k_l, defined as

$$k_l = \frac{j}{\Delta c} = \frac{j}{C_s - C_b} \tag{2.1}$$

where j is the gas flux across the interface, C_b is the dissolved gas concentration in the bulk region and C_s is the equilibrium gas concentration

C.S. Garbe, R.A. Handler, B. Jähne (eds.): Transport at the Air Sea Interface
pp. 25-41, 2007,

given by Henry's law. Earlier studies have developed conceptual models that relate the transfer velocity k_l to the properties of the dissolved gas and the liquid in order to describe the gas transfer process. The simplest one is the the stagnant film-model $k_l = D/\delta$ [13] and more elaborated ones are the renewal models [5, 10]. Many other researchers attempted to find a practical relation between k_l and measurable flow properties (stream parameters), such as the mean velocity, slope or water depth [e.g. 6, 18].

For gases with low solubility, including oxygen, the transfer across the air-water interface is controlled by resistance on the liquid side. Assuming that the gas transfer process is horizontally homogeneous, the total gas flux across the air-water interface can be written as

$$\bar{j} = -(D\frac{\partial \bar{c}}{\partial z} - \overline{c'w'}) \tag{2.2}$$

where z denotes the vertical direction. The interaction of molecular diffusion (first term on the right hand side) and turbulent transport (second term on the right hand side) governs the process, with the latter being generally much more effective. Comparing (2.2) and (2.1) shows that the turbulent mass transport given by the correlation term $c'w'$ represents part of the transfer velocity in the term k_l. The quantification of k_l alone does not suffice in order to fully understand the gas transfer process. For that purpose, it is necessary to resolve separately in time and space both the molecular diffusive transport ($D\partial c/\partial z$) and turbulent transport ($c'w'$) terms in (2.2). The largest obstacle, which has hampered the study of gas transfer process in such turbulent environments is due to the fact that the aqueous boundary layer, which controls the gas transfer process, is only tens to hundreds μm thin. With the aid of optical measurement techniques, detailed near surface concentration measurements were made possible [e.g. 19, 26, 28, 29]. Other researchers conducted numerical simulations in order to understand the gas transfer process such as Magnaudet and Calmet [14], Pan and Banerjee [21]. However, the turbulent mass transport term $\overline{c'w'}$, which combines the information from the fluctuating turbulent part of velocity and concentration, is up to now typically not known and has to be modelled. Direct quantification of $c'w'$ is difficult, but still possible through simultaneous measurements of the gas concentration fluctuations and velocity fluctuations (known as the eddy correlation method).

The purpose of this study is to investigate the gas transfer process through detailed laboratory experiments with advanced non-intrusive optical measurement techniques that enable simultaneous and spatially synoptic measurements with high resolution of two-dimensional velocity and concentration fields, allowing detailed mapping of the concentration and velocity fields and direct quantification of the turbulent mass flux $c'w'$.

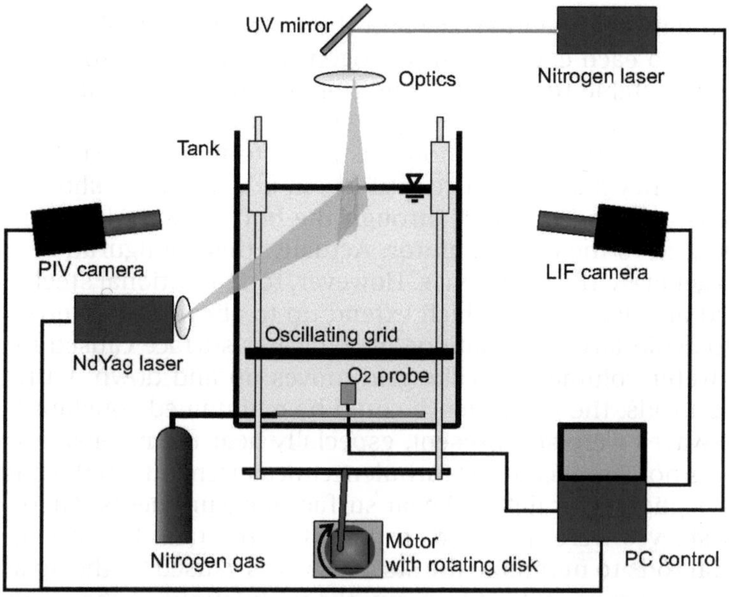

Figure 2.1. Schematic illustration of the experimental setup showing the oscillating grid system and the optical setup for the simultaneous PIV-LIF measurement techniques.

2.2 Experimental Set-Up

2.2.1 Tank

The experiments were performed in a tank made of perspex with an inner dimension of $50 \times 50 \times 65$ cm^3. The tank was equipped with a vertically oscillating grid near the bottom (Figure 2.1). In such a system, the turbulence generated by the grid decays as it diffuses towards the water surface. Despite the differences of turbulence source, the interaction between the interface and the turbulence has similar characteristics as that occurring at the water surface in rivers. Earlier studies [3, 4] showed that the grid-stirred tank is suitable for investigation in the area of gas transfer. The characteristics of grid-stirred turbulence have been studied in depth by Hopfinger and Toly [11], Thompson and Turner [25] and recently also by Matsunaga et al. [15] and McKenna and McGillis [16].

The oscillating grid consisted of an aluminium plate perforated to form a 7×7 bar grid, with a centre to centre mesh size M of 62.5 mm. The hole size was 50×50 mm^2, resulting in a mesh solidity of 36 %, which is optimal to avoid secondary motions and inhomogeneity [11].

The interfacial gas transfer process in all series was induced by lowering the dissolved oxygen concentration so that a concentration gradient is forced between the interface and the water in the bulk region. Therefore, prior to each experiment, nitrogen was bubbled into the test tank. To ensure a stable turbulence condition, the measurements commenced at least 20-30 minutes after the grid started to oscillate.

In addition, special attention was given to minimize surface undulations since they may introduce artifacts in w'. Figure 2.1 shows that four steel rods entering the tank through the bottom were installed to connect the grid to the driving motor. Actually, this configuration was sufficient to generate the turbulence. However, four additional steel rods were mounted above the grid, which extend up to above the water surface, in order to avoid large fluctuations of the water surface caused by the displaced water volume when the grid moves up and down. With the four extension rods, the water depth could be maintained constant but small surface waves were still present, especially near the extension rods. The effect was not significant for turbulence measurements in the bulk region but not negligible for detailed near surface measurements. Therefore, stationary sleeves were mounted to the lid of the tank to cover the upper extension rods to minimize the surface waves caused by the up and down motion of the rods. With this setup, the undulations at the water surface were typically smaller than 30 μm so that the water surface can be considered as flat at least in the present range of grid frequencies and stroke. The water depth varied between different experiments from 480 to 490 mm. For the analysis, we chose the following coordinate system. The horizontal axis x indicates the distance from the wall, whereas the vertical axis z indicates the depth from the water surface.

2.2.2 Simultaneous PIV-LIF

Two optical measurement techniques, Particle Image Velocimetry (PIV) and Laser Induced Fluorescence (LIF), were employed to measure planar velocity and concentration fields near the interface, respectively. The LIF technique based on the oxygen quenching phenomenon that was developed by Vaughan and Weber [27] was chosen to measure the gas concentration. This method does not involve any chemical reactions and is independent of external parameters like pH value.

Both the PIV and LIF techniques deliver instantaneous field information with high spatial resolution that is an advantage for elucidating the profiles near the boundary with its limited thickness. Moreover, these techniques can be coupled providing a simultaneous measurement of velocity and concentration which allows direct quantification of the turbulent mass flux ($c'w'$ in (2.2)).

The PIV-LIF system is schematically illustrated in Figure 2.1. Polyamid particles with nominal diameter of 10 μm were used as PIV tracer and a pyrenebutyric acid (PBA) concentration of $2 \cdot 10^{-5}$ Molar was used as LIF tracer. The area of interest was illuminated with two different lasers, one for PIV and the other for LIF. The PIV laser was a dual-cavity Q-switched Nd:YAG laser which emits pulse-pairs with an energy level of 25 mJ per pulse at a wavelength of 532 nm (green). The LIF laser was a pulsed nitrogen laser (MNL 801) with a mean energy power of 0.4 mJ. The laser emission was at 337.1 nm which was ideal since the absorption peak of PBA is at 340 nm [27]. Both laser beams were expanded into a light sheet which were then guided into the centre of the test tank. Two SensiCam charge coupled device (CCD) cameras, which have a resolution of 1280×1240 pixels and a grey scale depth of 12 bits, were employed to capture the illuminated flow fields. The PIV camera was positioned at the opposite side of the LIF camera. The measurement window of the PIV camera was approximately 20×20 mm^2 which covered the 10×10 mm^2 area of the LIF camera. The recording rate was 4 Hz (double frames). The PBA fluorescence lies between 370 - 410 nm and the scattered light from the PIV laser was at 532nm. An optical bandpass filter was mounted in front of each camera to ensure that only the fluorescent light and scattered NdYag Laser light could reach the camera chips from the LIF and PIV cameras, respectively. The synchronization of the cameras and lasers through a programmable timing unit (PTU) board was managed using the DaVis software developed by LaVision. DaVis also provides the algorithm for computing the vector fields. In order to acquire quantitative concentration information from the recorded LIF images, several image processing steps must be performed. These include noise removal, water surface detection, correction of laser attenuation and correction of optical blurring near the interface. We have described the image processing procedure in greater detail in Herlina and Jirka [8].

2.3 Results

2.3.1 Grid-Stirred Turbulence

Before discussing the gas transfer measurements near the interface region, the turbulence characteristic produced in the present grid-stirred tank is described.

The turbulence intensities of the system are characterized using the turbulent Reynolds number Re_T [3], defined by

$$Re_T = \frac{u'_\infty 2L_\infty}{\nu} \tag{2.3}$$

with u'_∞ is the far-field turbulence intensity, L_∞ the longitudinal integral length scale and ν is the kinematic viscosity. The term u'_∞ is estimated by the relation $u' = 0.25 f S^{1.5} M^{0.5} z_s^{-1}$, with f being the frequency of the oscillation, S the stroke and z_s the distance from the centre of the stroke towards the water surface [11]. The length scale can be approximated by

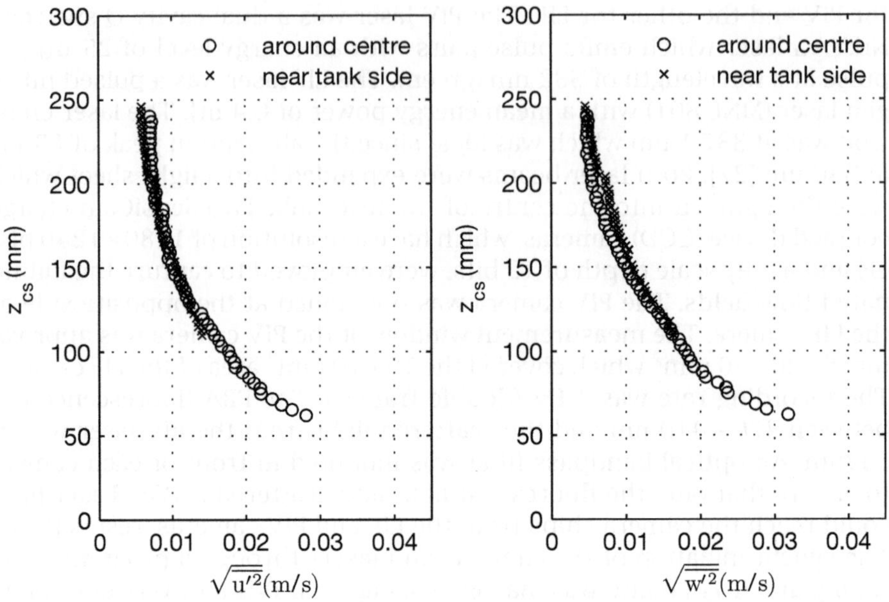

(a) horizontal component (b) vertical component

Figure 2.2. Temporally and spatially averaged turbulence velocities at the centre and near the side wall of the tank, with $Re_T = 330$.

$L_\infty = 0.1z_s$ [25]. The stroke of the grid was 5 cm for all experiments. The different levels of turbulence intensities were controlled by varying the frequency of the oscillating grid from 2 to 6 Hz which corresponded to Re_T from 260 to 780.

The grid-stirred tank system was chosen because it is characterized by zero-mean velocity and homogeneous turbulence. However, as pointed out by McKenna and McGillis [16], the mean flows are characteristics of grid-stirred tanks. It is extremely difficult to avoid secondary circulation flows in grid-stirred tank systems. The estimated mean flow in our tank is around 0.001 m/s - 0.004 m/s in the present range of turbulent Reynolds numbers. To examine the horizontal homogeneity in the tank, the temporally and then spatially averaged $(\overline{u'^2})^{1/2}$ and $(\overline{w'^2})^{1/2}$ profiles obtained from additional PIV experiments in the bulk region performed at the centre and near the side wall of the tank are plotted in Figure 2.2. The figures show that the distribution of the turbulence fluctuations from both experiments are in very good agreement. This ensures that the degree of spatial non-uniformity in the present tank is minimal.

Asher and Pankow [1] as well as McKenna and McGillis [17] have shown that grid-stirred tank systems are very sensitive to the presence of surface

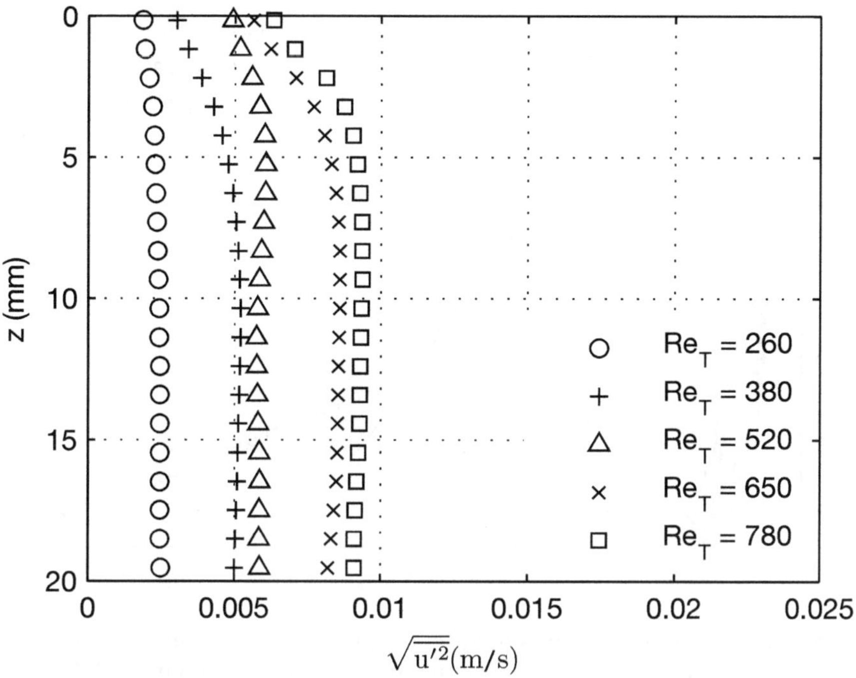

Figure 2.3. Horizontal turbulence fluctuations near the interface

films. The two extreme boundary conditions at the interface are specified by either zero velocity fluctuations (solid interface) or zero shear-stress (mobile interface). A solid interface represents a dirty surface whereas a mobile interface represents a clean surface. At a perfectly clean surface, only the vertical velocity fluctuations have to be zero while the horizontal components may be non-zero. In this study, the surface was cleaned from dust particles using a suction device prior to each experiment. Despite the careful precaution taken, it is impossible to maintain a perfectly clean surface in our laboratory. However, the non-zero horizontal velocity fluctuations at the interface, as shown in Figure 2.3, indicate that the surface in the present experiments is relatively far from the extreme case of solid interface and can be categorized as a mobile interface.

Although the source of turbulence is different, it is still interesting to make a rough comparison between the turbulence intensity level generated by the grid-stirred tank with the one occurring in open-channel or stream flows. For that purpose, the turbulent kinetic energy seems to be the most appropriate measure for the energy produced by the oscillating grid since no mean shear is present in the grid-stirred tank system. The turbulent kinetic energy for the present measurements (k =

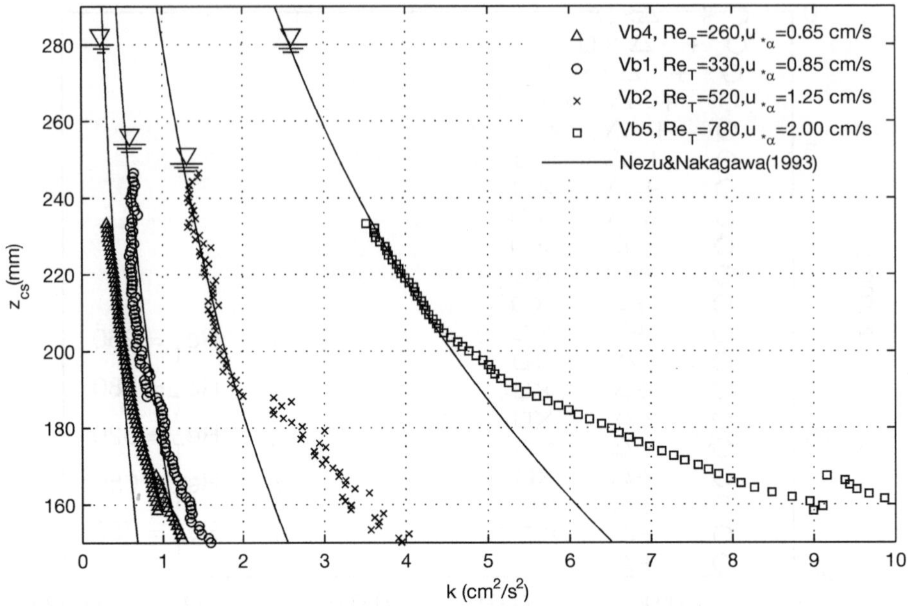

Figure 2.4. Measured turbulent kinetic energy k for four turbulence conditions in grid-stirred tank. The solid lines describe turbulence in open channel flow Nezu and Nakagawa [20].

$0.5\,[2u'^2 + w'^2]$) is plotted in Figure 2.4. The horizontal and vertical turbulence fluctuations (u' and w') were obtained from additional PIV measurements conducted in the bulk region, with the actual purpose of evaluating the grid-stirred tank system. The solid lines in the figure are the relation following Nezu and Nakagawa [20] for open channel flows. By allowing the comparison between the magnitude of the friction velocity $u_{\star\alpha}$ generated in the tank with $u_\star$ occurring in streams or open-channel flows, the turbulence levels produced in the tank (Re_T = 260 to 780) correspond to stream flows with $u_\star$ ranging approximately between 0.65 to 2 cm/s. Although this is just a crude comparison, it may help to give a sense of how strong the turbulence levels generated in the grid-stirred tank are. Of course, higher values than 2 cm/s may occur in nature. However, natural streams with this range of $u_\star$ do exist. For example, the Eaton Nord River in Canada has a typical $u_\star$ value ranging from 0.5 to 1.3 cm/s [22]. The levels of turbulence generated in this study spanned both low and high turbulence flow regimes suggested by Theofanus [24] so that a sufficiently wide spectrum of turbulent eddies could be investigated.

2.3.2 Velocity and Concentration Fields

Figure 2.5 shows a flow sequence of 12 instantaneous velocity and concentration fields for a near-interface image area of 8×12 mm. The light and dark colors indicate the regions with low and high dissolved oxygen concentration, respectively. In all images, a very thin dark layer, immediately below the interface, indicating the boundary layer in which the oxygen concentration decreases rapidly from saturated to the bulk concentration can be observed. Below this boundary layer, the images are dominated with light color corresponding to the low oxygen concentration in the bulk region that is constantly being mixed by the turbulence generated by the oscillating grid.

The transport of fluid with higher oxygen concentration into the bulk region could be visualized. The presence of turbulence obviously introduces constant changes to the boundary layer thickness and as well aid the oxygen absorption process into the water body. An earlier step of this study has been published in Herlina and Jirka [8]. In that paper, detailed discussion on the qualitative observations of the LIF sequences were presented, including the discussion on the eddy sizes initiating the transport processes.

It should be noted that the sequences shown above clearly demonstrate the logical correlations of the flow-field with the concentration distribution. Thus, it visually convinces and proves that the data acquisitions of the velocity and concentration fields were indeed simultaneous both in time and space.

2.3.3 Mean and Fluctuation Profiles

For better interpretation, it is convenient to present the concentration profiles in a normalized form. Generally, the concentration is presented as $(\overline{c} - C_b)/(C_s - C_b)$ and the depth is scaled with the boundary layer thickness δ_e. Here, δ_e is defined as the depth where the value of $(\overline{c} - C_b)/(C_s - C_b)$ is equal to $1/e$. The mean boundary layer thickness measured in this study decreased from 800 μm at $Re_T = 260$ to 300 μm at $Re_T = 780$ as shown in Figure 2.6.

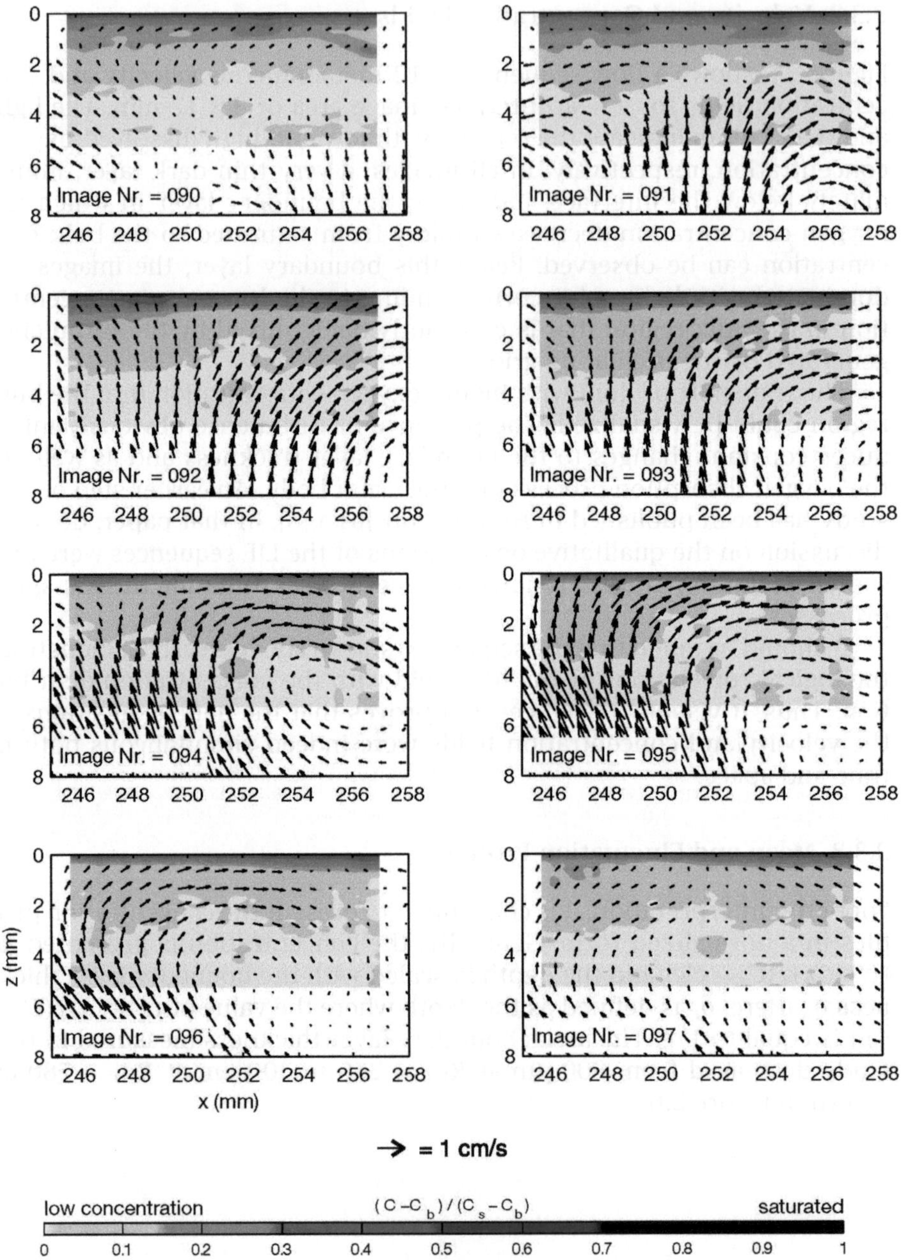

Figure 2.5. A sequence of oxygen concentration and velocity maps. Time interval between shown images is 0.25 s and $z = 0$ denotes the water surface.

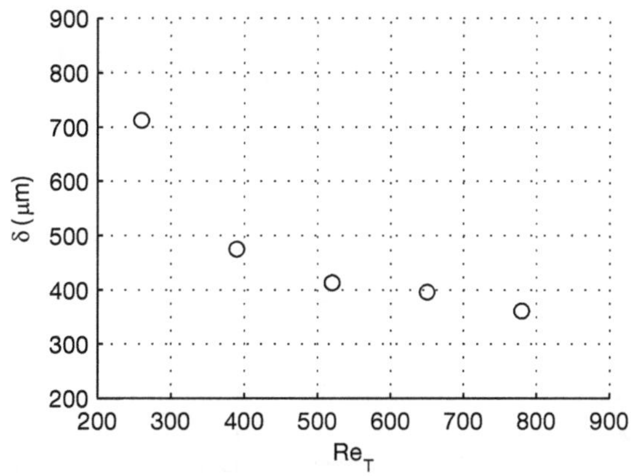

Figure 2.6. Variation of boundary layer thickness with different Re_T.

The normalized mean concentration profiles are shown in Figure 2.7a. The solid and dashed lines included in the figure are the solution of the mass flux equation for mobile interface based on the surface renewal model assuming the large scale are dominant and a small eddy, multistage model, respectively [12]. The measured profiles fall between these two theoretical profiles. From a practical point of view, the results can be interpreted that the first model holds better for low Re_T, whereas the small eddy, multistage model appears preferable for high Re_T. This supports the idea of the two-flow regime model suggested by Theofanus [24].

Also shown in the figure are the data from Chu and Jirka [4] with $Re_T = 360$, who performed concentration measurements using polarographic oxygen microprobe. The figure obviously highlights the advantage of the LIF quenching method over their invasive point-based measurement technique which was probably limited near the interface. The profile obtained by Chu and Jirka [4] agreed relatively well with the surface renewal model. It should be noted that their Re_T measurement range was mostly below 500 which was classified as low turbulent intensity by Theofanus [24].

The variation of the concentration fluctuation (rms values) with the depth measured in this study is shown in Figure 2.7b, with $(\overline{c'^2})^{1/2}$ is normalized with $C_s - C_b$ and the depth with δ_e. Figure 2.7b shows that the fluctuations increase from smaller values near the interface to a maximum around the boundary layer thickness ($z/\delta_e = 1$). The maximum peaks in the range between 0.15 and 0.2. After reaching a maximum, the fluctuation values decrease with further submergence. Theoretically, the $(\overline{c'^2})^{1/2}$ values in the bulk region should be equal to zero. The margins of error for an instantaneous profile was estimated to be ±5% [7]. For the mean concentration profiles, it is expected that the noise levels contained in the instantaneous profiles would cancel out each other. In contrast, the

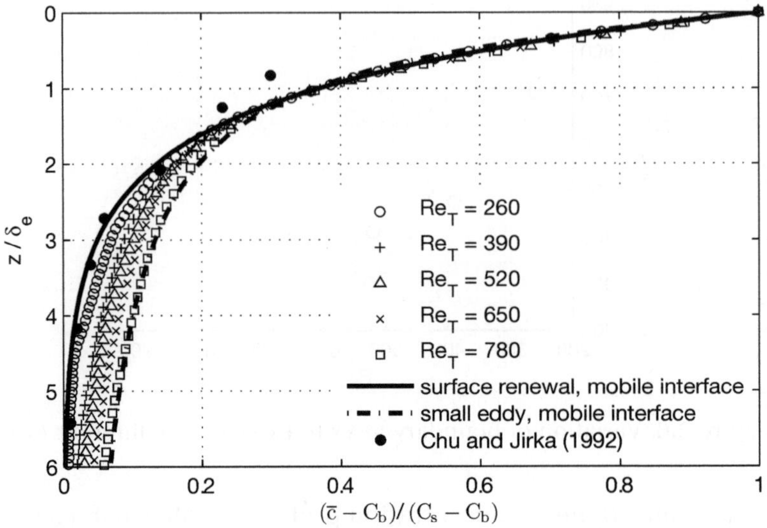

(a) mean

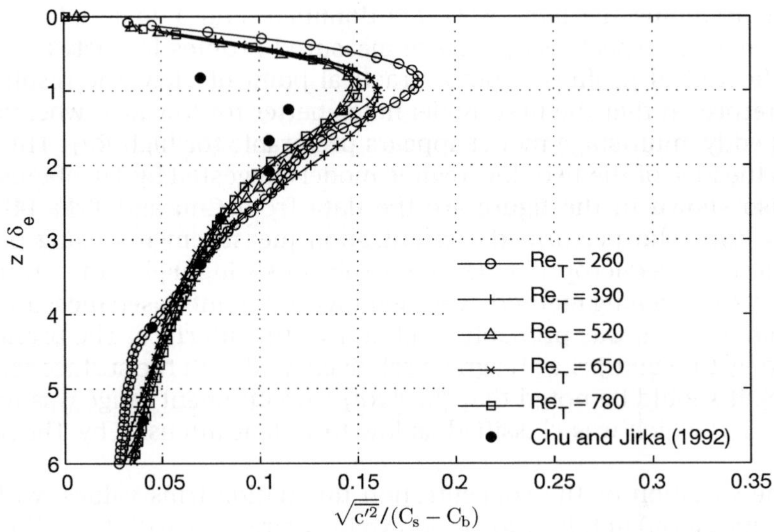

(b) fluctuation

Figure 2.7. Mean and fluctuation concentration profiles for five turbulence conditions. For clarity, only every seventh data point is shown. The figure highlights the advantage of the present LIF technique over the invasive point-based measurement technique used by Chu and Jirka [4] which was limited near the interface.

Table 2.1. Total mean flux values determined from the bulk measurements($\overline{j} = k_l(C_s - C_b)$). The k_l values are at 20°C reference temperature

Exp. Serie	Re_T	$k_{l,t}$ (cm/s)	C_s (mg/l)	C_b (mg/l)	$\overline{j}$ (mg/l · cm/s)
CV1	260	$4.16 \cdot 10^{-4}$	10.12	1.02	$33.8 \cdot 10^{-4}$
CV2	390	$5.58 \cdot 10^{-4}$	10.10	0.97	$45.9 \cdot 10^{-4}$
CV3	520	$6.82 \cdot 10^{-4}$	10.09	0.98	$54.2 \cdot 10^{-4}$
CV4	650	$7.93 \cdot 10^{-4}$	9.96	0.77	$64.7 \cdot 10^{-4}$
CV5	780	$9.50 \cdot 10^{-4}$	9.90	0.89	$75.7 \cdot 10^{-4}$

noise level in the c' is maintained. Therefore, the non-zero value (2-5%) in the bulk region is due to the noise level present in the instantaneous concentration profiles. Similar to the mean profiles, the data points from Chu and Jirka [4] show discrepancies with the present data in the near interface region. This is most probably due to their invasive measurement techniques.

2.3.4 Transfer Velocity k_l and Total Mean Flux

In addition to the gas transfer measurements directly at the interface, long time oxygen concentration measurements in the bulk region were also performed. From this reaeration experiments, the transfer velocity k_l could be estimated, the obtained values are summarized in Table 2.1. As expected, the transfer velocities increase with higher turbulent Reynolds numbers.

It has been long recognized that gas transfer processes are sensitive to the presence of surfactants. Recent study by McKenna and McGillis [17] showed that the presence of surfactants may reduce the transfer velocities significantly compared to the cases with cleaned surface conditions. The k_l values measured in this study are lower than the values measured by McKenna and McGillis [17] with cleaned surface conditions. This indicates that the surface conditions in the present study were not perfectly clean as already pointed out in Section 2.3.1. The surface pressures in the present study were not measured so that the degrees of cleanliness could not be quantified.

It should be noted that researchers have found that the surface divergence provides a relationship for the gas transfer that takes into account the effect of surface conditions [2, 17, 23]. Unfortunately, direct comparison with their data is difficult since their measurements focused on the surface divergence (horizontal velocity fields) whereas our study focuses on elucidating the vertical plane.

Based on (2.1), the total mean fluxes $\overline{j}$ can be calculated using the k_l values obtained from these bulk measurements ($\overline{j} = k_l(C_s - C_b)$). The values of C_s and C_b as well as the computed $\overline{j}$ are listed in Table 2.1.

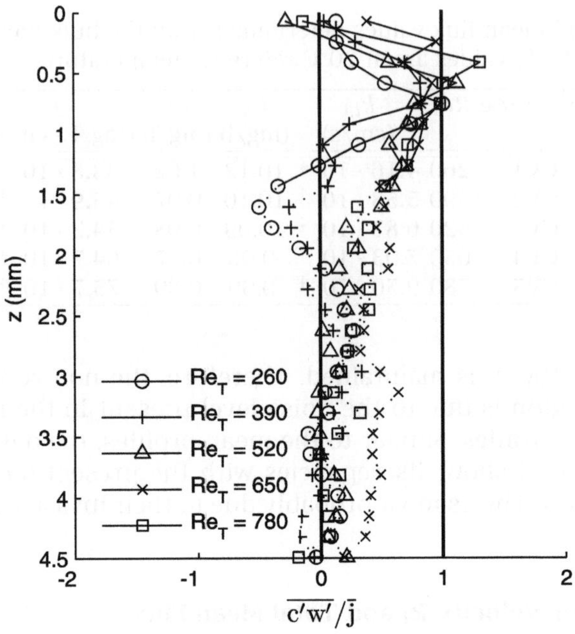

Figure 2.8. Variation of measured normalized turbulent mass flux. Values below 1.5 mm (dashed lines) are unreliable due to insufficient LIF accuracy as far as the eddy-correlation method is concerned.

In the next section, these $\overline{j}$ values are used to normalize the turbulent mass flux values ($c'w'$) obtained from the direct eddy-covariance method (simultaneous PIV-LIF measurements).

2.3.5 Turbulent Mass Flux

The multiplication of the velocity fluctuations with its coinciding concentration fluctuations (see Figure 2.5) gives the instantaneous turbulent mass flux $c'w'$ quantities. The variation of the measured mean turbulent mass flux $\overline{c'w'}$ over the depth is presented in Figure 2.8, with $\overline{c'w'}$ normalized by the total mean flux $\overline{j}$ determined from the reaeration (bulk) measurements (see Table 2.1).

The figure shows that the mean turbulent mass fluxes increase from around 0 at the interface to about the total mass flux values within approximately 0.5 to 0.8 mm, which is typically twice the boundary layer thickness ($2\delta_e$). This result is in line with the fact that near the interface any turbulent transport vanishes and the transfer should be dominated by molecular diffusion. As it goes deeper into the bulk, the turbulent transport becomes more effective and dominates the transfer process. Below $z = 1.5$ mm, the normalized turbulent fluxes tend to go to zero, which of course should not be the case since pure turbulent flux should occur in that region. Figure 2.7b shows that the mean concentration fluctuations tend to reach zero in the deeper bulk region. Contrary, the velocity fluctuations would of course reach large values as they get closer to the turbulence source. The multiplication of c' and w', however, should still results in something finite. Due to the decay of the laser light intensity as it travels through the water column, the LIF accuracy becomes more and more obscure with the depth (estimated accuracy $\pm 5\%$). This leads, unfortunately, to the fact that the detection of minute fluctuation in the bulk region is impossible and thus the LIF accuracy is insufficient in this deeper bulk region, as far as the eddy-correlation method is concerned.

The results from the simultaneous PIV-LIF measurements allow direct quantification of the total mass flux across the interface through summation of the molecular diffusive transport term $D\partial\overline{c}/\partial z$ and the turbulent mass transport term $\overline{c'w'}$ (2.2). The term $D\partial\overline{c}/\partial z$ can be easily computed from the vertical concentration profile $\overline{c}(z)$, whereas the term $\overline{c'w'}$ is the measured value presented in Figure 2.8. A deeper discussion on this is presented in Herlina and Jirka [9].

Furthermore, Figure 2.8 also shows that the $\overline{c'w'}$ profiles are of the same order as the total mean flux determined from the bulk measurements (i.e. $\overline{c'w'}/\overline{j} \approx 1$). This indicates that the contribution of the turbulent mass flux to the total mass flux is significant and should be taken into account when performing numerical simulations.

2.4 Conclusion

The gas transfer across the air-water interface induced by oscillating grid turbulence was investigated using a combined PIV-LIF technique. The turbulent mass flux could be quantified directly and it could be shown that the turbulent flux contribution increases from around 0 at the interface to about the total flux as it goes deeper into the bulk. These new results should be very useful for refining numerical models and developing more accurate models for the prediction of the transfer velocity.

Acknowledgement. The authors wish to thank the German Science Foundation (Deutsche Forschungsgemeinschaft) for the financial support through Grants No. DFG Ji 18/7 and Ji 18/13.

References

[1] W. E. Asher and J. Pankow. The interaction of mechanically generated turbulence and interfacial films with a liquid phase controlled

gas/liquid transport process. *Tellus: a bi-monthly journal of geographics*, 38B:305–318, 1986.

[2] S. Banerjee, D. Lakehal, and M. Fulgosi. Surface divergence model for scalar exchange between turbulent streams. *Intl J. Multiphase Flow*, 30:963–977, 2004.

[3] B. H. Brumley and G. H. Jirka. Near-surface turbulence in a grid-stirred tank. *J. Fluid Mech.*, (183):236–263, 1987.

[4] C. R. Chu and G. H. Jirka. Turbulent gas flux measurements below the air-water interface of a grid-stirred tank. *Intl J. Heat Mass Transfer*, 35(8):1957–1968, 1992.

[5] P. V. Danckwerts. Significance of liquid-film coefficients in gas absorption. *Indl Engng Chem.*, 43:1460–1467, 1951.

[6] J. S. Gulliver and M. J. Halverson. Air-water gas transfer in open channels. *Water Resources Research*, 25(8):1783–1793, 1989.

[7] Herlina. Gas transfer at the air-water interface in a turbulent flow environment. *Doctoral Thesis, Inst.Hydromechanics, Univ.Karlsruhe, University Press*, ISBN 3-937300-74-0, 2005.

[8] Herlina and G. H. Jirka. Application of LIF to investigate gas transfer near the air-water interface in a grid-stirred tank. *Exps Fluids*, 37(3), 341–349, 2004.

[9] Herlina and G. H. Jirka. Experiments on gas transfer at the air-water interface induced by oscillating grid turbulence. *submitted to J. Fluid Mech.*, 2006.

[10] R. Higbie. The rate of absorption of a pure gas into a still liquid during short periods of exposure. *AIChE Transactions*, 31:365–390, 1935.

[11] E.J. Hopfinger and J.A. Toly. Spatially decaying turbulence and its relation to mixing across density interfaces. *J. Fluid Mech.*, 78:155–175, 1976.

[12] B. Jähne and H. Haussecker. Air-water gas exchange. *Ann. Rev. Fluid Mech.*, 30:443–468, 1998.

[13] W. K. Lewis and W.G. Whitman. Principles of gas absorption. *Indl Engng Chem.*, 16, 1924.

[14] Jacques Magnaudet and Isabelle Calmet. Turbulent mass transfer through a flat shear-free surface. *J. Fluid Mech.*, 553:155–185, 2006.

[15] N. Matsunaga, Y. Sugihara, T. Komatsu, and A. Masuda. Quantitative properties of oscillating-grid turbulence in a homogeneous fluid. *J. Fluid Dyn. Res.*, 25:147–165, 1999.

[16] S. P. McKenna and W.R. McGillis. Observations of flow repeatibility and secondary circulation in oscillating grid-stirred tank. *Phys. Fluids*, 16(9):3499–3502, 2004.

[17] S. P. McKenna and W.R. McGillis. The role of free-surface turbulence and surfactants in air-water gas transfer. *Int.J.Heat Mass Transfer*, 47:539–553, 2004.

[18] D. B. Moog and G. H. Jirka. Air-water gas transfer in uniform flows with large gravel-bed roughness. *Geophysical Monograph 127, Gas Transfer at Water Surfaces*, pages 371–376, 2002.

[19] T. Münsterer, H.J. Mayer, and B. Jähne. Dual-tracer measurements of concentration profiles in the aqueous mass boundary layer. *Air-Water Gas Transfer : 3rd International Symposium on Air-Water Gas Transfer*, pages 637–648, 1995.

[20] I. Nezu and H. Nakagawa. *Turbulence in Open Channel Flows.* Balkema, Rotterdam, 1993.

[21] Y. Pan and S. Banerjee. A numerical study of free surface turbulence in channel flow. *Phys. Fluids*, 7:1649–1664, 1995.

[22] A.G. Roy, T. Buffin-Bélanger, H. Lamarre, and A.D. Kirkbride. Size,shape and dynamics of large-scale turbulent flow structures in a gravel-bed river. *J. Fluid Mech.*, 500:1–27, 2004.

[23] Y. Sugihara and H. Tsumori. Surface-renewal eddies at the air-water interface in oscillating-grid turbulence. *Environmental Hydraulics and Sustainable Water Management-Lee and Lam(eds)*, pages 199–205, 2005.

[24] T. G. Theofanus. Conceptual models of gas exchange. *Gas transfer at air-water surfaces. Reidel.*, pages 271–281, 1984.

[25] S. M. Thompson and J. S. Turner. Mixing across an interface due to turbulence generated by an oscillating grid. *J. Fluid Mech.*, 67:349–368, 1975.

[26] E. Variano and E. A. Cowen. Quantitative imaging of CO_2 transfer at an unsheared free surface. In C.S. Garbe, R.A. Handler, and B. Jähne, editors, *Transport at the Air Sea Interface - Measurements, Models and Parameterizations.* Chapter 3, pages 43–57, Springer Verlag, 2007. This volume.

[27] W. M. Vaughan and G. Weber. Oxygen quenching of pyrenebutyric acid fluorescence in water, a dynamic probe of the microenvironment. *Biochemistry*, 9:464–473, 1970.

[28] L. M. Wolff, Z. C. Liu, and T. J. Hanratty. A fluorescence technique to measure concentration gradients near an interface. *Proceedings 2nd International Symposium on Gas Transfer at Water Surfaces*, pages 210–218, 1990.

[29] P. T. Woodrow and S. R. Duke. Laser-induced fluorescence studies of oxygen transfer across unsheared flat and wavy air-water interfaces. *Indl Engng Chem. Res.*, 40:1985–1995, 2001.

3

Quantitative Imaging of CO_2 Transfer at an Unsheared Free Surface

Evan A. Variano and Edwin A. Cowen

DeFrees Hydraulics Laboratory,
School of Civil and Environmental Engineering,
Hollister Hall, Cornell University, Ithaca, NY 14853, USA
{ev42,eac20}@cornell.edu

Abstract We present laboratory measurements of simultaneous velocity and concentration fields for the transfer of CO_2 across a free surface. The interface is subject to the effects of free shear turbulence generated far beneath the surface, exhibiting low mean flow and excellent homogeneity. From measurements of the spatio-temporal mass flux we examine coherent structures below the free surface, as well as one-point statistics to better understand the fundamental physics of turbulent transport at a free surface in the absence of mean shear. We observe surface penetration events caused by bulk fluid impacting the interface from below, as well as downwelling events in which the near-surface fluid is injected into the bulk in narrow filaments. Both types of events contribute to the turbulent mass flux, and we measure that downwelling events are responsible for at least as much mass transfer as the upwellings on which existing models are based. Our measurements indicate that the dominant length and time scales are different for upwellings and downwellings; the quantification of these will be important to modeling efforts.

3.1 Introduction

Gas transfer at a turbulent free surface without mean shear serves as a base case for understanding the physics that operate in more complicated gas transfer situations such as wind shear and waves. The past century has seen a steady progression of measurement techniques, which have been used to verify and motivate an evolving series of gas transfer models. Most models are intended to provide a spatio-temporal average interfacial transfer rate, and do so based on a simplified model of the instantaneous turbulent mass transfer phenomena. Key examples are the models based on renewal (more properly penetration) events [3, 8], surface divergence [4, 12], or idealized eddies of a specific size interacting with the free surface [11, 14]. Each of these models includes assumptions about what part of the physics dominates transfer, based on physical reasoning

C.S. Garbe, R.A. Handler, B. Jähne (eds.): Transport at the Air Sea Interface
pp. 43-57, 2007,

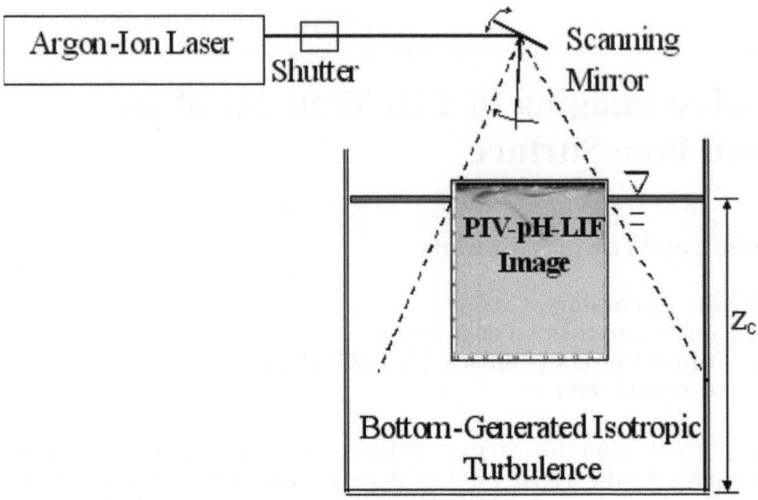

Figure 3.1. Experimental setup. Water depth z_c=34 cm in the small tank (width 15.4 cm) and 80 cm in the large tank (width 80 cm). In the large tank, the laser light sheet is delivered from below the tank

and measurements. Here we employ a quantitative imaging technique to investigate the basic physics of gas transfer, to aid in the continued development of effective models. This technique extends the work of Asher and Pankow [1], and is similar to measurements taken with a different technique by Herlina and Jirka [13].

3.2 Experimental Setup

3.2.1 Experimental Setup in Brief

We perform a series of laboratory experiments in which pure CO_2 gas fills the headspace above a tank of distilled water with a low concentration of dissolved CO_2. The CO_2 gas is added to the headspace at a flowrate set to maintain a pure CO_2 layer extending from the water surface to the top of the tank, where excess CO_2 overflows out of the facility. Both the air-side and water-side of the interface are turbulent, and the invasion of CO_2 into the water-side is measured via a pH-sensitive fluorescent dye. Turbulence is generated on the air-side by a single round jet which supplies the CO_2 gas, and on the water-side by a novel method (see Sect. 3.2.2) designed to minimize mean flows. This latter feature is important, as mean flows can have a large effect on gas transfer measurements by steadily drawing the surface layer into the bulk [22].

We take simultaneous measurements of the 2D velocity and concentration fields in a plane orthogonal to and intersecting the free surface (see Fig. 3.1). This is done in a non-invasive manner via high-resolution digital imaging of an area illuminated with a planar laser light sheet (Innova-90 Argon Ion, Coherent Inc.) with a wavelength of 488 nm. Velocity fields are determined from particle image velocimetry, PIV, in which we correlate the position of effectively passive (Stokes number < 0.05) tracer particles in successive images. Concentration fields are from laser induced fluorescence, LIF, in which a fluorescent dye, well-mixed in the tank, fluoresces in proportion to the local CO_2 concentration [2]. Concentration is denoted herein by square brackets, e.g. [CO_2]).

Data are reported here from two different facilities, both at the DeFrees Hydraulics Laboratory [20, 22]. The "small tank" is 15.4 cm $\times$ 15.4 cm with 34 cm water depth. The turbulent Reynolds number is ≈ 120, and the Taylor microscale Reynolds number $R_\lambda \approx 40$ [20]. The "large tank" is 80 cm $\times$ 80 cm with 80 cm water depth. The turbulent Reynolds number is 3220, and the Taylor microscale Reynolds number $R_\lambda = 314$ [22].

3.2.2 Turbulence Generation

Turbulence is driven in these experiments by an array of randomly actuated synthetic jets at the tank bottom [20, 22]. These synthetic (zero net mass flux) jets introduce momentum and strong shear at the bottom of the tank in a spatio-temporally random manner. With increasing distance from the jets, the driven flow evolves into turbulence that is homogeneous and isotropic in the horizontal direction, and turbulent kinetic energy decays in the vertical direction. We measure this turbulence near the free surface, where the kinematic constraint causes a strong anisotropy in which vertical fluctuations are redistributed to the horizontal directions. We prefer this turbulent forcing method to the grid-stirred tank traditionally used for studies of unsheared free surfaces [6, 9] because it exhibits lower mean flows as well as better homogeneity in the surface-parallel planes [22].

3.2.3 Water Chemistry

Distilled water (0.25 MΩ cm) is stripped of CO_2 by bubbling with Helium through a diffuser stone. NaCl is added to the water at 0.05 molar, which will provide a known and effectively constant ionic strength, which will be of use when calculating bulk [CO_2] from pH electrode readings. The fluorescent dye, *2',7' dichlorofluorescein* (Acros Organics, distributed by Fisher Scientific), is added to 0.04 μmolar and mixed throughout the tank to allow the LIF technique. Hollow glass spheres (Sphericel, Potters Industries #110P8) with mean diameter of 11 μm are placed in solution and allowed to sit for > 2 hours, after which only those which are neutrally buoyant are siphoned off and added to the tank at a concentration < 1 ppm so as not to alter the flow. This also serves to rinse manufacturing-related chemicals from the particles, which McKenna and McGillis show can be a significant source of surfactants [16].

Any surface contamination, especially a surfactant layer, can radically change the gas transfer physics. To keep the surface as clean as possible, we skim the water surface immediately before the experiment. This is achieved in the small tank by adding water to the tank bottom while an overflow drain at the free surface level allows the surface layer, and any contaminant on it, to leave the tank. The large tank features a sharp-crested skim drain with adjustable height and a filter system that returns the purged surface water to the tank. During skimming, surface cleanliness is qualitatively monitored via digital imaging – a contaminated free surface appears as a bright line regardless of camera angle, and this line is removed by skimming or when upwelling fluid pushes surface contaminants aside. Skimming is turned off for the CO_2 invasion experiments, during which surfactants will slowly leach from the submerged equipment and materials. As a result, we keep our experiment short (< 15 minutes) to avoid significant changes in surfactant concentration.

3.2.4 Imaging and Illumination

PIV and LIF images are captured simultaneously by a single camera, increasing the accuracy of spatial coordination. Image size is chosen to balance fine resolution (allowing us to resolve the concentration and momentum boundary layers, which exhibit sharp gradients near the surface) with enough spatial coverage to include some of the bulk flow region. We achieve this with image areas ~ 2 cm and pixel resolutions of 25 μm in the small tank and 67 μm in the large tank.

Illumination is from a continuous Argon Ion laser running in a light-limited mode. The beam is turned into a planar light sheet via a scanning mirror (Cambridge Technologies, Cambridge MA, galvanometer model 6650). Image pairs capture successive same-direction sweeps from the mirror's sawtooth pattern. An electronic shutter (NM Laser Products, Sunnyvale CA, model LS200FNC) upstream of the scanning mirror lets the beam pass (thereby exposing the image) only during the linear portion of the mirror's scan. This exposure time is 3 ms for the small tank, with 7.92 ms between image pairs (measured from the start of the first to the start of the second image), and 4.4 ms for the large tank, with 6 ms between image pairs. In the small tank, the laser (running at 0.25 Watts) enters the tank through the free surface. This allows the free surface motions to distort the light sheet. Fortunately this effect is quite small, due to the low Reynolds number and the absence of facility-induced capillary waves. The large tank has a transparent bottom through which the light sheet (laser power 5.5 Watts) is delivered, minimizing surface disruptions. However, this introduces a long beam path over which laser power is absorbed and passing particles can cast randomly timed shadows. Thus both illumination techniques introduce light sheet distortions which appear as noise (measured to be $\approx$ 10% of signal intensity) in measured concentration fluctuations.

3.2.5 Quantitative Imaging Procedure

The fluorescent dye mixed throughout the water provides an initially bright image. Before adding CO_2 gas to the system, a series of images are taken to define a reference value at each pixel. During a CO_2 invasion experiment, images will show some darker regions corresponding to areas of high CO_2 concentration and brighter spots corresponding to images of tracer particles. Because of this, the signals for PIV and LIF are separable - everything brighter than the reference value is a tracer for PIV and everything darker is our [CO_2] measurement by LIF.

After a CO_2 invasion experiment, we calibrate the LIF technique by measuring each pixel's intensity while the entire water volume is set to a known value of [CO_2]. These known values are obtained by removing the air-side forcing and gradually adding CO_2 by bubbling. We keep the tank well mixed so that the CO_2 concentration at any given point is equal to the bulk concentration (a concentration boundary layer at the free surface would violate this assumption, this is why we stop the air-side forcing by filling the headspace with air from the ambient environment). The bulk concentration is measured via two pH electrodes (Ross Orion combination electrode, Corning gel electrode) mounted 5 cm beneath the free surface. We use these electrodes to compute bulk [CO_2] from the electroneutrality equation and initial and boundary conditions [21]. Each pixel can be calibrated individually (from its intensity timeseries during calibration) or based on a universal calibration curve. This universal curve is based on the timeseries of several pixels throughout the image area, non-dimensionalized by the reference value and averaged. The individual pixel calibration method is used in the large tank experiments, and the universal method is used in the small tank experiments.

During calibration, we vary [CO_2] so that fluorescence intensity covers the entire range observed in the experiments. In the both tanks, this dynamic range is pH = [6.5,4.0], [CO_2]=[0, 270 mg/L], which is equivalent to pixel values down to 40% of maximum image intensity. The [CO_2] range we measure in is ideal, because it corresponds closely to the pH range in which the dye response is strongest and reaction kinetics are not prohibitively complex [1].

The coordinate system used here is fixed with respect to the free surface; this reference frame allows for proper accounting of interfacial fluxes. In the small tank we identify the free surface location manually in each image using the centerpoint of the reflection caused by the free surface, plus any particles on the free surface as indicators. In the large tank we identify the free surface location using a second camera viewing the interface at a slight downward angle as in Banner and Peirson [5].

Velocity measurements are made using the PIV technique of Cowen and Monismith [7]. In the data reported here, each reported vector is based on a subregion of 32 pixels, the measurement in this region being informed by a series of iterative passes using larger subregions. Vectors are interrogated on a grid of 16 pixels, so that vectors are independent from their second-nearest neighbors. Spurious vectors are identified by local median, adaptive Gaussian, or correlation based filters, and are neither replaced by interpolation nor included in calculations.

Data is collected in the small tank at 1.25 Hz (5 image pairs every 4 seconds) for $\approx$ 6 minutes, in an image area of 2.1 cm by 2.1 cm, with a pixel resolution of 25.4 μm. LIF resolution is equal to the three times the pixel resolution, as the reported concentration measurements are the median concentrations over 3x3 pixel regions centered on the velocity field grid points. In the large tank, data is collected at 25 Hz for 10 minutes, in an image area of 3.3 by 4.4 pixels, with a pixel resolution of 67 μm.

3.3 Results

3.3.1 Surface Renewals and Injections

Our coupled PIV-LIF technique gives us a timeseries of instantaneous mass flux values in a 2D plane. This affords a window into the fundamental physics of how turbulent mass transport occurs. One major feature we observe is visually striking mass injection events, such as those in Fig. 3.2. In such events the concentration boundary layer is pulled down into the fluid and then sheared away. Similar events are seen in the small tank experiment and examples can be found in [21].

These mass injection events represent turbulent mass flux from the surface into the bulk, which can be seen by considering the fluctuating vertical velocity W' and the fluctuating concentration C' (defined below). The product of these is the turbulent mass flux, and a negative value indicates flux into the bulk. Points inside an injection event show $W' < 0$ and $C' > 0$.

Turbulent mass flux into the fluid is also caused by upward-moving fluid with low [CO_2]. Such events are the surface renewal eddies hypothesized by Danckwerts [8] and have been a major focus of gas transfer research. Their presence in our images is not as visually striking as the mass injection events, as their visible signal is a thinning of the already thin concentration boundary layer. However, their presence is observed clearly in animated image series and in conditional timeseries such as those in Fig. 3.5.

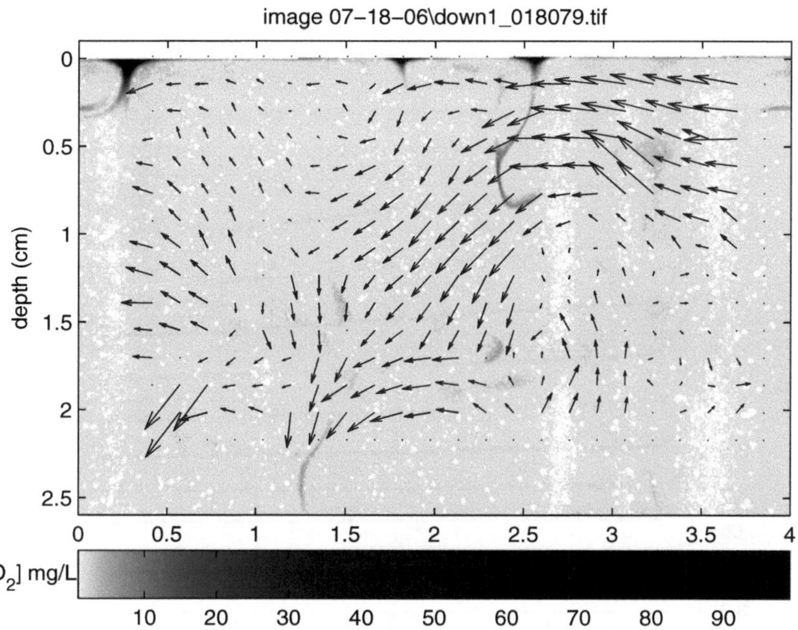

Figure 3.2. Instantaneous concentration and velocity fields showing both mass injection events and surface penetration events (downwellings and upwelligns, respectively) from the large tank dataset. For ease of viewing in grayscale, color scale is logarithmic

Quantifying the role of injection events in turbulent mass transfer is the focus of this paper, but the presence of those events is by no means a new result. Fortescue and Pearson [11], in their model of idealized eddies at a surface, state that "the absorbed gas is carried into the bulk of the fluid largely by the layers of 'surface' fluid that plunge downwards near the edge of the eddy." Recent studies of oxygen transfer across an air-water interface by Takehara and Etoh [19], Herlina and Jirka [13] and Falkenroth et al. [10] all observe injection events. Simulations by Nagaosa [17] and Magnaudet and Calmet [15] also reveal injection motions for scalars of varying diffusivities.

Because mass injection events resemble the "fingers" created in a layer of negative buoyancy, we must confirm that they are really signatures of turbulent flux and not buoyancy flux. A density difference in this tank could be thermal, due to evaporative cooling of the surface, or chemical, from the increase in dissolved CO_2. The chemical effect on density is negligible – the highest measured CO_2 concentration corresponds to a density change of $< 0.1\%$ [21]. We test the thermal effects by running an exper-

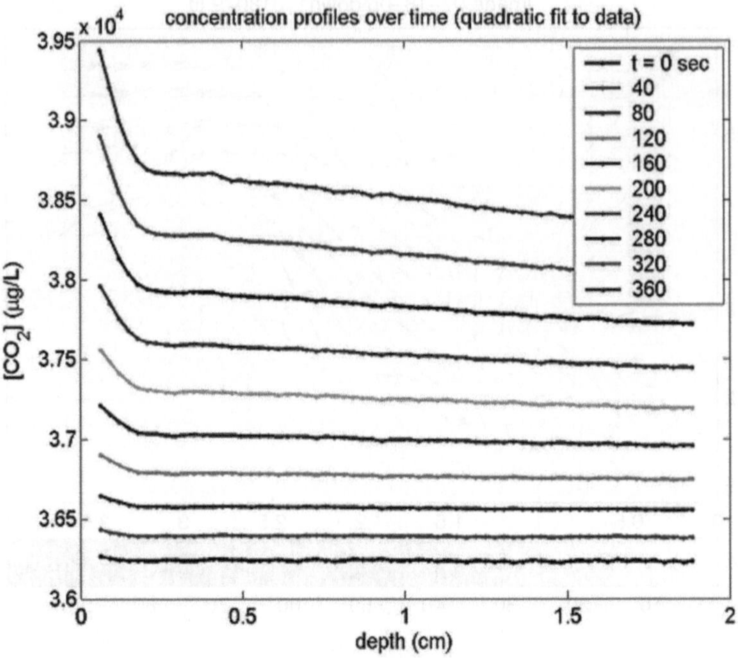

Figure 3.3. Concentration profiles over time (start of the experiment is the lowest curve). Curves are smooth because they are polynomial fits to the data

iment with and without turbulence generation, and find that injections cease when turbulence does. Furthermore, Schladow et al. [18] specifically study plunging plumes from a cooled surface layer, and to obtain a gas transfer velocity similar to that measured here they must create a temperature difference of 18 degrees Celsius between the surface and bulk (a density change of 0.2%). This is much larger than the temperature gradient in our system (< 5 degrees Celsius). Thus we conclude mass injections in our system are not due to density effects.

3.3.2 Single Point Statistics

Single point timeseries of velocity, concentration, and turbulent mass flux are discussed here, using the small tank dataset. Results from the large tank are currently in preparation.

The mean velocity (denoted $\langle U \rangle$) is found at each PIV grid point by ensemble averaging across our 500 samples. The instantaneous fluctuating values are defined as $U' = \langle (U - \langle U \rangle)^2 \rangle^{1/2}$. Because the concentration timeseries at any point is non-stationary ($[CO_2]$ is increasing throughout the experiment, see Fig. 3.3) we cannot find the fluctuating values by simply

Table 3.1. The distributions of W' and mass flux conditional on W' show that the predominance of mass flux during downwelling events are not due to a predominance of downwelling flow in general. Recall positive skewness indicates a distribution peak "skewed" to the left of a normal distribution

	Mean	Median	Skewness
W'	-0.04	-0.09	0.61
$W' \mid W'C' < 0$	-0.03	-0.12	1.24

subtracting a time-averaged mean value. Rather, we consider the horizontally averaged concentration at each depth over time. We curve-fit these timeseries and use the curves ($\overline{C}(z,t)$) as a proxy for the expectation value of concentration. Thus for any instantaneous pixel value $C(x,z,t)$ we define $C'(x,z,t) = C(x,z,t) - \overline{C}(z,t)$.

$C'(x,z,t)$ can be multiplied by $W'(x,z,t)$ to find the instantaneous turbulent CO_2 flux $F_t(x,z,t) \equiv W'C'$. A negative value indicates transfer of CO_2 towards the bulk of the water-side. The horizontally averaged instantaneous turbulent flux near the surface $F_t(z = 0.06\ \text{cm}, t)$ is seen in Fig. 3.4. At the free surface, molecular processes are responsible for the CO_2 flux, and F_t should go to zero, as W' is kinematically constrained to zero there. Beneath the viscous boundary layer, the temporal or horizontal average of F_t should equal the total flux F. We can measure this total flux independently from the time rate of change of the bulk concentration, $F(t) = (V/A) \times \mathbf{d}C_{bulk}(t)/\mathbf{d}t$, where V = tank volume and A = free surface area. Comparing the measurements of F and F_t (which are independent other than sharing the [CO_2] calibration curve) in Fig. 3.4 shows excellent agreement.

The events discussed above – surface injection and surface renewal – are both evidenced by negative values of $W'C'$, yet have oppositely signed values of W'. We can thus divide the observed instances of negative turbulent mass flux into those with $W' > 0$ and those with $W' < 0$, i.e. the mass flux due to renewals and injections, respectively. Figure 3.5 shows timeseries of instantaneous mass flux from the small tank experiment, horizontally averaged across the depth closest to the surface and conditioned on the sign of W'. The conditional timeseries in Figure 3.5b and c indicate that there are many more injection events than renewals. Summing each of these timeseries to find the total turbulent mass flux due to each event type shows that the injection events (those with downwelling flow) are responsible for $\approx$ 8 times more mass flux than the surface renewals (upwellings).

We determine that the dominance of downwelling mass flux is not due simply to a predominance of downwelling flow. We do so by examining the skewness of the distribution of W' for the entire dataset and the distribution of W' conditional on mass flux into the fluid ($W'C' < 0$). The

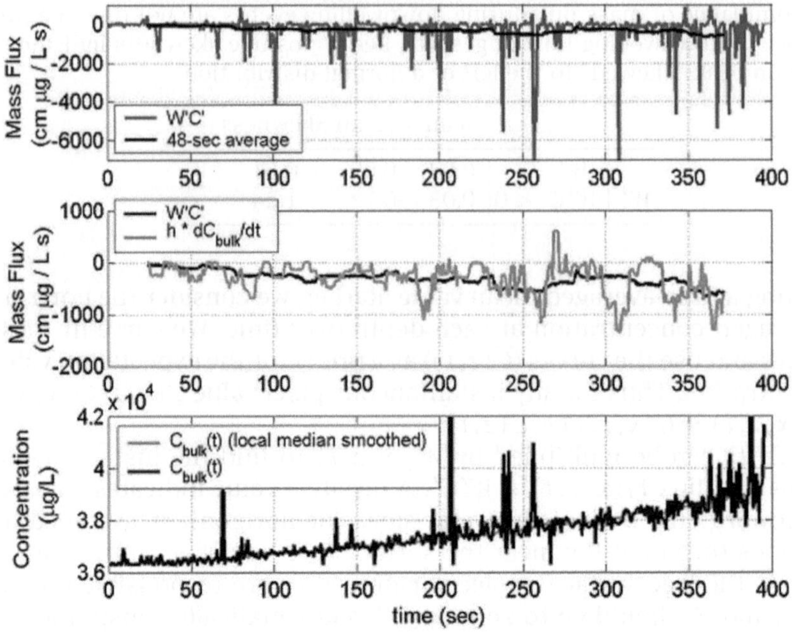

Figure 3.4. (a) $W'C'$ horizontally averaged at the measurement depth closest to the surface (0.06 cm), peaks are smoothed away with a 48-second running average to show the underlying trend. (b) Two measures of mass flux: smoothed $W'C'$ curve from (a) and time rate of change of smoothed bulk [CO_2] curve from (c). The latter is calculated with a simple finite difference method, $dC_{bulk}/dt = C(t+4 \text{ sec}) - C(t - 4 \text{ sec})/8$ sec. (c) Bulk [CO_2] from median LIF values over 3x3 pixel window plotted on top of the temporally smoothed (local median filter) version of this timeseries

results of this analysis, seen in Table 3.1, show that downwellings are slightly more common than upwellings in this flow overall, but are much more common during periods of mass flux into the water.

Importantly, when this analysis is repeated for the large tank dataset, we find a different result for the relative strength of turbulent flux due to upwellings and turbulent flux due to downwellings. The ratio is close to 1:1, as opposed to the 8:1 ratio measured in the small tank. Further analysis shows that this ratio is extremely sensitive to the value of $\overline{C}(z,t)$ which is used to define the concentration fluctuations. Therefore we expect variation between experiments, as any individual dataset will have some deviation between the calculated $\overline{C}(z,t)$ and the true expectation value for which it is a proxy. Interestingly, initial results suggest that while the relative strength of upwelling and downwelling fluxes is sensitive to the value of $\overline{C}(z,t)$, the combined value of these fluxes (i.e. the total flux

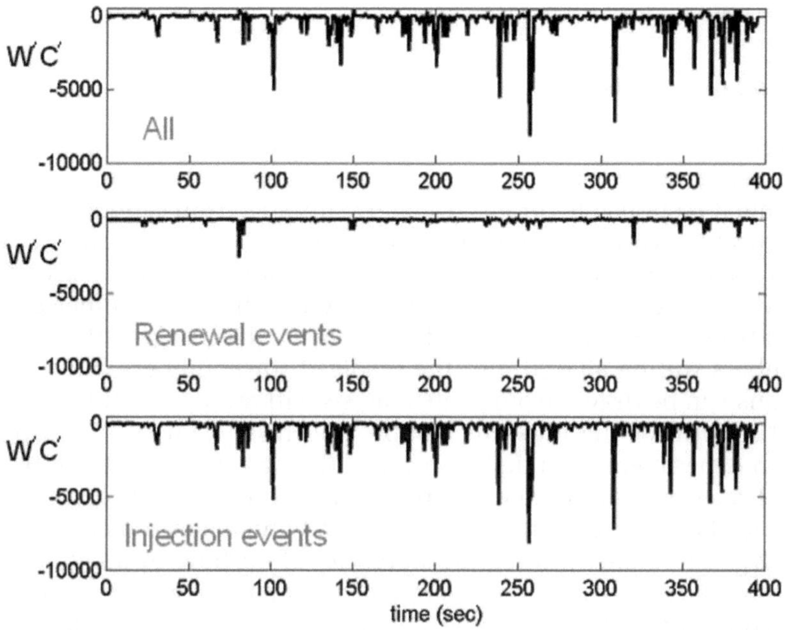

Figure 3.5. (a) Timeseries of instantaneous mass flux, horizontally averaged across a line parallel to the free surface at the smallest depth measured. (b) and (c) show this timeseries conditional on the sign of W'

into the bulk) is not. Analysis is underway to further quantify and explain this trend.

The two experiments compared here have further differences which could also contribute to the difference in flux ratios. The two tanks have very different Reynolds numbers, and small tank's size causes wall effects to cover a greater fraction of the tank volume than in the large tank. Finally, differences in the flux ratio could be due to the fact that the large tank dataset was recorded for a longer time period, thereby sampling more of the intermittent powerful upwelling events.

While the difference in turbulent mass flux ratios is noteworthy, we would like to emphasize that results from both tanks show that the downwellings' contribution is within an order of magnitude of the upwellings' contribution. This shows that downwellings are not simply a companion process to the upwellings, but are of direct importance to turbulent mass transfer. This understanding will be important when creating models for turbulent mass transfer in cases other than homogeneous and isotropic turbulence. For example, when upwelling is weak and covers a large spatial area, and downwelling is strong and covers a small spatial area, parameterizations based solely on upwellings would likely be inferior to those incorporating downwellings.

3.3.3 Two Point Statistics

The power spectrum (in wavenumber space) of the turbulent mass flux field can help resolve the longstanding question as to the lengthscale of motions most responsible for gas transfer. This spectrum cannot be measured accurately without resolving the largest scales of the flow. That is, the small measurement region used here does not accurately resolve the longest wavenumbers. If we attempt to calculate the wavenumber spectrum over the limited spatial extent of this dataset, the energy in the longer wavenumbers will be aliased into the resolved wavenumbers. No useful conclusions can be drawn from such analysis without preexisting knowledge of the form of the low-wavenumber spectrum. Thus we have performed an experiment intended to resolve the mass flux spectrum from the lowest relevant wavenumbers to the highest resolvable wavenumbers. We do so in the large tank (which is statistically homogeneous over a region much greater than the integral lengthscale of the flow) by imaging two identical measurement regions, separated by a distance r and measured simultaneously using the technique described above. Varying r allows us to accurately resolve the full spatial autocorrelation function of turbulent mass flux, and from this determine the turbulent mass flux power spectrum. At the time of this writing, this analysis is underway and results will be reported when available.

In the absence of more formal two-point statistics, we perform a quick measurement of the relevant size scales for injection events. We define the length scale to as the horizontal width at the point closest to the free surface. We measure the distribution of event sizes by manually inspecting the images from the small tank dataset, and find a unimodal distribution with a long tail which continues to the largest events observable in our field of view ($\approx$ 2 cm). This tail may follow a power law, but further data and a more robust size metric is needed before this can be verified.

3.4 Analysis

The above results demonstrate that injections (of the concentration boundary layer into the bulk) are at least as important to turbulent mass transfer as surface renewal events. Upwellings and downwellings clearly are related, though they need not be identical in size, strength, or net effect on gas flux. We observe differences in their spatial and temporal distributions, such that the elusive question of "what length and time scales are most important for turbulent interfacial mass transfer" may be more complicated still – the question is best answered by identifying the dominant scales for each phenomenon directly responsible for turbulent mass transfer, i.e. considering upwellings and downwellings separately.

Qualitative observations from the above datasets show that there are intermittent large and energetic (high velocity) upwellings. We did not observe similarly large injection events. Thus the injections may provide the dominant mass transfer at smaller scales, while upwellings could dominate at larger scales. An investigation of this requires long records (to capture intermittent large events) as well as two-point statistics. Such data were obtained from our large tank experiments, to be reported in the near future.

We note that existing models parameterizing gas transfer based on the frequency, size, and strength of eddies approaching the free surface have been successful, even if the models do not include the existence and possible dominance of injections. Thus the sum total of gas transfer phenomena (injections, upwellings, and others) responds similarly to the model parameters as do the simplified phenomena used to create these models. As these models of gas transfer at a shear free surface are adapted to describe other cases (e.g. wind shear and waves), this similarity may break down, and a more accurate view of the physics may prove helpful.

3.5 Conclusions, Extensions, and Acknowledgements

Examination of the turbulent mass flux field reveals events in which fluid is injected to the bulk from near the free surface. Analysis shows that these contribute to the turbulent mass transfer at least as much as the upwelling events that are typically discussed as the mechanism for turbulent mass transfer. In the near future, analysis of data from the large tank experiments will be used to further investigate this phenomenon.

Acknowledgement. We would like to sincerely thank the editors of this volume for organizing an educational and productive workshop, and those at the Ruprecht-Karls-University of Heidelberg's Interdisciplinary Center for Scientific Computing for being gracious hosts. This work was supported in part by NSF grants CTS-0093794, DGE-9870631, and GK12-0231913.

References

[1] Asher W.J., Pankow J. (1986) The interaction of mechanically generated turbulence and interfacial films with a liquid phase controlled gas/liquid transport process. *Tellus* 38B: 305-318

[2] Asher W.J. (1987) An examination of the hydrodynamics governing a liquid-phase rate controlled gas/liquid mass transport process at

clean and film-covered liquid surfaces. PhD Thesis, Environmental Science and Engineering, Oregon Graduate Center, Oregon

[3] Asher W.J., Jessup A.T., Atmane M.A. (2004) Oceanic application of the active controlled flux technique for measuring air-sea transfer velocities of heat and gases *J. Geophys. Res.* 109

[4] Banerjee S. (1990). Turbulence Structure and Transport Mechanisms at Interfaces. *9th Int. Heat Transfer Conference*, Hemisphere Press, NY

[5] Banner, M.L., Peirson W.L. (1998) Tangential stress beneath wind-driven air–water interfaces. *J. Fluid Mech.* 364: 115-145

[6] Brumley B., Jirka G. (1987). Near-Surface Turbulence in a Grid-Stirred Tank. *J. Fluid Mech.* 183: 235-263

[7] Cowen E.A., Monismith S. (1997). A hybrid digital particle tracking velocimetry technique. *Exp. Fluids* 22: 199-211

[8] Danckwerts P.V. (1951) Significance of liquid-film coefficients in gas adsorption Industrial and Engineering Chemistry 43(6): 1460-1467

[9] DeSilva I., Fernando H.J.S. (1994). Oscillating grids as a source of nearly isotropic turbulence. *Phys Fluids* 6(7): 2455-2464

[10] Falkenroth A., Degreif K., Jähne B. (2007) Visualisation of Oxygen Concentration Fields in the Mass Boundary Layer by Fluorescence Quenching. In: Garbe C.S., Handler R.A. and Jähne B. (eds) Transport at the Air Sea Interface - Measurements, Models and Parameterizations. Chapter 4, pages 59–72, Springer Verlag. This volume.

[11] Fortescue G.E., Pearson J.R.A. (1967) On Gas absorption into a turbulent liquid. *Chemical Engineering Science* 22: 1163-1176

[12] Hanratty T.J. (1990). Effect of Gas Flow on Physical Absorption. In: *Air-Water Mass Transfer: Selected Papers from the Second International Symposium on Gas Transfer at Water Surfaces*, American Society of Civil Engineers, NY

[13] Herlina, Jirka G.H. (2004) Application of LIF to investigate gas transfer near the air-water interface in a grid-stirred tank *Exp. Fluids* 37: 341-349

[14] Lamont J.C., Scott D.S. (1970) An eddy cell model of mass transfer into the surface of a turbulent liquid *AIChE J* 16:513-519

[15] Magnaudet J., Calmet I. (2006). Turbulent mass transfer through a flat shear-free surface. *J. Fluid Mech.* 553: 155-185

[16] McKenna S.P., McGillis W.R. (2004). The role of free-surface turbulence and surfactants in air-water gas transfer. *International Journal of Heat and Mass Transfer.* 47(3):539-553

[17] Nagaosa R. (1999). Direct numerical simulation of vortex structures and turbulent scalar transfer across a free surface in a fully developed turbulence. *Phys. Fluids* 11(6): 1581-1595

[18] Schladow S.G., et al. (2002) Oxygen transfer across the air-water interface by natural convection in lakes. *Limnology and Oceanography* 47(5): 1394-1404

[19] Takehara K., Etoh G. (2002) A Direct Visualization Method of CO2 Gas Transfer at Water Surface Driven by Wind Waves. In: Donelan et al. (eds) *Gas Transfer at Water Surfaces.* Geophysical Monograph 127, 2002, American Geophysical Union
[20] Variano E.A., Bodenschatz E., Cowen E.A. (2004). A random synthetic jet array driven turbulence tank. *Exp. Fluids.* 37(4):613-615
[21] Variano E.A. (2005) Quantitative Visualization of Carbon Dioxide gas transfer at a turbulent free surface. MS Thesis, Cornell University, New York
[22] Variano E.A., Cowen E.A. (2006). A turbulent stirred tank using randomized jets. *J. Fluid Mech.* (submitted)

4

Visualisation of Oxygen Concentration Fields in the Mass Boundary Layer by Fluorescence Quenching

Achim Falkenroth[1], Kai Degreif[2], and Bernd Jähne[1,2]

[1] Institute for Environmental Physics, IUP
University of Heidelberg, Germany
Achim.Falkenroth@iup.uni-heidelberg.de
[2] Interdisciplinary Center of Scientific Computing, IWR
University of Heidelberg, Germany
Bernd.Jaehne@iwr.uni-heidelberg.de

Abstract Laser-Induced Fluorescence (LIF) is applied to observe directly the mechanism of gas exchange in the aqueous viscous boundary layer at a free water surface. In order to make dissolved oxygen (DO) visible, a new class of dyes with a long phosphorescent lifetime in the order of microseconds is used. This property makes the quenching constant for DO sufficiently high for sensitive measurements. Depth profiles of the O_2 concentration near the water surface are obtained by a vertical laser light sheet at a rate of 185 frames per second. This technique is capable of visualising a measurement window of some centimetres down from the water surface with a resolution in the order of 50–100 µm. For a small circular wind-wave facility a correlation between wind speed and gas-transfer velocities calculated from the extracted mean boundary-layer thickness are presented and compared to the results of parallel measurements with a mass balance method for other gases with given Schmidt numbers.

4.1 Introduction

Gas transfer across gas–liquid interfaces is of importance in natural environments and technical applications. The exchange of gases between the atmosphere and the oceans determines the global distribution of many gaseous and volatile chemical species. Reaeration of lakes and rivers across the water surface is a critical process for the ecology of these environments, especially if they show high biological activity. In technical systems the mass transfer across gas–liquid interfaces is an essential process in various gas–liquid reactors such as bubble columns and falling film columns.

The difficulty in studying the mechanisms of air–water gas exchange is due to the small thickness (30–300 µm) of the aqueous mass boundary

C.S. Garbe, R.A. Handler, B. Jähne (eds.): Transport at the Air Sea Interface
pp. 59-72, 2007,

layer at a free interface and an overlaying undulation by waves. Measuring the mean flux by mass balance techniques results only in mean values for the transfer coefficient and mass boundary layer thickness. Only techniques that visualise concentration fields of the dissolved gases give direct insight into the turbulence close to the interface.

In the literature two different kinds of visualisation techniques for gases dissolved in water are reported: the pH-indicator technique and the oxygen-quenching technique with the pH-indicator technique being the older one. The key point of this technique is the conversion of the flux of an acid or alkaline gas from the air to the water surface into a flux of fluorescence intensity by a chemical reaction with a dye.

40 years ago this technique was used to investigate the gas transport mechanisms in falling films [5, 10, 11, 12]. In 1989, the pH-indicator technique was first used in a grid-stirred tank by Asher and Pankow [1, 2]. The authors measured time series of the CO_2 concentration fluctuations close to a gas-liquid interface at a fixed position using dichloro fluorescein. The first successful measurements of vertical concentration-profiles within the aqueous mass boundary layer at a free interface in a wind–wave flume were reported by Jähne [13] on a symposium in Minneapolis using HCl gas. More detailed studies using fluorescein followed later (Münsterer and Jähne [17], Münsterer [19], Variano and Cowen [21]).

At the same symposium the first successful measurements from wind-wave flumes using the oxygen-quenching technique were reported by Wolff et al. [23]. Pyrenebutyric acid (PBA) was used, a dye known from measurements of dissolved oxygen concentrations in cells (cf. Vaughan and Weber [22]). PBA fluorescence was stimulated by a N_2 laser in the UV at 337 nm. Later other authors used the same dye in wind-waves flumes (Münsterer et al. [18], Münsterer [19], Woodrow and Duke [24]) and grid-stirred tanks (Herlina [7], Herlina and Jirka [8, 9]) for oxygen exchange studies.

The rather qualitative character of the previously published results indicates that the currently used fluorescent dyes still show significant disadvantages. The use of PBA to measure oxygen dissolved in water gives rise to several problems. Firstly, the quenching effect is rather weak causing a poor signal-to-noise ratio of the concentration measurements. Secondly, it is difficult to solve PBA in water. Thirdly, PBA is a surface active chemical species. This means that the hydrodynamic boundary conditions at the air–water interface (surface tension and surface elasticity) are altered by PBA.

Thus, the search for a more suitable fluorescent dye seemed promising, and a better luminescent dye for the oxygen-quenching technique could be found. Herein the properties and advantages are described and demonstrated by some visualisation experiments in a wind–wave flume with a surfactant film suppressing wind waves.

4.2 Visualisation by Oxygen Quenching

The presence of oxygen in the water is unveiled by its ability to quench luminescence, i.e. to induce the energy dissipation of the excited state without the emission of a photon by a collision with the a chromophore. A dye with a long lifetime of its excited state is needed as an oxygen sensible luminophore in order to increase the probability of a successful hit.

The Stern–Volmer equation [20] describes the quenching of the luminescence as a function of the concentration of a quencher:

$$I(c)/I_0 = \frac{1}{(1 + K_{SV} \cdot c)} \tag{4.1}$$

where $I(c)/I_0$ is the fraction of the actual luminescence related to the luminescence in absence of the quencher; c is the concentration of the quencher in the liquid; and K_{SV} is the Stern–Volmer quenching constant.

As the quenching process is fast compared to mass fluxes, the concentration of the quencher O_2 can be evaluated reliably at any point by measuring the intensity of the light emission from that point. Other influences, such as intensity of the exciting light and temperature, have to be the same as in the calibration procedure.

An ideal dye for oxygen-gas transfer measurements should have a high quenching constant, be readily soluble in water and show no significant surface activity. A high quenching constant, and thus, high sensitivity according to Equation (4.1), requires a long lifetime of the excited state in the order of some µs. A family of diimine ruthenium complexes, especially sodium tris(4,7-diphenyl-1,10-phenanthroline disulfonic acid)ruthenate(II) complex ($Na_4[Ru(dpp\,ds)_3]$), has these desired properties [3, 15].

The following is a list of the advantages of $Ru(dpp\,ds)_3$ as compared to the widely used pyrenebutyric acid (PBA):

- PBA requires a UV laser for excitation (e.g. a N_2 laser at 337 nm) leading to some bleaching, while the ruthenium complex absorbs in the visible blue where cheap and handy light sources for the stimulation of the fluorescence are available. Fig. 4.1.a shows 465 nm as a maximum of absorption in the visible.
- The Stokes shift of $Ru(dpp\,ds)_3$ for stimulation with 473 nm and an emission maximum of 610 nm (cf. Fig. 4.1.a) is 137 nm. Above 530 nm the absorption is low enough to have no significant self absorption of the fluorescent light. This is a much larger Stokes shift than for PBA when employing a N_2 laser with a stimulation at 337 nm, an emission maximum at 375 nm and a Stokes shift of 38 nm (cf. Vaughan and Weber [22]) resulting in self absorption.
- The quenching constant K_{SV} of $Ru(dpp\,ds)_3$ is about 8700 ± 1300 L/mol (cf. Fig. 4.1.b), while the quenching constant for PBA is more than

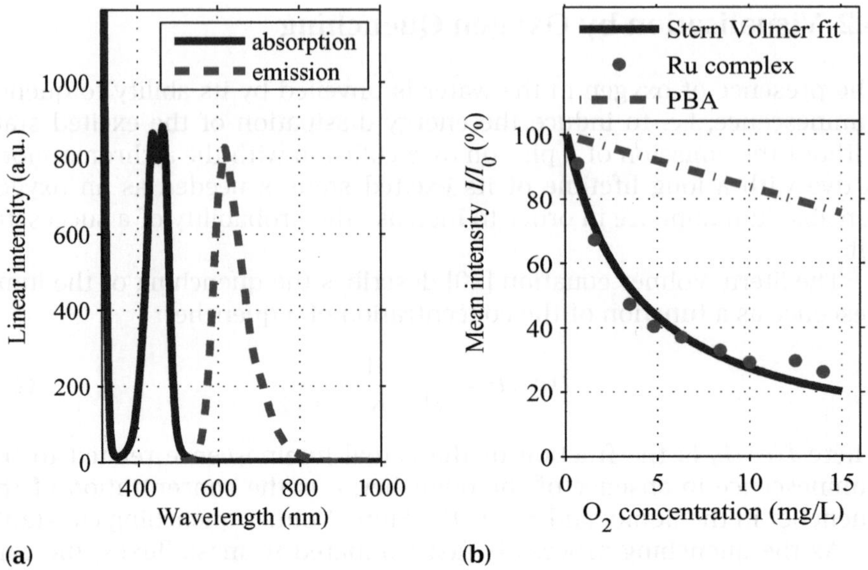

Figure 4.1. Characterisation of the luminescence of the dye Ru(dpp ds)$_3$. (**a**) Spectra for absorption and emission stimulated at 473 nm. (**b**) Fluorescence quenching as a function of the oxygen concentration for Ru(dpp ds)$_3$ and PBA (8 mg/L O_2 at equilibrium with ambient air corresponds to 250 µmol/L)

ten times lower (683 ±70 L/mol in Münsterer [19] or 645 ±79 L/mol in Vaughan and Weber [22], Fig. 4.1.b). Therefore, at small oxygen concentrations the measurements with the Ru-complex are more than ten times more sensitive (cf. Equation (4.1)). At the saturation concentration of 8 mg/l the phosphorescence quenching for Ru(dpp ds)$_3$ is 63%, while it is only 13% for PBA.

- The six sulfonic groups make Ru(dpp ds)$_3$ excellently soluble in water, and the dye shows no surface activity. In contrast, PBA is almost insoluble in water. It can only be solved in NaOH before mixing with water resulting in much lower concentrations.

4.3 Experimental Set-up

The set-up shown in Fig. 4.2.a uses a vertical laser sheet generated by a cylindric convex mirror and is suitable for a measurement footprint of 1–2 cm at the water surface. The pixel resolution hereby was 25 µm/pixel. Illumination by a 50 mW laser at 473 nm is sufficient to acquire image sequences from the side with the maximum frame rate of 185 Hz of a Dragonfly CCD-camera (480×640 pixel, Point Grey Research, Canada). The camera was tilted by 10 degrees to avoid occlusion by the moving surface.

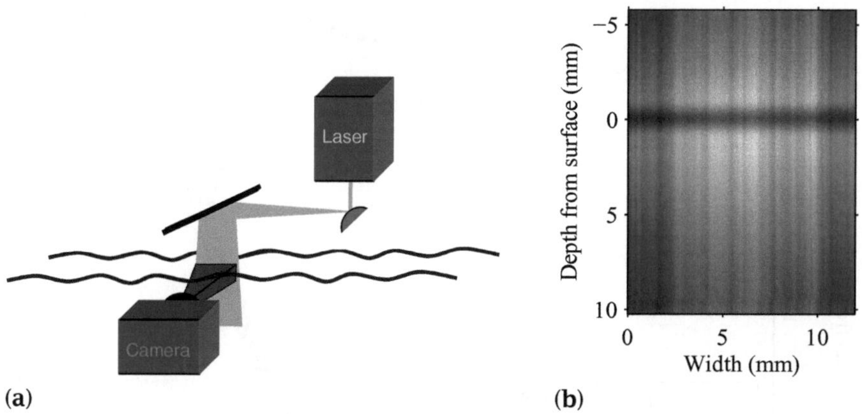

Figure 4.2. (**a**) Set-up for measuring the concentration profile in a small annular wind-wave facility. (**b**) Image captured by the camera. N.b.: The upper part of the image is a total reflection at the surface

This generates a mirror image above the surface line by total reflection at the water surface as seen in Fig. 4.2.b.

The measurements are carried out in a circular wind-wave flume with 12 cm water height and a water volume of 70 L. The channel is 20 cm in width, 41 cm in height with a diameter of 1.2 m. Wind is generated by a rotating paddle wheel. The concentration of the dye was about 10^{-5} mol/L.

In order to strip dissolved oxygen from the water, a "Jostra Quadrox" gas-exchange module from Maquet is installed containing membrane capillaries with a total membrane surface of 1.8 m^2 and a water volume of 250 ml. This device is normally used in cardio-pulmonary machines to load blood with oxygen and extract carbon dioxide. The gas exchanger can be used for other gases as well. In the work described herein we employed it with vacuum (60 mbar) to degas the water to oxygen concentrations as low as 0.8 mg/L. The oxygen concentration is measured constantly by a commercial oxygen probe sensor.

4.4 Calculation of the Boundary-Layer Thickness

Images of the luminescence profiles, like the one in Fig. 4.2.b, demonstrate the good contrast in the boundary layer. In the middle of the dark horizontal line lies the air-water interface. Here oxygen penetrates the water surface and quenches the phosphorescence in the water. In the depth profile of Fig. 4.3 this distinct decrease of the emitted light intensity near the surface of the degassed water is even more evident.

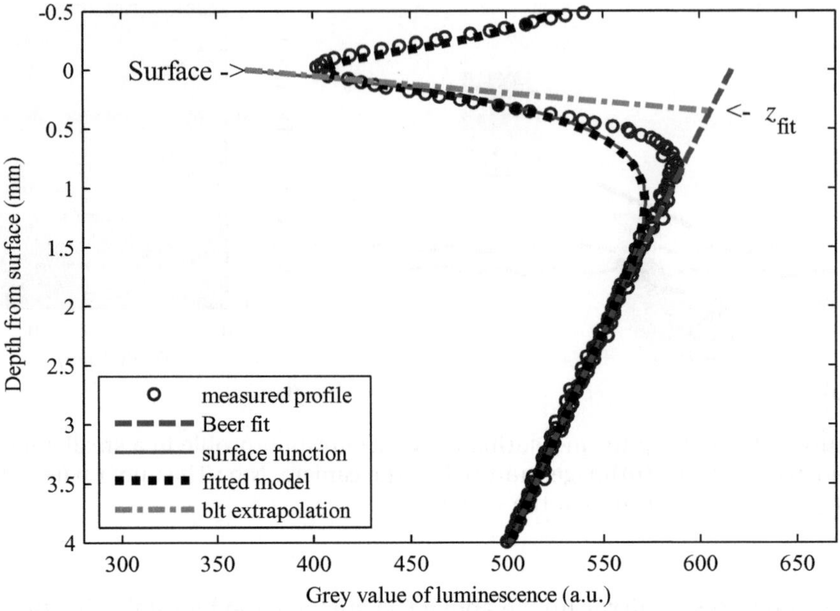

Figure 4.3. Measured depth profile (detail) at a wind speed of 3 m/s and modelling. To analyse the mean of the measured data (*circles*), a modelled function is fitted to it by a least-squares method. The base is a Beer–Lambert fit (*broken line*). A multiplied exponential surface function model (*solid line*) describes the quenching and a convolution with a Gaussian representing the optical blurring leads to the model (*dots*) that is fitted to the mean of the measured data. Extrapolation of the surface-function gradient to the base line (*dash-dot line*) results in the boundary-layer thickness z_{fit}

To extract quantitative values some steps of image processing are needed. The boundary-layer thickness $z^{\star}$ is defined as a mean value so that all vertical lines of a measurement image like the one in Fig. 4.2.b were averaged to get a mean depth profile as shown in Fig. 4.3.

Above the surface the mirror effect is visible. The calculation of the actual concentration via a complex calibration is omitted because the effect turned out to be not significant.

Towards greater deepness a decrease of phosphorescence is due to the Beer–Lambert extinction of the incident light by the known absorption of the ruthenium dye. This exponential function represents the base for the modelling function.

In the boundary layer the signal decreases because of the quenching by the solving oxygen. This can be modelled by an exponential decay of the luminescence at the surface. It shows the steepest gradient at the surface. The extrapolation of this gradient to the base line of the Beer–

Lambert function results in the extracted boundary-layer thickness z_{fit} as also mentioned by Jähne et al. [14] in this volume.

Because of an optical blurring the function of the light intensity appears not to be pointed that much at the surface as a flat mirror would suggest. This blurring is a result of the quality of the optical path at the high magnification used. It reduces the effective resolution of the imaging. To account for it, the modelled boundary-layer function was convoluted with a Gaussian representing the blurring. In order to determine the unknown parameters of the different functions, the resulting model function was fitted to the measured mean profile.

For comparison, $z^\star$ was also calculated using a second evaluation method. Here the distance from the surface was measured where the luminescence decreased to 1/e between the luminescence at the surface by reconstruction of the unblurred surface intensity starting from the steepest gradient of the vertical profile in the way as proposed by Woodrow and Duke [24] and Herlina [7].

This thickness z_{e} is overestimated systematically because the steepest gradient is lower than the gradient of an exponential function at the surface. Only for an exponential function the distance at 1/e of a certain value is the same as the extrapolation of its gradient at the same value.

4.5 Gas-Transfer Velocity

The presented method is most powerful without waves. Waves lead to occlusions of the surface when observed from the side. In an experiment in the circular channel, different wind speeds are generated over a water surface on which a surfactant was applied to generate a stagnant film that suppresses wind waves.

Figure 4.4 shows time series of one vertical image line in the laser sheet observed with 185 Hz for 27 s giving 5000 frames in one row. The images shown here are taken after a constant wind speed was established for half an hour. The distance from the surface to the bottom of the image is around 12 mm.

In the first four images shown in Fig. 4.4.a–d the boundary layer is visible and gets thinner with increasing wind speeds. The time series with bulk turbulences in Fig. 4.4.e demonstrates the penetration of the boundary layer by eddies. Most obvious are injection events of packages of water with high oxygen concentrations that are transported in filaments quickly away from the surface where they dilute in the well mixed bulk (described also by Variano and Cowen [21] in this volume).

From theory the boundary-layer thickness $z^\star$ is directly related to the transfer velocity k (see also Jähne et al. [14]):

$$k = D_{\mathrm{ox}}/z^\star \tag{4.2}$$

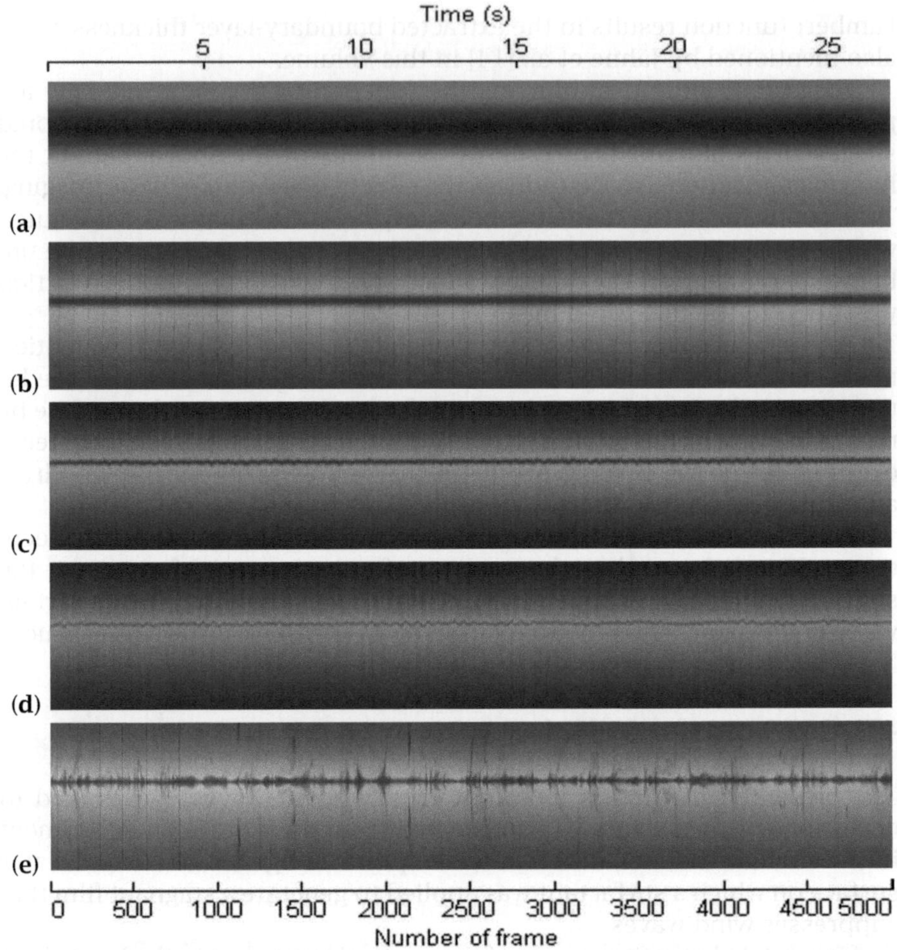

Figure 4.4. Time series: one vertical line (640 pixel) of the laser sheet. (**a**) 20 min. after stopping the O_2-stripping pump and without wind. (**b**) Wind speed of 0.8 m/s. (**c**) 3 m/s. (**d**) 6 m/s. (**e**) Turbulences in the bulk generated by a pump. N.b.: The dark vertical lines are occlusions by the four wind paddles.

where D_{ox} is the molecular diffusivity of oxygen in water. For a wind speed of 0.8 m/s the thickness of the boundary layer z_{fit} with the fitting method of Sec. 4.4 is 18.7 pixel corresponding to 0.48 mm. With D_{ox} of 2.37 10^{-5} cm^2/s at 25.3°C (from Mayer [16]) the gas-transfer velocity k is 1.8 cm/h what is reasonable for a slow gas exchange with little turbulence.

The values of the boundary-layer thickness extracted with the methods described are plotted against the wind speed in Fig. 4.5.a. For higher wind speeds $z^\star$ decreases just as expected. The estimation fails for wind speeds

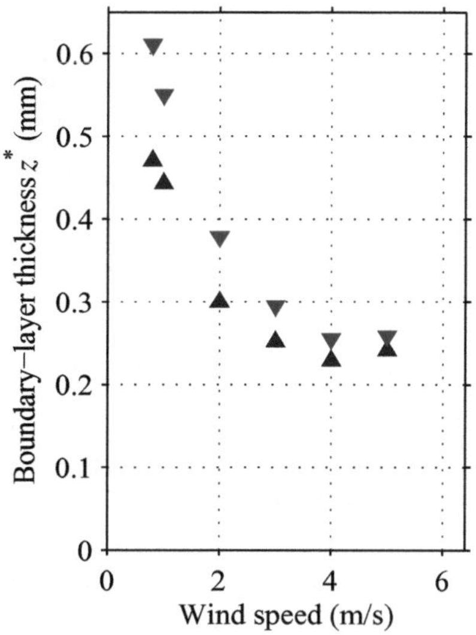

(a)

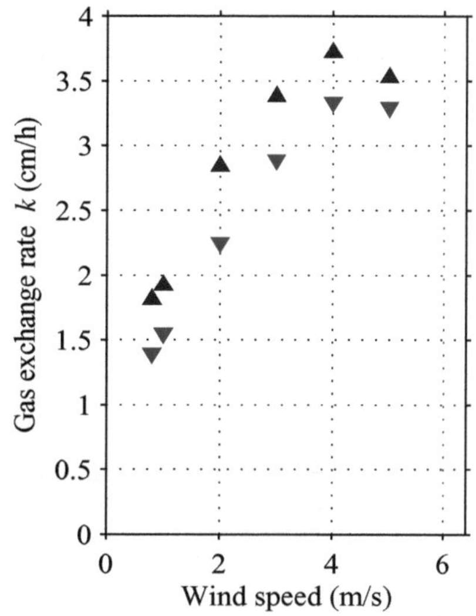

(b)

Figure 4.5. Parameters for oxygen invasion with a stagnant film (*blue upwards triangles*: z_{fit}, *green downwards triangles*: z_e). (**a**) Boundary-layer thickness depending on the wind speed. (**b**) Gas-transfer velocity calculated from the boundary-layer thickness. Method is less accurate for small $z^\star$ at high wind speeds

beyond 5 m/s because here $z^\star$ gets in the range of the optical resolution. After three hours of increasing wind speeds, the oxygen concentration in the bulk got high, and the intensity of luminescence got low, so no computation of $z^\star$ was done for higher wind speeds than 5 m/s. The gas-transfer velocity of oxygen calculated with Equation (4.2) for the different wind speeds is shown in Fig. 4.5.b.

4.6 Comparison with Other Measurement Techniques

To give an estimate for the quality of the achieved data, parallel measurements of dissolved oxygen concentration and trace gases were performed permitting the calculation of reference transfer velocities.

Using an oxygen sensor, the water side increase in O_2 concentration was recorded during the experiment (cf. Fig. 4.6.a). Assuming that a concentration change in the water is only due to the gas exchange over the water surface, we can formulate the mass balance as follows:

$$V_w \, \dot{c}_w = -k A \,(c_w - \alpha c_a) \tag{4.3}$$

where the variables denote:

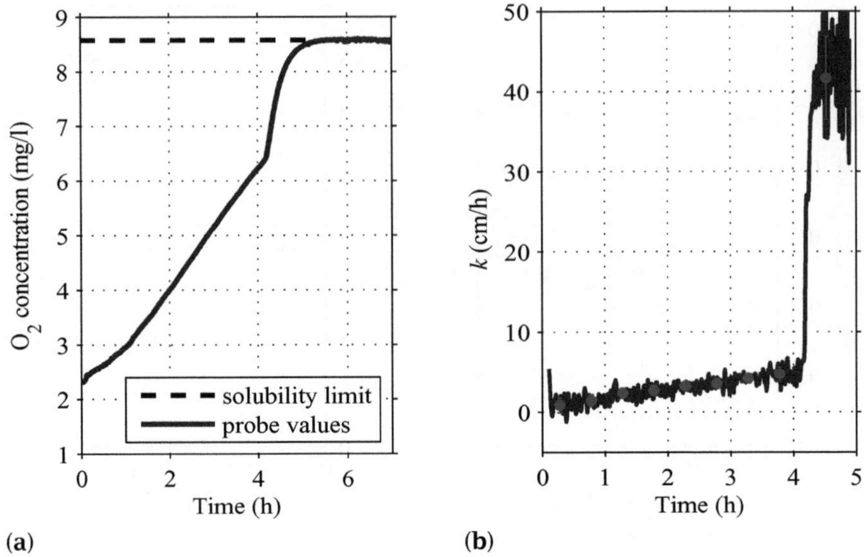

Figure 4.6. (**a**) Water side measurements of O_2 concentration. (**b**) Gas-transfer velocities calculated from Equation 4.3. Mean values for different wind speeds are indicated by red dots

c_a, c_w : air- and water-side concentration
$\dot{c}_w$: temporal derivative of water-side concentration
V_w : water volume (m^3)
α : dimensionless solubility (Ostwald's solubility)
k : transfer velocity (m/s)
A : water surface (m^2)

From this equation the transfer velocity k can be calculated if we determine the temporal derivative of the water-side concentration directly from the measured data. The equilibrium water concentration $c_{eq} = \alpha c_a$ can be determined from concentration convergence in Fig. 4.6.a because the air side concentration did not change during the measurement. The resulting transfer rate of this calculation is shown in Fig. 4.6.b. As the temporal resolution of this calculation method is of the order of minutes, the variance of the calculated values is very high. Averaged values are represented by red points. The abrupt increase in the transfer velocity after four hours is due to the formation of waves when the stability of the surfactant film breaks beyond a critical wind speed that lies between 7 and 8 m/s for the circular channel.

Using a similar mass balancing method, gas-exchange rates were calculated from air-side concentration measurements for the evading trace gases H_2 and N_2O (for further details see Degreif [4]). For comparison,

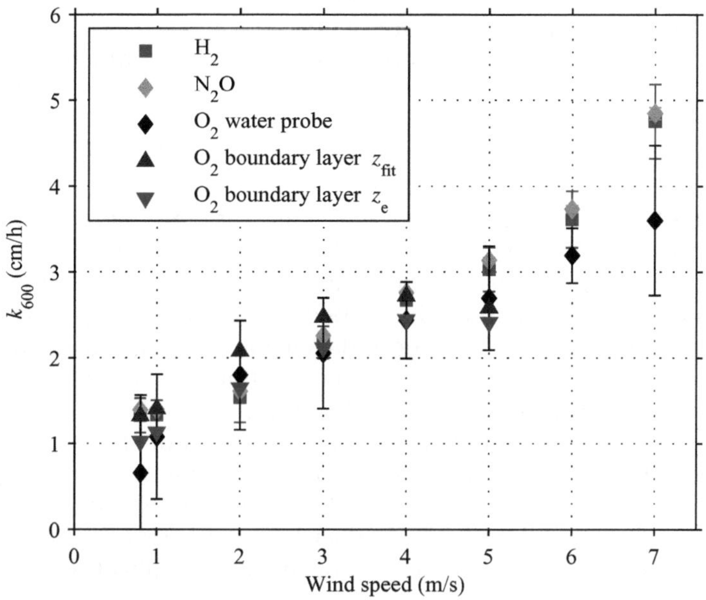

Figure 4.7. Comparison of the normalised gas-transfer velocities k_{600} for a film covered surface derived from different measurement techniques. *Blue and green triangles*: calculated from the boundary-layer thickness for oxygen transport (cf. Fig. 4.5). *Black diamonds*: calculated from the temporal change in O_2 concentration. *Red squares* and *green diamonds*: mass balancing calculation from the measured air-side concentration change in H_2 and N_2O

these transfer velocities are shown as normalised to a Schmidt number of 600 in Fig. 4.7. This figure shows that the transfer velocities derived from the boundary layer are in agreement with the reference measurements.

4.7 Discussions and Outlook

The new phosphorescent dye for the quantitative visualisation of concentration fields within the aqueous mass boundary layer shows significant advantages over the previously used dyes. Thus, much better signal-to-noise ratio can be expected than with previous studies.

The three different approaches for calculating the transfer velocities are in good agreement proving the applicability of the used techniques. There is no fundamental difference in the results for the evasion of the trace gases N_2O and H_2 from the invasion of oxygen measured with a mass balance method and using the boundary-layer thickness. The measurement of oxygen concentration profiles turned out to be an accurate method for gas-exchange measurements.

The method for fitting a boundary-layer function is much less accurate or even fails for thin boundary layers that are in the range of the optical blurring or the pixel resolution of the camera. For the calculation of the gas-transfer velocity, the mean of several boundary-layer thicknesses and the mean of line profiles were taken.

This method enhances the signal to noise ratio but has two drawbacks: firstly a further blurring happens in the case that the surface is moving from one line to the other; secondly, taking the mean underestimates systematically the gas-transfer velocity because the latter is proportional to the mean of the inverse of the boundary-layer thickness as mentioned in Jähne et al. [14] in this volume. Having low fluctuations from one profile line to the other, the effect is negligible here.

The publication of details about the synthesis of the ruthenium dye, more about the characteristics and on further experiments is in preparation [6]. Also another evaluation method using other functions than an exponential for modelling the boundary layer are presented that yield results that are more accurate after concentration calibration and surface detection.

Currently a new inert and chemically clean linear wind–wave facility is being constructed that is specifically designed for the visualisation technique described in this paper. The air space and the water channel will be coated with Teflon and the whole facility is gas-tight. Therefore, acid and alkaline gases can be used in this facility. The water channel will be 4 m long, about 0.4 m wide and 0.1 m high and can be filled with ultraclean water. This facility will provide an improved optical access allowing for imaging with a higher optical resolution.

Acknowledgement. We gratefully acknowledge financial support by the Research Training Group (Graduiertenkolleg 1114) (`http://www.grk1114.de`) of the German Science Foundation (DFG). The luminescence spectra of Fig. 4.1.a were taken with the support of Pia Heinlein of the Institute of Physical Chemistry, University of Heidelberg. The measurements of the wide range of oxygen concentration in Fig. 4.1.b were performed by Felix Vogel. We would like to thank Ingo Klimant from the Analytical Chemistry Department, University of Graz, for providing the organic ruthenium complex and Maquet Cardiopulmonary AG, Hirrlingen, for providing the "Jostra Quadrox" gas-exchange module at no charge.

References

[1] W. E. Asher and J. F. Pankow. The interaction of mechanically generated turbulence and interfacial films with a liquid phase controlled gas/liquid transport process. *Tellus*, 38B:305–318, 1986.

[2] W. E. Asher and J. F. Pankow. Direct observation of concentration fluctuations close to a gas-liquid interface. *Chemical Engineering Science*, 44:1451-1455, 1989.

[3] F. N. Castellano and J. R. Lakowicz. A water-soluble luminescence oxygen sensor. *Photochemisty and Photobiology*, 67(2):179-183, 1998.

[4] Kai Degreif. *Untersuchungen zum Gasaustausch - Entwicklung und Applikation eines zeitlich aufgelösten Massenbilanzverfahrens.* PhD thesis, University of Heidelberg, 2006. URL http://www.ub.uni-heidelberg.de/archiv/6120.

[5] H. Fahlenkamp. *Zum Mechanismus des Stofftransports im laminar-welligen Rieselfilm.* PhD thesis, Technical University of Achen, 1979.

[6] Achim Falkenroth. *(in preparation).* PhD thesis, University of Heidelberg, 2007. URL http://archiv.ub.uni-heidelberg.de/volltextserver/index.php?la=en.

[7] Herlina. *Gas Transfer at the Air-Water Interface in a Turbulent Flow Environment.* PhD thesis, University of Karlsruhe, 2005. URL http://www.uvka.de/univerlag/volltexte/2005/71/.

[8] Herlina and G. H. Jirka. Application of LIF to inverstigate gas transfer near the air-water interface in a grid-stirred tank. *Experiments in Fluids*, 37:341-348, 2004.

[9] Herlina and G. H. Jirka. Turbulent gas flux measurements near the air-water interface in a grid-stirred tank. In C. S. Garbe, R. A. Handler, and B. Jähne, editors, *Transport at the Air Sea Interface - Measurements, Models and Parameterizations.* Chapter 2, pages 25-41, Springer Verlag, this volume, 2007.

[10] J. W. Hiby. Eine Fluoreszenzmethode zur Untersuchung des Transportmechanismus bei der Gasabsorption im Rieselfilm. *Wärme- und Stoffübertr.*, 1:105-116, 1968.

[11] J. W. Hiby. The chemical indicator: a tool for the investigation of concentration fields in liquid. *Ann.N.Y.Acad.Sci.*, 404:348-349, 1983.

[12] J. W. Hiby, D. Braun, and K. H. Eickel. Eine Fluoreszenzmethode zur Untersuchung des Stoffübergangs bei der Gasabsorption im Rieselfilm. *Chemie-Ing.-Techn.*, 39:297-301, 1967.

[13] B. Jähne. From mean fluxes to a detailed experimental investigation of the gas transfer process. In S. C. Wilhelms and J. S. Gulliver, editors, *Air-Water Mass Transfer, selected papers from the 2nd Int. Symposium on Gas Transfer at Water Surfaces*, Minneapolis, MN, 1991. ASCE.

[14] Bernd Jähne, C. Popp, U. Schimpf, and C. S. Garbe. Analysis of the heat transfer process across the aqueous heat boundary layer by active thermography: Mean transfer velocities and intermittence. In C. S. Garbe, R. A. Handler, and B. Jähne, editors, *Transport at the Air Sea Interface - Measurements, Models and Parameterizations.* Chapter 18, pages 255-274, Springer Verlag, this volume, 2007.

[15] R. A. Krause. Synthesis of ruthenium(II) complexes of aromatic chelating hererocycles. In *Structure and Bonding*, volume 67. Springer-Verlag, 1987.

[16] Hans Jürgen Mayer. Entwicklung einer laserinduzierten Fluoreszenz-Technik zum Messen von Konzentrationsprofilen und Diffusionskonstanten. Master's thesis, University of Heidelberg, 1995.

[17] T. Münsterer and B. Jähne. LIF measurements of concentration profiles in the aqueous mass boundary layer. *Experiments in Fluids*, 25: 190–196, 1997.

[18] T. Münsterer, H. J. Mayer, and B. Jähne. Dual-tracer measurements of concentration profiles in the aqueous mass boundary layer. In B. Jähne and E. C. Monahan, editors, *Air-Water Gas Transfer, selected papers from the 3rd Int. Symposium on Air-Water Gas Transfer*, Hanau, 1995. Aeon.

[19] Thomas Münsterer. *LIF Investigation of the Mechanisms Controlling Air-Water Mass Transfer at a Free Interface.* PhD thesis, University of Heidelberg, 1996.

[20] O. Stern and M. Volmer. Über die Abklingungszeit der Fluoreszenz. *Phys.Z*, 20:183–188, 1919.

[21] Evan Variano and E. A. Cowen. Quantitative imaging of CO_2 transfer at an unsheared free surface. In C. S. Garbe, R. A. Handler, and B. Jähne, editors, *Transport at the Air Sea Interface - Measurements, Models and Parameterizations.* Chapter 3, pages 43–57, Springer Verlag, this volume, 2007.

[22] W. M. Vaughan and G. Weber. Oxygen quenching of pyrenebutyric acid fluorescence in water. *Biochemistry*, 9:464, 1970.

[23] L. M. Wolff, Z. C. Liu, and T. J. Hanratty. A fluorescence technique to measure concentration gradients near an interface. In S. C. Wilhelms and J. S. Gulliver, editors, *Air-Water Mass Transfer, selected papers from the 2nd Int. Symposium on Gas Transfer at Water Surfaces*, pages 210–218, Minneapolis, 1991. ASCE.

[24] P. T. Woodrow and S. R. Duke. LIF measurements of oxygen concentration gradients along flat and wavy air-water interfaces. In M. A. Donelan, W. M. Drennan, E. S. Saltzman, and R. Wanninkhof, editors, *Gas Transfer at Water Surfaces, selected papers from the 4th Int. Symposium on Gas Transfer at Water Surfaces*, volume 127 of *Geophysical Monograph*, pages 83–88, 2002.

5

Visualization of 2-D Divergence on the Free Surface and its Relation to Gas Transfer

Aldo Tamburrino[1], Claudio Aravena[2], and John S. Gulliver[3]

[1] Associate Professor
Department of Civil Engineering
University of Chile
Santiago, Chile
atamburr@ing.uchile.cl

[2] Civil Engineer
Public Work Ministry
Santiago, Chile
claudio.aravena.r@moptt.gov.cl

[3] Joseph T. and Rose S. Ling Professor
Department of Civil Engineering
University of Minnesota
Minneapolis, MN 55455
gulli003@umn.edu

Abstract An experimental study designed to determine free-surface characteristics of a turbulent flow, generated by jets that emerge from the bottom of the pool, is described. Particle Tracking Velocimetry at the free-surface was used to measure the velocity field, vorticity field and two-dimensional divergence (Hanratty's β). While the high magnitudes of β did occur at similar locations to high surface vorticity, the scale of these maximal and minimal are approximately 1/3 of the scales of surface vorticity. This indicates that the scales of vorticity and β may not be correlated, or at least are correlated at a substantial multiplicative reduction. The relationship between β and liquid film coefficient is discussed. For flows where transfer is dominated by all frequencies of β, $K_L \sim (D\beta_{rms})^{1/2}$ is sufficient. However, there are flows where the large eddies are more important, and a spectra of β needs to be computed, according to McCready et al's [12] relation $K_L \sim (DS_{\beta max})^{1/2}$. In this case, a methodology is needed for transferring wave number spectra of β to frequency spectra in tanks without a significant flow, in order to estimate K_L. In our experiments, 1% of the turbulent kinetic energy was chosen to simulate similar results in a flume.

C.S. Garbe, R.A. Handler, B. Jähne (eds.): Transport at the Air Sea Interface
pp. 73-86, 2007,

5.1 Introduction

The structure of the flow on the free surface plays an important role in mass and heat transfer through the air-water interface. A first principal approach was performed by Chan and Scriven [3] and Sikar and Hanratty [15]. Chan and Scriven reduced the important components of the mass transport equation very near a free surface to:

$$\frac{\partial C}{\partial t} + w\frac{\partial C}{\partial z} = D\frac{\partial^2 C}{\partial z^2} \tag{5.1}$$

where C is concentration, w is velocity normal to the interface, D is diffusion coefficient, t is time and z is dimension normal to the interface. They solved this equation for a continuous upwelling of an irrotational flow near a free surface. Simultaneously, Sikar and Hanratty [15] solved the above equation for fixed surfaces without the irrotational flow assumption by dividing turbulence into high frequency and low frequency components and performing an order of magnitude analysis. Both of these articles identified $\beta(t) = \partial w/\partial z$ as the important parameter to describe the influence of turbulence on mass transfer near an interface. The challenge was then to measure β close to an interface. This was accomplished by Lau [10] who used special optics on a Laser doppler velocimetry system to measure $\beta(t)$ near a fixed interface, in this case the inside of a clear pipe. Campbell and Hanratty [2] used Lau's measurements to solve Equation (5.1) for transfer from a solid surface, measured mass transfer from a solid surface, and found that the liquid mass transfer coefficient could be characterized with the equation:

$$K_L = u_\star F(\beta(t)) Sc^{-7/10} \tag{5.2}$$

where $u_\star$ is a shear velocity at the interface, Sc is a Schmidt number and F is a function that uses the frequency spectra of $\beta(t)$:

$$F(\beta(t)) = 0.237 {S^+_{\beta m}}^{0.21} \tag{5.3}$$

where $S^+_{\beta m} = S_{\beta m}\nu/u_\star^2$ and $S_{\beta m}$ is the maximum value of the frequency spectrum of β with units of time^{-1}.

McCready, Vassiliadou, and Hanratty [12] assumed that the turbulence that was measured by Lau near a fixed interface could also be applied to a free interface as long as the free interface boundary conditions were applied, and developed the following expressions for the liquid-side mass transfer coefficient, K_L from measurements of gas transfer in a wind-wave flume:

$$K_L = \sqrt{a_1 D \beta_{rms}} \quad \text{large } \omega_c \tag{5.4}$$

$$K_L = \sqrt{a_2 D S_{\beta max}} \quad \text{small } \omega_c \tag{5.5}$$

where β_{rms} is the root mean square value of β and ω_c is a cut off frequency defining whether small or large frequencies dominate the mass transfer process. Since their wind-wave flume had a large cut-off frequency, the coefficient, $a_1 = 0.5$, could only be determined for Equation (5.4). Equations (5.2) through (5.5) show us that a free surface boundary condition, with no restriction to movement in the surface, has a $K_L \sim D^{1/2}$ relation, and a fixed surface boundary condition has a $K_L \sim D^{0.7}$ relationship.

Law and Khoo [11] measured β for stirred tanks and wind-wave flumes through PIV measurements of a vertical Laser sheet close to the free surface, finding that $a_1 = 0.05$ applied to both types of experiments. This value, however, was an order of magnitude below that of McCready, et al. [12]. Because of the predominant work of Hanratty and his students in the development of the theory, Law and Khoo also labeled it Hanratty's β.

Tamburrino and Gulliver [16] realized that, if the free surface was sufficiently normal to the camera, streaklines or PIV of the two-dimensional divergence of the free-surface turbulence and an application of continuity would result in a determination of β:

$$\beta = \frac{\partial w}{\partial z} = -\left(\frac{\partial u}{\partial x} + \frac{\partial v}{\partial y}\right) \tag{5.6}$$

Equations (5.5) and (5.6) were applied to free-surface turbulence measurements by Tamburrino and Gulliver [17] to determine the liquid film coefficient in a flume without wind, with $a_2 = 0.21$.

Equations (5.4) and (5.5) bear a remarkable resemblance to the Danckwerts' [4] surface renewal theory.

$$K_L = \sqrt{Dr} \tag{5.7}$$

where r is a mean surface renewal rate. The difficulty in applying Equation (5.7) has been its conceptual nature, i.e. there is no quantitative definition of what constitutes a surface renewal eddy. Prior experimental studies have struggled with this question, and used open spots that develop on a free surface sprinkled with flour or another surface tracer . Equations (5.4) and (5.5) are the first equations applied at a free surface that will enable direct characterization of surface renewal, or

$$r \sim \beta_{rms} \quad \text{large } \omega_c \tag{5.8}$$

$$r \sim S_{\beta max} \quad \text{small } \omega_c \tag{5.9}$$

Thus, identification of the vertical velocity gradient through Hanratty's β has the potential to experimentally characterize the mean surface renewal at a free surface that is exposed to turbulent flow.

This article presents experimental results on free-surface turbulence characteristics in a tank stirred by jets that emanate vertically from the

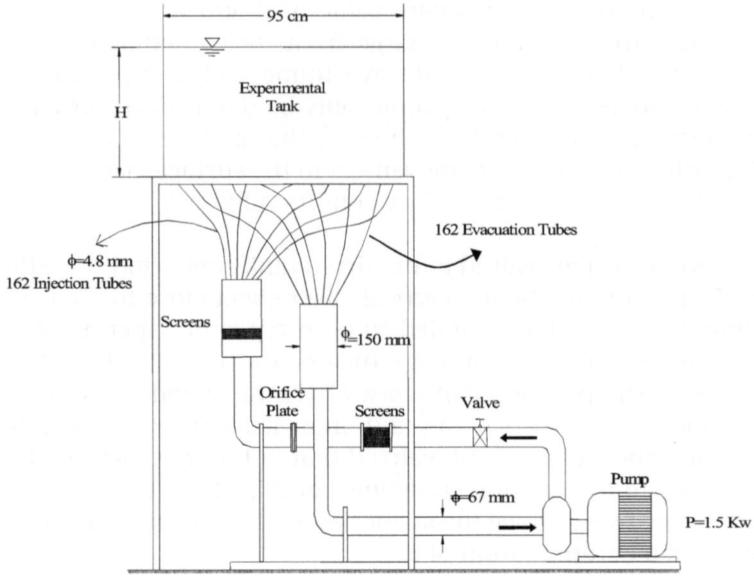

Figure 5.1. Experimental set-up. A 95 cm square tank was supplied by 162 jets and 162 evacuation tubes

bottom, and to compare them with the measurements from other experimental setups, such as grid-stirred tanks and open channels. Attempts to relate liquid film coefficient to Equation (5.8) indicate some restrictions to the use of this convenient relation.

5.2 Experiments and Analysis

The experimental installation consists of an acrylic tank, with a 95 cm square base and height of 70 cm. The mixing was supplied by 18×18 jets of 2.9 mm inner diameter, orthogonally spaced 5 cm and 5 cm from sidewalls of the pool. Half of the jets were used for injection of water, and half for evacuation, with the injection and evacuation ports equally spaced. The water circulated through a closed circuit, impelled by one 1.5 kW stainless steel pump, as outlined in Fig. 5.1.

The free surface velocity field was obtained from the trajectory of photographed tracer particles or particle tracking velocimetry (PTV). In order to be able to distinguish the direction of tracer movement, a flash was instigated at the beginning of each exposure, so that the particles exhibited a head followed by a tail in the direction of movement. The images were expanded to $1 \text{ m} \times 1 \text{ m}$ and digitized for analysis. If the surface is essentially normal to the camera [17], the images will give the velocity vec-

Table 5.1. Experimental Conditions. H is water depth, V is the mean jet velocity, ν is kinematic viscosity, β_{rms} is the root mean square value of Hanratty's β, k is the kinetic energy of turbulence on the free surface and Re_T is the free-surface turbulent Reynolds number, $\sqrt{k}H/\nu$

Exp.	H	V	ν	β_{rms}	k	K_L	Re_t
	(cm)	(cm/s)	(cm^2/s)	(1/s)	(cm^2/s^2)	(cm/s)	
A1	44	0.72	1.088×10^{-2}	0.06	0.058	3.64×10^{-4}	970
A2	44	1.17	1.017×10^{-2}	0.21	0.185	4.67×10^{-4}	1860
A3	44	1.69	0.999×10^{-2}	0.40	0.306	5.75×10^{-4}	2440
B1	33	0.71	1.127×10^{-2}	0.16	0.153	3.81×10^{-4}	1140
B2	33	1.14	1.042×10^{-2}	0.27	0.331	5.42×10^{-4}	1820
B3	33	1.72	1.061×10^{-2}	0.33	0.565	7.50×10^{-4}	2340
C1	22	0.69	1.035×10^{-2}	0.43	0.564	8.67×10^{-4}	1200
C2	22	1.16	0.986×10^{-2}	0.61	1.200	1.32×10^{-3}	2440

tors on the water surface. The velocity magnitude was determined from the length of the tracer image divided by the exposure time. The velocity vectors are distributed over the surface, as shown in Fig. 5.2. They were interpolated to the nodes of a mesh using the Kriging method because it takes into account the presence of clusters of data and regions with low density of information in the interpolation process [1], with a separation of 5 mm between nodes. The experimental conditions are summarized in Table 5.1.

5.3 Results

5.3.1 Free Surface Vorticity

The velocity vector at each node was calculated and, from this, the vorticity ω:

$$\omega = \frac{\partial u}{\partial y} - \frac{\partial v}{\partial x} \tag{5.10}$$

where u and v are the x and y components of velocity on the free surface and z is the axis normal to the free surface. The distribution of vorticity can reflect the presence of flow structures on the free surface, as shown in Fig. 5.3. Experiment A1, with a vertical/horizontal aspect ratio of 0.46, indicates zones of greater vorticity near the corners of the tank. This occurs because the presence of the non-slip walls is still significant and a mean movement across the free surface towards the walls creates vorticity in the corners. The formation of pairs of circulation cells in the corners and center of the tank are observed, as shown in Fig. 5.2. Experiment C1, however, with an aspect ratio of 0.23 has a more uniform distribution of vorticity maxima and minima.

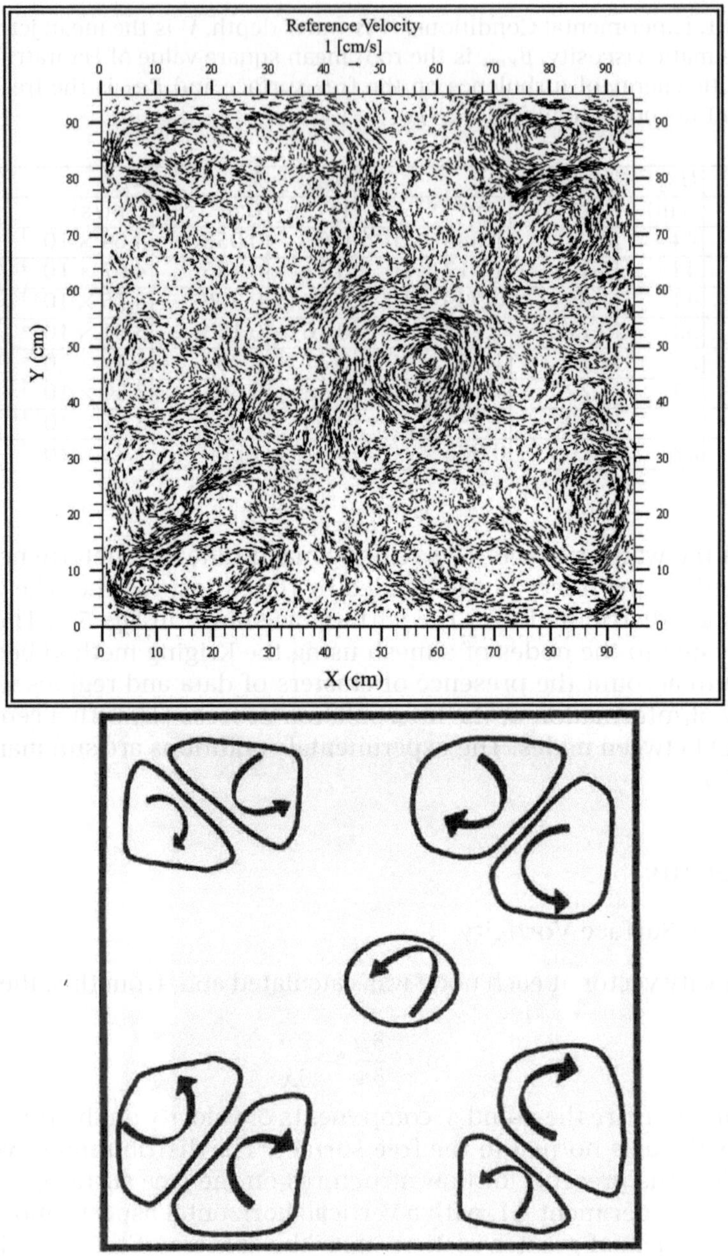

Figure 5.2. Particle tracer velocities at the surface with 46 cm of depth, experiment A1, indicate large structure in the turbulence

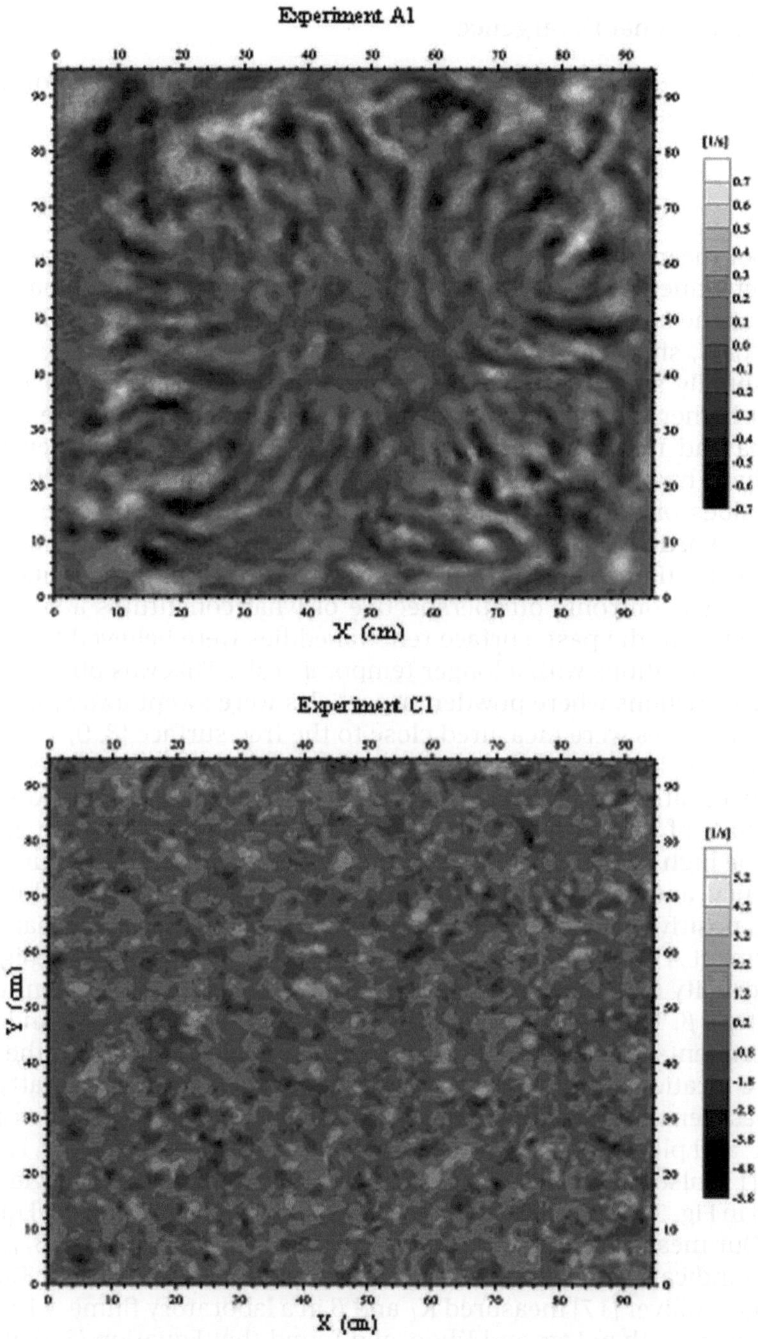

Figure 5.3. Vorticity on the free surface of the tank. Top: 46 cm depth indicates the presence of large flow structures. Bottom: 23 cm depth indicates minimal large flow structures.

Two-Dimensional Divergence

The velocity vectors at each node were also used to compute two-dimensional divergence (Hanratty's β):

$$\beta = -\left(\frac{\partial u}{\partial x} + \frac{\partial v}{\partial y}\right) \tag{5.11}$$

The root mean square (RMS) value of β are given in Table 5.1, as are the turbulent kinetic energy of the free surface turbulence per unit mass, computed as the square of surface velocity scale: $k = \frac{1}{2}(u^2 + v^2)$. The spatial scales of β, shown in Fig. 5.4 for the same experiments as Fig. 5.3, are substantially smaller than those of vorticity. While Experiment A1 does exhibit higher maxima and minima in Fig. 5.4, the spatial scale of these maxima and minima are approximately 1/3 of the scales given in Fig. 5.3. There has been speculation that the high values of β occur where high values of vorticity exist [17]. Figures 5.2 and 5.4 indicate that the scales of vorticity and β may not be correlated, or at least are correlated at a substantial multiplicative reduction. The information contained in Fig. 5.4 also confronts our perspective of what constitutes a surface renewal eddy. In the past, surface renewal eddies were believed to occur at upwelling locations with a longer temporal scale. This was observed with photos of regions where powder or particles were swept away [5, 6, 7], or where velocities were measured close to the free-surface [8, 9, 14]. A similar region is known to occur in the center of Experiment A1, because of the aspect ratio. Figure 5.4 indicates that this region does not exhibit the magnitude of β maxima and minima that occur in the corners. It appears to be the high velocity gradients of the surface vortices that occur at high frequency, and not low frequency upwelling, that constitute the locations of high positive values of Hanratty's β or surface renewal. A comparison of Experiment A1 in Figs. 5.2 and 5.4 indicates that the locations of high surface vorticity (i.e., the corners) are associated with a higher magnitude of Hanratty's β. This indication was also seen in all of the series A and series B experiments. In series C, there is little discernable pattern to the water surface locations of high magnitudes of either vorticity or Hanratty's β.

Measurements of liquid film coefficient in the tank are given in Table 5.1, and plotted to test the validity of Equation (5.4) in Fig. 5.5. Law and Khoo [11] also measured K_L and β_{rms} in a tank stirred by a single jet, as shown in Fig. 5.5. For their data, $K_L \sim (D\beta_{rms})^{1/2}$, as indicated by Equation (5.4). Our measurements had a lower value of K_L, with a lower β_{rms}, and did not indicate the dependence given in Equation (5.4). Further, Tamburrino and Gulliver [17] measured K_L and β in a laboratory flume, with lower values of each than Law and Khoo, and found that Equation (5.4) also did not fit their measurements. Finally, McKenna and McGillis [13] measured K_L and β in a tank with an oscillating grid. A slope of approximately 0.5 is seen below $\beta_{rms} = 2\,\mathrm{s}^{-1}$. Thus, it appears that Equation (5.4) applies to

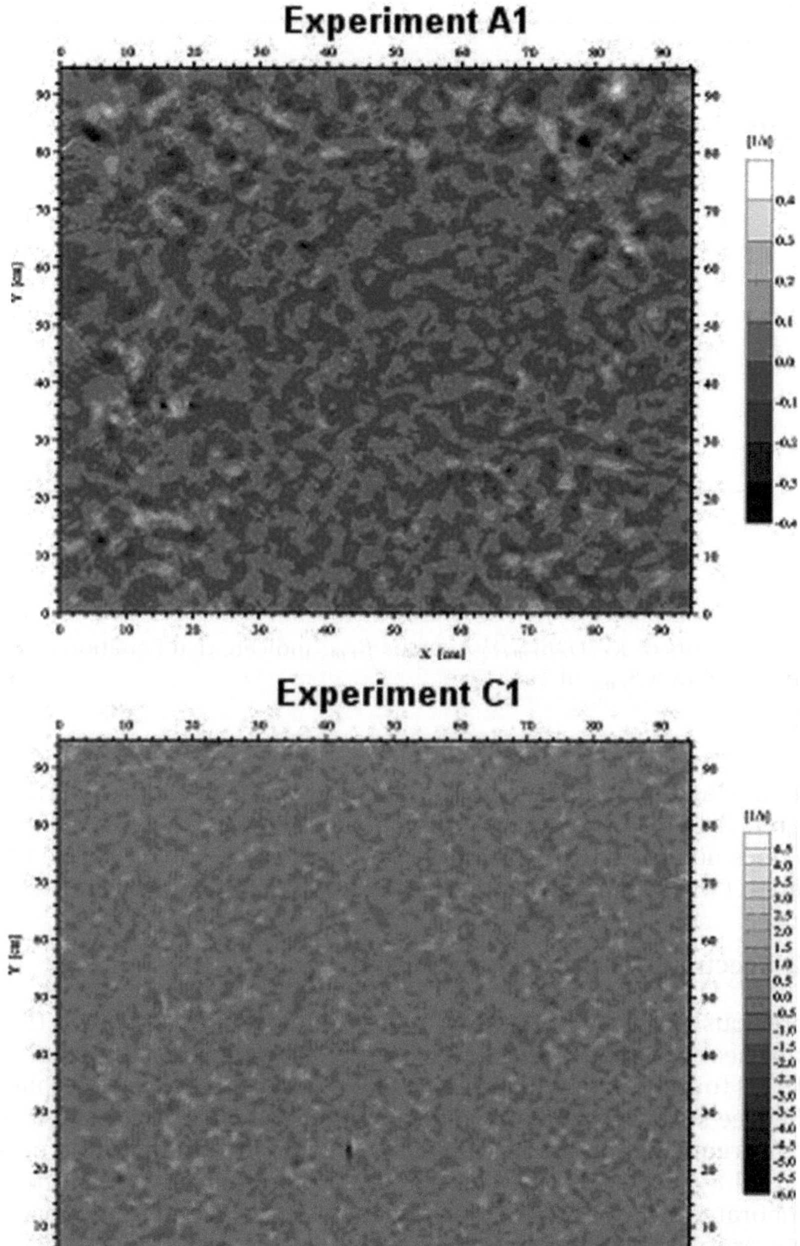

Figure 5.4. Two dimensional divergence (Hanratty's β) for experiments A1 and C1.

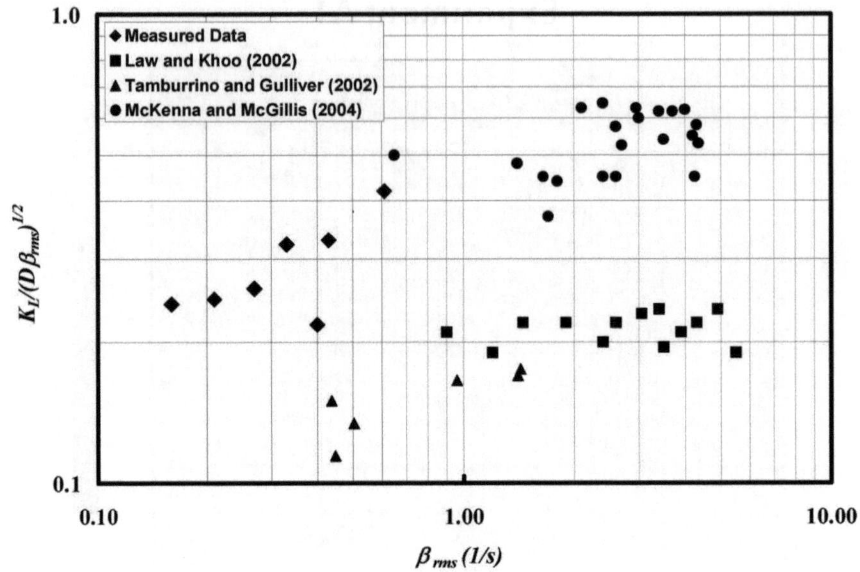

Figure 5.5. Plots of $K_L/(D\beta_{rms})^{1/2}$ versus β_{rms} indicate that Equation (5.4) does not apply below a β_{rms} of $1\mathrm{s}^{-1}$.

water surfaces with a moderate to high wind and strongly agitated tanks. For mildly agitated tanks, such that the deviations of the free surface are not greater than approximately 1 mm and with $\beta_{rms} = 1 - 2\,\mathrm{s}^{-1}$, Equation (5.4) does not describe the gas transfer process. We will therefore turn to Equation (5.5), which means that a spectra of Hanratty's β is needed.

5.3.2 Spectra of Two-Dimensional Divergence

As previously indicated, Hanratty's β is an important variable in the process of heat and mass transfer across the air-water interface. McCready, et al. [12] found, in applying the turbulence measured close to a solid wall to the free-surface boundary condition, that the β spectra has a -2 slope at high frequencies. Tamburrino and Gulliver [17] modified the parameters used by McCready, et al. to develop a more universal β-spectra, and corroborated its validity for open channels. In wave number space, the Tamburrino and Gulliver spectrum of β, $S_{\beta\beta}$, is described by:

$$\frac{S_{\beta\beta}}{L \cdot \beta_{rms}^2} = \frac{a}{L(1 + b(\kappa - \kappa_0)^2)} \tag{5.12}$$

where L is some integral turbulence length scale, a and b are constants and κ_0 is the wave number corresponding to a maximum in $S_{\beta\beta}/\beta_{rms}^2$. The

Table 5.2. Best-fit parameters of the β spectrum in the jet-stirred tank.

Exp.	a (cm)	b (cm^2)	κ_0 (1/cm)
A1	0,051	2,35	0,60
A2	0,028	0,56	0.98
A3	0,026	0,43	1,09
B1	0,025	0,40	0,95
B2	0,026	0,45	0,94
B3	0,032	0,75	0,80
C1	0,030	0,64	0,82
C2	0,031	0,70	0,88

values of parameters a, b and κ_0 for the different experiments appear in Table 5.2. Samples of the β spectra of the measurements are given in Fig. 5.6. The solid, continuous line in Fig. 5.6 corresponds to Equation (5.12), with the parameters provided in Table 5.2. The spectra proposed by McCready et al. is successfully fit to the data, with a slope of -2 at high frequencies. The spectra of McCready, et al. was developed for a sheared, free-surface boundary. Tamburrino and Gulliver found that, with some modification, this spectra could also be applied to an open-channel flow, with shear generated away from the free-surface. Now, the spectra has also been applied successfully to a jet-stirred tank.

Equation (5.5) uses only the peak value of the spectrum, meaning that low frequency renewal events are more important than the high frequency events in determining K_L. The peak of the wave number spectra, a in Table 5.2, must also be transformed into the peak value of a frequency spectra. Tamburrino and Gulliver used the mean surface velocity in their flume, with Taylor's hypothesis of frozen turbulence, to transform their

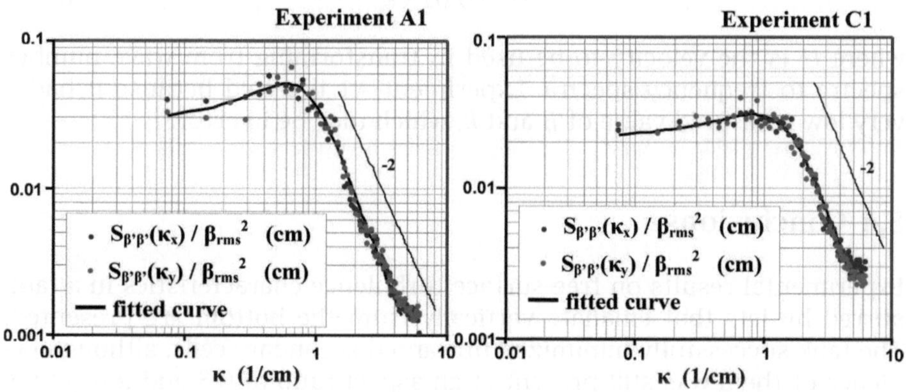

Figure 5.6. Spectra of Hanratty's β for experiments A1 and C1.

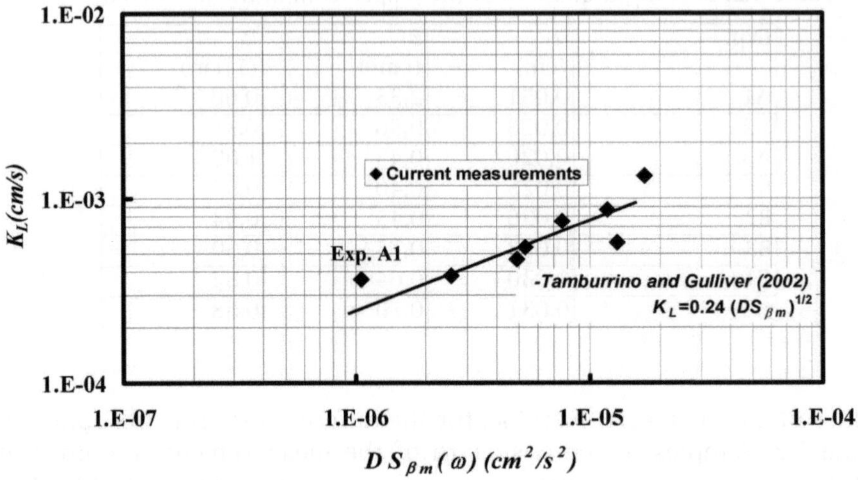

Figure 5.7. Comparison of Tamburrino and Gulliver's [17] relationship with the current measurements when $U = 0.01k^{1/2}$ was used to convert wave number spectra to frequency spectra.

wave number into a frequency. They found the following relation fit their flume data:

$$K_L = 0.24\,(DS_{\beta m})^{1/2} \tag{5.13}$$

where $S_{\beta m}$ is a transform of a from a wave number spectra. There is no mean velocity in the tank, however, so we used the kinetic energy of turbulence as a surrogate, and fit the data to Equation (5.13), as shown in Fig. 5.7, with the following wave number to frequency transformation:

$$U = 0.01\sqrt{k} \tag{5.14}$$

where U is the velocity to be used in transforming from wave number spectra to frequency spectra. Experiment A1 is noted because it had a very low measured value of β and k, which may be in error.

5.4 Conclusions

Experimental results on free-surface turbulence characteristics in a tank stirred by jets that emanate vertically from the bottom are presented. The tank successfully minimized unwanted secondary cells, although evidence of them was still present at an aspect ratio of 0.5 and low jet velocities. The measurements revealed the following:

1. Two-dimensional divergence of the free-surface turbulence was computed for the eight experiments and β-spectra were developed. The spectra proposed by McCready et al. [12] is successfully fit to the data, with a log-slope of -2 at high frequencies for all experiments.
2. Views of Hanratty's β on the free surface, computed from measurements, changes the perspective of what constitutes a surface renewal eddy. In the past, surface renewal eddies were believed to occur at upwelling locations with a longer temporal scale. Although these regions existed in the tank experiments, they did not exhibit the magnitude of β maxima and minima that occur in the corners. This is an indication that the high velocity gradients of the surface vortices that occur at high frequency, and not low frequency upwelling, that constitute the locations of high positive values of β or surface renewal.
3. While the high magnitudes of β did occur at similar locations to those of high surface vorticity, the scale of these maxima and minima are approximately 1/3 of the scales of surface vorticity. This indicates that the scales of vorticity and β may not be correlated, or at least are correlated at a substantial multiplicative reduction.
4. Our measurements, at a lower β and liquid film coefficient than Law and Khoo's, did not seem to follow the $K_L \sim (D\beta_{rms})^{1/2}$ relationship. The implications are that the free surface turbulence was less intense, the lower frequency renewal events are dominant, and a $K_L \sim (DS_{\beta m})^{1/2}$ may be preferable.
5. The wave number to frequency transformation using $0.01\sqrt{k}$ needs to be investigated further.

Acknowledgement. Funding for this project was provided by the Foundation for Scientific Research and Technology of Chile, through projects FONDECYT No. 1990025 and 1020822. The last author would also like to thank the Fulbright Program of the U.S. State Department and the Commission for Educational Exchange between the United States and Chile for the Fulbright Scholar support.

References

[1] J. C. Agüi and J. Jimenez. On the performance of particle tracking. *J. Fluid Mech.*, 185:447–468, 1987.

[2] J. A. Campbell and T. J. Hanratty. Mass transfer between a turbulent fluid and a solid boundary: Linear theory. *AIChE J.*, 28:988, 1982.

[3] W. C. Chan and L. E. Scriven. Absorption into irrotational stagnation flow. *Industrial and Engineering Fundamentals*, 9(1):114–120, 1970.

[4] P. V. Danckwerts. Significance of liquid-film coefficients in gas adsorption. *Ind. Eng. Chem.*, 43(6):1460–1467, 1951.

[5] J. T. Davies and W. Khan. Surface clearing by eddies. *Chem. Eng. Sci.*, 20:713, 1965.

[6] J. T. Davies and F. J. Lozano. Turbulence and surface renewal at the clear surface in a stirred vessel. *AIChE Journal*, 30:502, 1984.

[7] J. S. Gulliver and M. J. Halverson. Air-water gas transfer in open channels. *Water Resourc. Res.*, 25:1783, 1989.

[8] S. Komori, Y. Murakami, and H. Ueda. The relationship between surface renewal and bursting motions in an open-channel flow. *J. Fluid Mech.*, 203:103, 1989.

[9] S. Komori, R. Nagosa, and Y. Murakami. 'mass transfer into a turbulent liquid across the zero-shear gas-liquid interface. *AIChE Journal*, 36:957, 1990.

[10] K. K. Lau. *Study of Turbulent Structure Close to a Wall Using Conditional Sampling Techniques.* PhD thesis, Univ. of Illinois, Urbana, 1980.

[11] C. N. S. Law and B. C. Khoo. Transport across a turbulent air-water interface. *AIChE Journal*, 48(9):1856–1868, 2002.

[12] M. McCready, E. Vassiliadou, and T. Hanratty. Computer simulation of turbulent mass transfer at a mobile interface. *AIChE Journal*, 32(7):1108–1115, 1986.

[13] S. P. McKenna and W. R. McGillis. The role of free-surface turbulence and surfactants in air-water gas transfer. *Int. J. Heat Mass Transfer*, 47(3):539–553, 2004.

[14] M. Rashidi and S. Banerjee. Turbulence structure in free-surface channel flows. *Phys. Fluids*, 31(9):2491–2503, 1988.

[15] K. K. Sikar and T. J. Hanratty. Relation of turbulent mass transfer to a wall at high schmidt numbers to the velocity field. *J. Fluid Mech.*, 44:589–603, 1970.

[16] A. Tamburrino and J. S. Gulliver. Free-surface turbulence measurements in an open-channel flow. *E. P. in Rood and J. Katz (Eds.) ASME*, FED Vol. 181:103–112, 1994.

[17] A. Tamburrino and J. S. Gulliver. Free-surface turbulence and mass transfer in a channel flow. *AIChE Journal*, 48(12):2732–2743, 2002.

6

The Air-Water Interface: Turbulence and Scalar Exchange

Sanjoy Banerjee

Department of Chemical Engineering,
Department of Mechanical Engineering,
Bren School of Environmental Science and Management,
University of California, Santa Barbara, CA 93105
banerjee@engineering.ucsb.edu

Abstract Transport processes at gas–liquid interfaces play a central role in many industrial and environmental systems. For example, such processes at the air–water interface impact evaporation of water from reservoirs, control of regional climate, evasion of carbon in tropical river systems, and the aeration of hypoxic water, to name a few. There has also been recent intense interest in this area associated with the transport of greenhouse gases and moisture between the atmosphere and terrestrial water bodies. But despite all this activity, recent estimates of oceanic carbon dioxide uptake still vary by factors ~2. Since carbon dioxide is thought to be sequestered in the oceans at a rate equivalent to ~40% of its generation rate from man-made sources, such uncertainties have obvious policy implications.

This poor state of knowledge arises primarily due to interfacial scalar exchange being controlled by near-surface turbulence, which is difficult to measure and simulate as interfaces move, deform, and sometimes break. However, recent advances in particle imaging velocimetry and numerical approaches have led to some progress in this area, which is discussed. In low winds, the large-scale turbulence structure on the liquid side of the surface is found to be quasi two-dimensional, consisting primarily of attached spiral eddies (whirlpools). With increasing wind, streaks and bursts appear, much like in wall turbulence, but with detailed differences in structure, e.g. interface-parallel intensities peak right at the surface. At even higher wind speeds, short length, $\sim \mathcal{O}(10\,\text{cm})$, interfacial waves start to break, qualitatively enhancing turbulence and scalar exchange rates. It is shown that over this whole range of conditions the mean surface-velocity divergence field is expected, for theoretical reasons, to be highly correlated with scalar transfer rates. Direct numerical simulations and experimental measurements appear to support this hypothesis. It is shown that the surface divergence field is related to the mean square wave slope, which may be remotely observed. Such remote measurements may then serve as scalar exchange surrogates, enabling estimation of reliable regional and global budgets.

C.S. Garbe, R.A. Handler, B. Jähne (eds.): Transport at the Air Sea Interface
pp. 87-101, 2007,

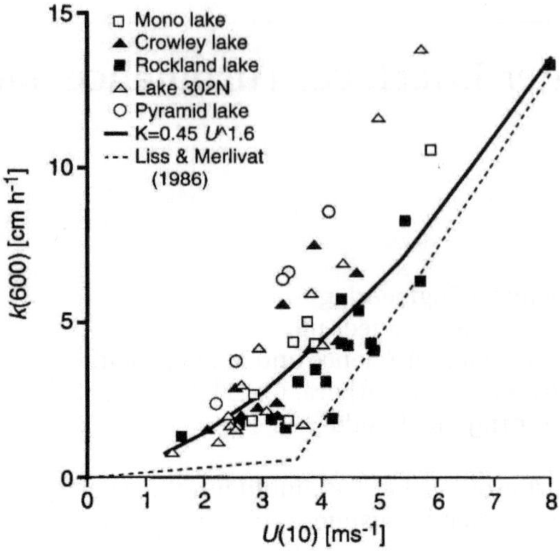

Figure 6.1. Gas transfer coefficients vs. wind speed from lakes [16]. Liss and Merlivat [14] fit shown dotted.

6.1 Introduction

Exchange of heat and mass across deformable fluid–fluid interfaces is central to many industrial and environmental processes but is still poorly understood, despite the considerable effort that has gone into its study. For example, such processes are important in condensers, evaporators, and absorbers, as well as in many environmental problems, like transfer of greenhouse gases at the air-sea interface.

Focusing on environmental applications, Figure 6.1 compares experimental air-water gas transfer coefficients, versus the wind speed 10 meters above the water surface, U_{10}, to the widely used parameterization by Liss and Merlivat [14]. The scatter in the data is due to several reasons discussed later. The effect of such uncertainties is shown in Table 6.1, where estimates of global carbon uptake by the oceans vary by factors of 3, even if other estimates [9, 10] suggest somewhat reduced variabilities, but still differing by a factor of 2. depending on the choice of parameterization. As a recent study of dissolved oceanic carbon concludes that virtually all the anthropogenic CO_2 released from 1800 to 1994 has partitioned between the ocean and atmosphere [22], such uncertainties obviously impact policies related to greenhouse gas emissions, which has major economic implications.

Table 6.1. Global oceanic CO_2 uptake estimates using different gas exchange-wind speed relationships [7]

Source	Equation		Flux $\mathrm{Pg\,C\,yr^{-1}}$
Liss and Merlivat [14]	$k = 0.17U_{10}$	$U_{10} < 3.6\,\mathrm{m\,s^{-1}}$	
	$k = 2.85U_{10} - 9.65$	$3.6\,\mathrm{m\,s^{-1}} < U_{10} < 13\,\mathrm{m\,s^{-1}}$	-1.1
	$k = 5.9U_{10} - 49.3$	$U_{10} > 13\,\mathrm{m\,s^{-1}}$	-1.1
Wanninkhof [24]	$k = 0.39U_{10}^2$ (long term averaged winds)		-2.2
Wanninkhof & McGillis [25]	$k = 1.09U_{10} - 0.333U_{10}^2 + 0.078U_{10}^3$ (long term averaged winds)		-3.3
Nightingale et al. [19] NCEP 6 hour winds	$k = 0.333U_{10} + 0.222U_{10}^2$		-1.7

The low quality of predictions arises from lack of understanding of fluid motion near deforming interfaces, which usually controls scalar exchange. This, in turn, arises from difficulties in measuring and numerically simulating velocity fields in the vicinity of moving surfaces. The problem is made even more complex by the flows being turbulent in most situations of practical interest. Understanding of turbulence near solid boundaries (like wall turbulence) is still incomplete, in spite of the vast effort devoted to this end. It is not surprising, therefore, that much less is known about turbulence near moving, deforming, and in some cases breaking, fluid-fluid interfaces. Nonetheless, progress is being made in this difficult but important area largely due to relatively recent developments of nonintrusive measurement techniques such as digital particle imaging velocimetry (DPIV) and direct numerical simulation (DNS) of turbulent flows over complex boundaries.

The objective of this paper is to discuss recent results, particularly from laboratory experiments and direct numerical simulations (DNS), regarding fluid motion and scalar exchange across gas-liquid interfaces. The simplest case is when the gas is not moving and turbulence is generated somewhere else, perhaps at the bottom of a channel or in a shear layer, and then impinges on the interface, on which there is no wind shear imposed. A more complex situation arises when the gas phase is moving, in which case the interface develops waves due to wind forcing, and these waves affect interfacial transport processes. We will discuss in brief what can be learned for both these situations and induce from these results a model for liquid-side controlled mass and heat exchange processes, e.g. interphase transfer of a sparingly soluble gas. Some experimental results will also be presented to indicate the validity of the direct numerical simulations and the proposed model.

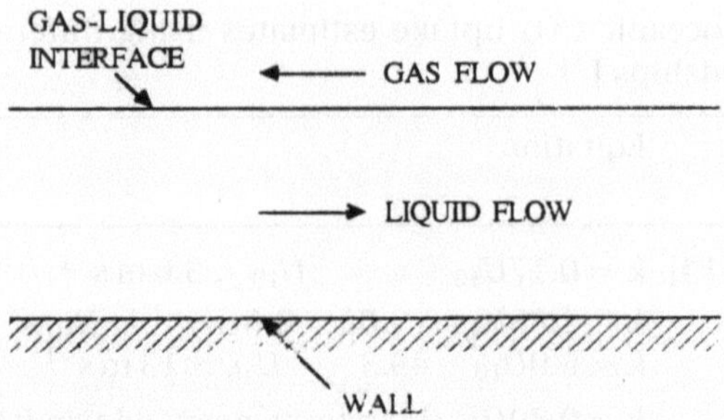

Figure 6.2. Schematic of the canonical problem studied here. The gas can flow cocurrent or countercurrent to the liquid stream, or not at all.

Of necessity, this review is rather brief and focuses on work in our own laboratory, but the work on experiments and accompanying DNS in Prof. Komori's laboratory at Kyoto University indicates very similar results [12, 13, 26].

Start by considering the canonical problem in Figure 6.2, which sketches the problem in general terms. A liquid layer flows over a bottom boundary, and the gas flow can either be cocurrent, countercurrent, or not flow at all. This is the situation that is to be simulated, and the gas–liquid interface can develop waves, which interact with the flows, which themselves can be turbulent. If one looks down at the free surface from the top when the flowing liquid is turbulent, then, in the absence of gas shear, the structures shown in Fig. 6.3 are obtained. These consist of upwellings, which are essentially bursts emanating from the bottom boundary, vortices associated with these upwellings, and downdrafts, which are not visible clearly in this figure but are nonetheless present [2]. The vortical structures can merge, if they have a like rotation direction, into larger vortices; sometimes these structures can annihilated by an upwelling. It is worth noting at this point that the upwellings arise from bursts that are generated at the boundary, as shown in this flow visualization from the side of the channel indicated in Fig. 6.2.

Turning now to turbulence phenomena that occur when wind shear is imposed, Figure 6.4 shows that the free surface region also generates burst-like structures, as well as bursts from the bottom boundary. The top three panels in Fig. 6.4 are for countercurrent gas flow and the bottom three for cocurrent gas flow [21] The interesting finding here is that, in spite of the boundary condition at the gas–liquid interface being quite different from that at the wall, the qualitative turbulence structure under shear conditions are rather similar, though of course different in detail. We will now discuss what happens at wind-sheared interfaces in more detail.

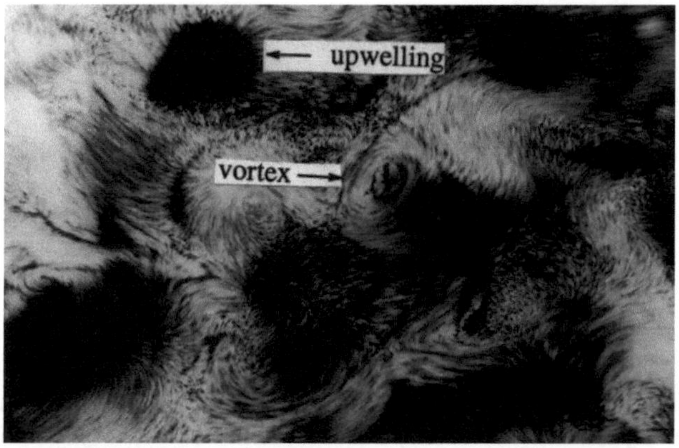

Figure 6.3. Flow structures at an unsheared gas-liquid interface with microparticles visualized looking at the liquid surface from above. The dark areas are upwellings emanating from the bottom boundary.

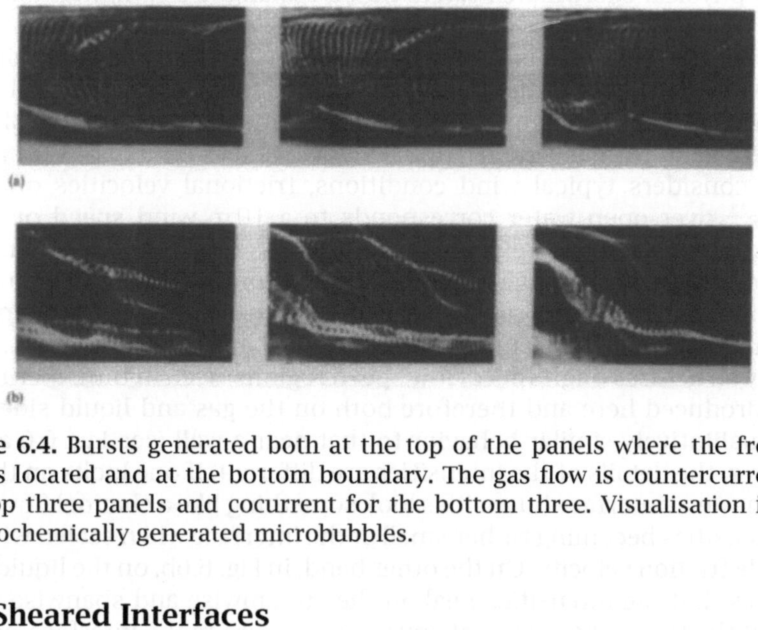

Figure 6.4. Bursts generated both at the top of the panels where the free surface is located and at the bottom boundary. The gas flow is countercurrent for the top three panels and cocurrent for the bottom three. Visualisation is with electrochemically generated microbubbles.

6.2 Sheared Interfaces

When wind stresses are imposed on the liquid surface, then the situation becomes more complex and interfacial waves start to have a significant effect. If the waves are not of high steepness and are in the capillary or capillary gravity range, then DNS can be conducted by using a fractional time-step method in which each of the liquid and gas domains are treated separately and coupled to interfacial stress and velocity continuity boundary conditions. In addition, at each time step the domains must be mapped

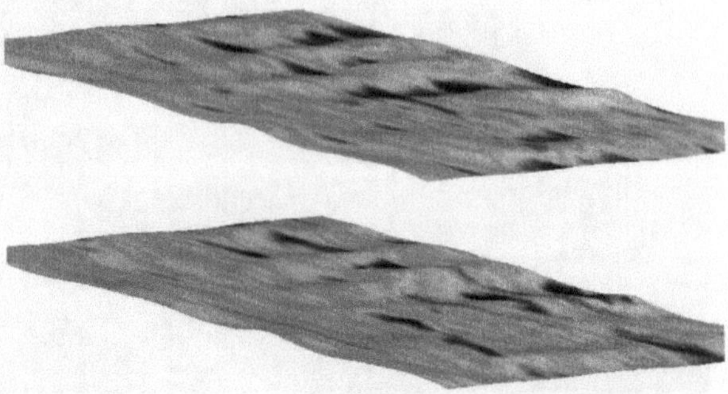

Figure 6.5. High-speed/low-speed structures near the gas-liquid from the DNS [6]. The top panel is just above the wavy surface, and the bottom just below. The blue regions have low streamwise velocity and the yellow/orange, high streamwise velocity.

from a physical coordinate system into a rectangular computational system.

While the procedure is somewhat laborious, De Angelis et al. [6] have done extensive testing validating convergence and accuracy, provided the steepness is small enough that the interface does not break. Such simulations are valid for gas-side frictional velocities less than about $0.1\,\mathrm{m\,s^{-1}}$. If one considers typical wind conditions, frictional velocities of about $0.1\,\mathrm{m\,s^{-1}}$ over open water corresponds to a 10 m wind speed of about $3.5\,\mathrm{m\,s^{-1}}$. Above these velocities, the interface starts to break and the DNS conducted in the manner described become inaccurate or infeasible. However, something can be learned from these DNS, and the typically structures seen in the vicinity of the interface are shown in Fig. 6.5. The streaky structures (high speed/low speed regions) seen in the experiments are reproduced here and therefore both on the gas and liquid sides one sees qualitatively similar behavior to that near a wall (see Fig. 6.6 a & b). However, the details of the intensities are different. If one looks on the gas side, the near-interface intensities look something like a channel flow, with the intensities becoming rather small at the interface when scaled with the gas-side friction velocity. On the other hand, in Fig. 6.6b, on the liquid side, one sees that the intensities peak in the streamwise and spanwise directions at the free surface and, of course, go to zero in the interface-normal direction, provided that the coordinate system is set relative to interfacial position. If scalar exchange such as gas transfer is studied with DNS, then, if the phenomenon is gas-side controlled, then the mass (or heat flux) is highly correlated with the interfacial shear stress[1]. However, on the liq-

[1] Typically, absorption of highly soluble gases such as ammonia from air would be gas-sid control. Another example would be water vapour transport at the air-water interface.

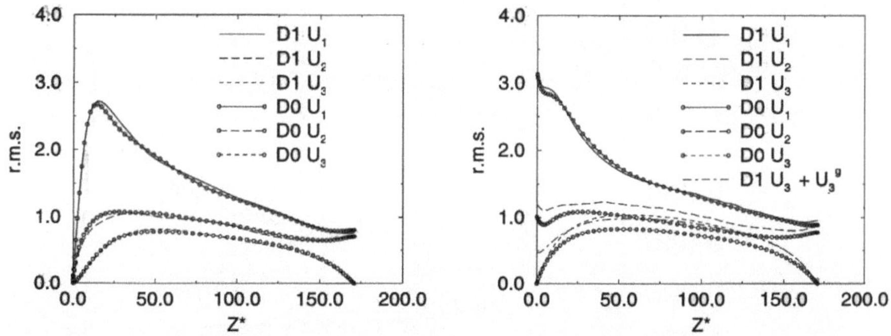

Figure 6.6. Turbulence intensities near the gas–liquid interface [6]. Ordinate is normalised with friction velocity and abscissa is in wall units. left = gas side; right = liquid side. z_* is the distance from the interface normalised by the interfacial shear stress, phase density and viscosity. D0 refers to cases run with a flat interface, D1 with a deformable interface and D1 $U_3+U_3^g$ to the result for the interface normal velocity on the liquid side, corrected for interfacial velocity.

uid side it is not. The DNS indicates, however, that ejections and sweeps, which can be discerned by a quadrant analysis, correlate with the shear stress on the gas side but do not on the liquid side.

Note here that the ejections on the gas side arise over the low shear regions, whereas the sweeps, which bring high-speed fluid towards the interface, are associated with high-sheer stress regions. This is to be expected, but the same phenomenon does not appear on the liquid side. This is due to the gas driving the shear-stress pattern. Therefore, on the gas side the high-shear stress regions may be expected to correspond to regions of high mass or heat transfer and the low-shear stress regions to low scalar transfer. However, on the liquid side the shear stress patterns are not an indicator of mass transfer rates, nor are they an indicator of ejections or sweeps. Therefore, one has to look at the DNS directly to determine what controls scalar exchange rates when it is liquid-side dominated. De Angelis et al. showed that they were also dominated by the sweeps, with high exchange rates occurring with the sweep, whereas low exchange rates occur with ejections. This is illustrated in Fig. 6.7. In this figure, a large negative value of u_1', u_2' ($-ve$ in the coordinate system used) indicates a sweep which brings high speed fluid towards this interface. This gives rise to a peak in β following which there is a decay that roughly corresponds to the predictions of surface renewal theory.

Since the frequency of sweeps and ejections scale with the frictional velocity on each side of the interface, De Angelis et al. [6] were able to develop a relationship between frictional velocity and scalar exchange rates, as shown in Equations (6.1) and (6.2) below.

$$\beta_w \mathrm{Sc}_w^{0.5} u_{\star,w} \sim 1 \tag{6.1}$$

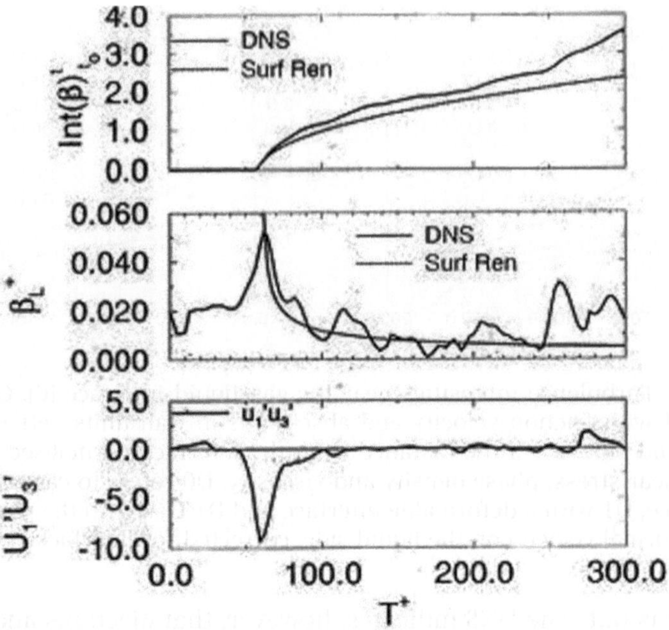

Figure 6.7. Panels showing that a sweep, with high $u_1 u_3$ gives rise to a high mass transfer rate β_T^* on the liquid side. Note that the surface renewal model predicts the instantaneous and integrated value of β^* quite well if the sweep is considered as a "renewing" event. β^* is the mass transfer velocity nondimensionalised by u_{fric}^*, and T^* is a time nondimensionalised by u_{fric}^*, and ν (the kinematic viscosity).

with the subscript w denoting the liquid side and $u_\star \sim D_f/\rho$ where ρ is the fluid density and D_f the frictional drag. $\text{Sc} = \nu/D$ is the Schmidt number, with ν being the kinematic viscosity. An equivalent expression was derived for the gas side, with subscript a denoting the gas side, in the form

$$\beta_a \text{Sc}_a^{2/3} u_{\star,a} \sim 1 \tag{6.2}$$

Here, Sc_a is the ratio of the kinematic viscosity on the air side divided by the molecular diffusivity of the species, such as water vapour. The mass transfer velocity β is in $\text{m}\,\text{s}^{-1}$ (also denoted as k in Fig. 6.10).

These correspond very well with experimental data at low wind velocities, i.e. $u_{10} < 3.5\,\text{m}\,\text{s}^{-1}$. However, there is another way to interpret the DNS, as discussed in the next section, which has a somewhat better chance of working under breaking conditions.

6.3 Surface Divergence Model and Microbreaking

Regions of high divergence occur in the upwellings which bring fresh bulk liquid to the interface, and regions of convergence form at the downdrafts, which are not apparent in the figures, but can be seen clearly in direct numerical simulation [20]. When the interface is sheared, bursts consisting of ejections (events which take interfacial fluid away) and sweeps (events which bring bulk fluid to the interface), much like in wall turbulence. Sweeps form regions of high surface divergence, and ejections, of high convergence. The surface divergence model is based on the realization that the region near the interface that controls scalar exchange is of very small interface-normal dimensions. Typically, the thickness of the "film" over which the resistance to transfer lies, is $\sim \mathcal{O}(100\,\mu\text{m})$ for transfer of sparingly soluble gases such as CO_2 and methane [6]. For the gas-side controlled processes it is $\sim \mathcal{O}(1\text{ mm})$. It is the motion in these very thin regions that dominate the transport processes. For a gas–liquid interface, the turbulent fluctuations parallel to the interface proceed relatively unimpeded (in the absence of surfactants), a fact validated by direct numerical simulations with the exact stress boundary conditions at a deformable interface [6, 8]. For example, even with mean shear imposed on the liquid by gas motion, the surface-parallel turbulent intensities peak at the interface [15], whereas in the gas it peaks some small distance away, as in wall turbulence. Therefore, to a first approximation, the liquid-side interface-normal velocity, w', can be written as

$$w' \sim \partial w'/\partial z|z_{\text{int}} + \text{HOT} \tag{6.3}$$

with z being the surface-normal coordinate and HOT standing for higher order terms. This can be related to the divergence of the interface-parallel motions at the surface, as

$$-\left.\frac{\partial w'}{\partial z}\right|_{z_{\text{int}}} = \left.\left(\frac{\partial u'}{\partial x} + \frac{\partial v'}{\partial y}\right)\right|_{z_{\text{int}}} = \gamma \tag{6.4}$$

where the quantity in parentheses is the surface divergence of the surface velocity field fluctuations, γ. The streamwise and spanwise coordinates tangential to the moving interface are x and y, and z is the normal. In a fixed coordinate system, Equation (6.3) requires an additional term due to surface dilation [3, 5].

For stagnation flows, interface-parallel motions and diffusion may be neglected, as pointed out by Chan and Scriven [5], who showed a similarity solution existed in this case. Clearly, a solution of the problem of turbulence giving rise to the surface divergence field, in which γ would vary as a function of position and time, is not possible analytically, but McCready et al. [17] in a landmark paper showed that the root mean square (rms)

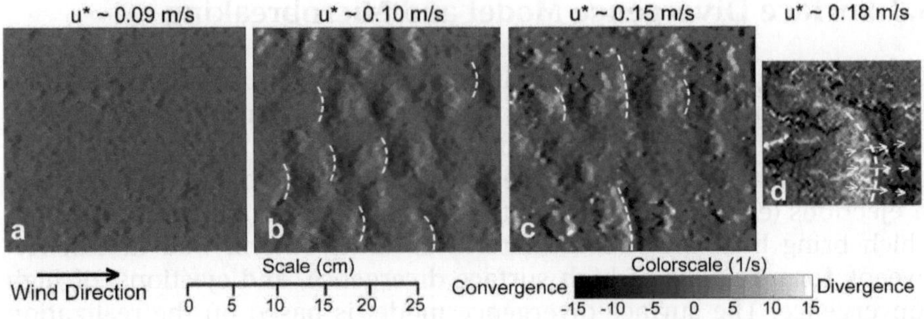

Figure 6.8. Plan-view image of the air-water interface with increasing wind speed - $u_\star$ is the wind side friction velocity. Grayscale represents velocity divergence. White dashed lines are wave crests. When microbreaking commences convergence is seen ahead of wave crests and divergence behind [23].

surface divergence was nonetheless approximately related to the average transfer coefficients by

$$\overline{\beta}/u \sim \mathrm{Sc}^{-1/2}\mathrm{Re}_t^{-1/2}\left[\overline{(\partial u'/\partial x + \partial v'/\partial y)^2}\right]_{\mathrm{int}}^{1/4} \tag{6.5}$$

Re_t is the turbulent Reynolds number based on far-field integral length scale, Λ, and velocity scale, u. Note that this is essentially a theoretical expression for $\overline{\gamma}$, and the constants arise by fitting a much more complex expression to this simplified form [3].

Hunt and Graham's theory [11] connects bulk turbulence parameters to those near the interface by superposing an image turbulence field on the other side, which impedes surface normal motions, redistributing the kinetic energy to surface parallel motions, which are enhanced. The predictions have been well verified in experiments [1]. The theory also allows a connection to be made between the bulk turbulence parameters and the surface divergence and leads to the correlation:

$$\overline{\beta}/u \approx C\mathrm{Sc}^{-1/2}\mathrm{Re}_t^{-1/2}\left[0.3\left(2.83\mathrm{Re}_t^{3/4} - 2.14\mathrm{Re}_t^{2/3}\right)\right]_{\mathrm{int}}^{1/4} \tag{6.6}$$

This last model has been used by researchers conducting gas transfer field experiments to correlate data and it applicability has been reviewed by Banerjee and MacIntyre [4], where it is called the surface divergence model. Forms of the surface divergence model were also tested for gas transfer at the surface of stirred vessels [18], as well as recently in direct simulations [3] - in both cases, with success. Most recently, Turney et al. [23] tested a form of the model for gas transfer across interfaces, with wind forcing sufficiently high to induce microbreaking, and found that it agreed with data in both microbreaking and non-microbreaking conditions.

We will now consider the accuracy of Equations (6.5) and (6.6) against both experiments and direct numerical simulations. First, turning to ex-

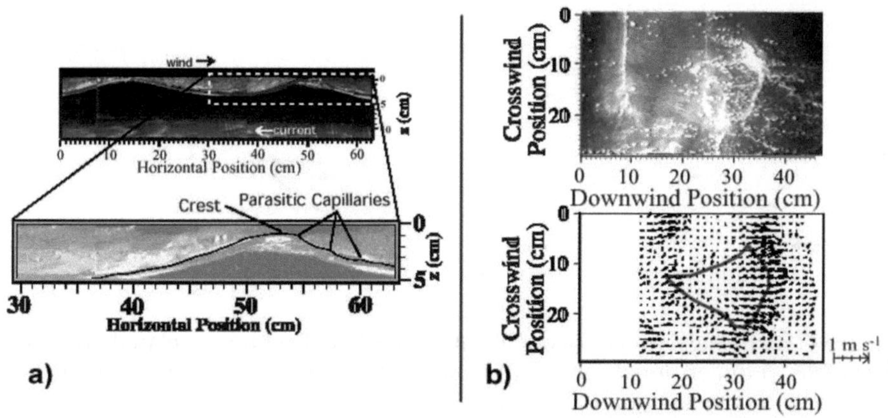

Figure 6.9. a) Side-view of a microbreaking wave, showing typical scale. b) Plan-view of a microbreaking wave, visualized with floating particles. Particles accumulate in the convergence zones ahead of the wave crests.

periments. A typical measurement of the surface divergence field made by top-view DPIV of glass micro balloons floating at the liquid surface is shown in Fig. 6.8. The wind conditions are indicated by $u_\star$ in the figure, and one can see the surface divergence patterns becoming more pronounced as the wind velocity is increased. Strong convergence zones form just upstream of microbreaking wave crests, and divergence zones form behind. Surface topography measured by a shadow-graph method [23], from which wave slopes can be calculated, and the surface is seen to roughen substantially as wind velocity is increased, coinciding with the onset of microbreaking – the operational definition of which is the movement of surface water at the same speed as the wave crest. For example, particles scattered in a liquid surface converge at a microbreaking wave crest and are swept along with it. A typical profile of a microbreaking wave is shown in Fig. 6.9, together with a top down view showing particle accumulation just ahead of the wave crests.

The surface divergence model presented in Equation (6.5) appears promising from the comparison in microbreaking conditions with measured gas transfer rates [23]. The connection between surface divergence and bulk turbulence parameters shown in Equation (6.6) has been tested in DNS of wind-forced flows (without breaking) [3]. While the model was expected to be applicable primarily to unsheared interfaces, it appears to work surprisingly well for wind-forced flows as well. The relationship between the rms surface divergence and the mass transfer coefficient is shown in Fig. 6.10a [23]. In Fig. 6.10b we also show the relationship between surface divergence and mean surface wave slope, which may be measured by satellite remote sensing methods [23]. The relationship be-

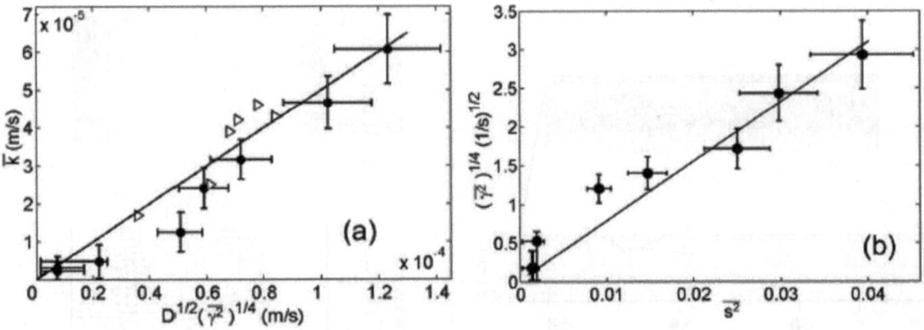

Figure 6.10. a) Mass transfer coefficient vs. surface divergence related group derived from a dimensional version of Equation (6.5). b) square root of rms surface divergence vs. mean-square wave slope.

tween gas transfer rates and wave slope has also been previously reviewed in several studies e.g. Glover et al. [9].

In Fig. 6.11, some results of a direct numerical simulation are shown for wind forcing without breaking. The connection between bulk turbulence parameters and surface divergence given in Equation (6.6) is checked and seen to be valid.

6.4 Conclusions

The discussion focussed on turbulence and gas transfer at interfaces sheared by the wind, and simulations were based on boundary fitting which successfully captured the behavior of low-steepness interfacial waves and the interaction with the underlying turbulence. The method worked at low wind stress imposed on the free surface, corresponding to $U_{10}<3.5\,\mathrm{m\,s^{-1}}$. Above these wind speeds, the DNS indicated interface breaking, which it could not capture. This also corresponds to what is observed in field experiments. Nonetheless, the DNS elucidated the structures that control scalar exchange rates on the gas and liquid side, indicating that sweeps were important in both. This then led to a parameterization that appears to be accurate at low wind speeds. When wind speeds exceeded these conditions, i.e. led to gas-side friction velocities of $\sim 0.1\,\mathrm{m\,s^{-1}}$ on the surface, waves of lengths ~ 10 cm, and amplitudes $\sim$ 1 cm started to "microbreak", leading to regions of convergence at the wave crests and divergence behind – as if the surface fluid were being sucked under the waves, rolling over the surface. These waves led to qualitative increases in scalar exchange rates on the liquid side, e.g. adsorption of sparingly soluble gases.

For conditions without microbreaking, simple parameterizations for scalar exchange rates in terms of friction velocity were derived. These

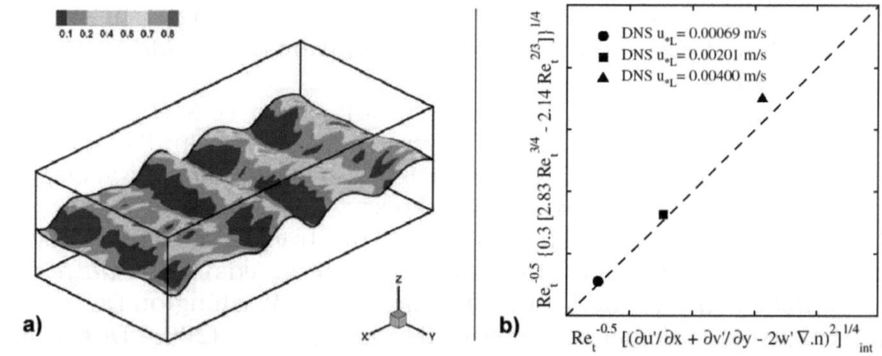

Figure 6.11. a) Interface configuration for the coupled gas–liquid DNS [3]. b) The surface divergence calculated from the DNS vs. the prediction in Equation (6.6) (shown as the ordinate).

applied only at low wind speeds. A more universal approach was needed, and the surface divergence model was assessed against simulations and experimental results, which correlated a broad range of effects, including those due to microbreaking. The surface divergence was also found to be related to mean square wave slope which might serve as a surrogate for gas transfer measurements, at least at typical oceanic wind speeds of $7.5\,\mathrm{m\,s^{-1}}$, and well beyond the limit where surfactants might affect transfer rates. Should this be proven in field experiments then it would be possible to obtain reliable regional and global estimates of CO_2 uptake by the oceans based on remote sensing of wave slope for lengths $< \mathcal{O}(1\,\mathrm{m})$. It is unusual that the laboratory studies and simulations, which are done at scales of a few cm, may contribute in this instance to better estimates of what might happen at scales of hundreds of kilometers.

References

[1] Banerjee S. (1990) Turbulence structure and transport mechanisms at interfaces. *Ninth Int Heat Transfer Conf.* Hemisphere Press New York

[2] Banerjee S. (1994) Upwellings, downdrafts, and whirlpools: Dominant structures in free surface turbulence. In: Kobayashi, AS (ed) Mechanics USA 47. Applied Mechanics Reviews, Seattle

[3] Banerjee S., Lakehal D., Fulgosi M. (2004) Surface divergence models for scalar exchange between turbulent streams. Int J Multiphase Flow 30:963

[4] Banerjee S., MacIntyre S. (2004) The air-water interface: Turbulence and scalar exchange. *Advances in coastal and ocean engineering* 9:181

[5] Chan W.C., Scriven L.E. (1970) Absorption into irrotational stagnation flow. A case study in convective diffusion theory. *Ind Eng Chem Fund* 9:114
[6] De Angelis V., Lombardi P., Andreussi P., Banerjee S. (1999) Microphysics of scalar transfer at air-water interfaces. In: Sajjadi S.G., Thomas N.H., Hunt J.C.R. (eds) *Wind-over-wave couplings: Perspectives and prospects.* Oxford Univ Press
[7] Donelan M.A., Wanninkhof R.H. (2002) Gas transfer at water surfaces – concepts and issues. In: Donelan MA et al. (eds) *Gas transfer at water surfaces.* American Geophysical Union Washington DC
[8] Fulgosi M., Lakehal D., Banerjee S., De Angelis V. (2003) Direct numerical simulation of turbulence in a sheared air-water flow with a deformable interface. *J. Fluid Mechanics* 482:319.
[9] Glover H., Frew D.M., McIne S.J., Bock E.J. (2002) A multi-layer time series of global gas transfer velocity from the TOPEX/POSEIDON dual frequency normalised backscatter algorithm. In: Donelan MA et al. (eds) *Gas transfer at water surfaces.* American Geophysical Union Washington DC
[10] Ho D.T., Law C.S., Smith M.J., Schlosser P., Harvey M., Hill P. (2002) Measurement of air–sea gas exchange at high wind speeds in the Southern Ocean. Implications for global parameterizations. In: Donelan MA et al. (eds) *Gas transfer at water surfaces.* American Geophysical Union Washington DC
[11] Hunt J.C.R., Graham J.M.R. (1978) Free-stream turbulence near plane boundaries. *J Fluid Mech* 84:209
[12] Komori S. (1990) Turbulence structure and mass transfer at a wind-driven air-water interface. In: Sajjadi S.G., Thomas N.H., Hunt J.C.R. (eds) *Wind-over-wave couplings: Perspectives and prospects.* Oxford University Press
[13] Komori S., Nagaosa R., Murukami Y. (1993) Turbulence structure and mass transfer across a sheared air-water interface. *J Fluid Mech* 249:161
[14] Liss P.S., Merlivat, L. (1986) Air–sea gas exchange rates: Introduction and synthesis. In: Buat-Menard P. (ed) *The role of air–sea exchange in geochemical cycles.* D Reidel Publishing Company
[15] Lombardi P., DeAngelis V., Banerjee S. (1996) Direct numerical simulation of near-interface turbulence in coupled gas–liquid flow. *Phys Fluids* 8;1643
[16] MacIntyre S., Wanninkhof R.H., Chanton J.P. (1995) Trace gas exchange across the air-water interface in freshwater and coastal marine environments. In: Matson P.A., Harris R.C. (eds) *Methods in ecology-biogenic trace gases: Measuring emissions from soil and water.* Blackwell Science New York

[17] McCready M.J., Vassiliadou E., Hanratty T.J. (1986) Computer-simulation of turbulent mass-transfer at a mobile interface. *AIChE J* 32:1108
[18] McKenna S.P., McGillis W.R. (2004) The role of free-surface turbulence and surfactants in air-water gas transfer. *Int J Heat Mass Tran* 47:539
[19] Nightingale P.D., Malin G., Law C.S., Watson A.J., Liss P.S. (2000) In situ evaluation of air–sea gas exchange parameterizations using novel conservative and volatile tracers. *Global Biogeochemical Cycles* 14:373
[20] Pan Y., Banerjee S. (1995) A numerical study of free-surface turbulence in channel flow. *Phys Fluids* 7:1649
[21] Rashidi M., Banerjee S. (1990) Streak Characteristics and Behavior Near Wall and Interface in Open Channel Flows. *J Fluids Eng* 112:164
[22] Sabine C.L., Feely R., Gruber N., Key R.M., Lee K. (2004) The oceanic sink for anthropogenic CO_2 *Science* 305:367
[23] Turney D., Smith W.C., Banerjee S. (2004) A measure of near-surface turbulence that predicts air–water gas transfer in a wide range of conditions. *Geophys Res Lett* 32:LO4607
[24] Wanninkhof R.H. (1992) Relationship between wind-speed and gas-exchange over the ocean. *J Geophys Res* 97:7373
[25] Wanninkhof R.H., McGillis W.R. (1999) A cubic relationship between air–sea CO_2 exchange and wind speed *Geophys Res Lett* 26:1889
[26] Zhao D., Toba Y., Suzuki Y., Komori S. (2003) Effect of wind waves on air-sea gas exchange. *Tellus B* 55:478

7

Turbulence Generation in the Wind-Driven Subsurface Water Flow

Guillemette Caulliez[1], Richard Dupont[1], and Victor I. Shrira[2]

[1] Institut de Recherche sur les Phénomènes Hors Equilibre
163 av. de Luminy, 13009 Marseille, France.
caulliez@irphe.univ-mrs.fr

[2] Keele University
Keele, Staffordshire, ST5 5BG, U.K.
v.i.shrira@maths.keele.ac.uk

Abstract The laminar-turbulent transition of the water surface boundary layer generated by a steady wind at the entrance of a wind-wave tank is investigated experimentally. Observations of the velocity field in water were made both by flow visualization techniques and laser Doppler velocimeter measurements. Two stages in the development of the perturbations have been clearly identified. First, the slow growth of streamwise longitudinal vortices embedded into the laminar flow is followed by a rapid development of secondary instabilities resulting in the pattern breakdown. The picture looks similar to the by-pass transition to turbulence in the boundary layers over rigid plate. At the second stage, peculiar to this free surface flow, an explosive deepening of the boundary layer and a fast development of inflexional instabilities occur inside localized areas. This phenomenon leads to an intense vertical mixing, which differs dramatically from the rigid plate scenario.

7.1 Introduction

Subsurface turbulent motions in wind-induced boundary layers in the oceans play a key role in heat and mass exchanges at the air-sea interface. Despite the increasing efforts aimed at investigating such flows in recent years (see e.g. [8], [7], [4]), the properties and the origin of the small-scale turbulence as well as its interaction with wind waves, mean shear current or other motions of larger scales present in water are still poorly known. This situation is largely due to the fact that all these processes coexist and overlap in space-time domain, making the observations extremely difficult to perform and to interpret. To shed light on the complex structure of the natural free surface boundary layers, a first step may consist in a thorough investigation of the individual mechanisms controlling their formation. In order to identify one of the fundamental processes

C.S. Garbe, R.A. Handler, B. Jähne (eds.): Transport at the Air Sea Interface
pp. 103-117, 2007,

at the origin of the turbulence development in subsurface water flows, the present work aims at describing the laminar-turbulent transition of the spatially-developing water surface boundary layer observed at the entrance of a wind-wave tank. Precise laboratory experiments performed in steady wind conditions thus enable us to establish the specific dynamical processes leading to the breakdown of the laminar shear flow and the rapid development of turbulent spots.

7.2 Experimental Procedure

The observations were carried out in the large and small IRPHE-Luminy wind-wave facilities, for wind speeds ranging from 2.5 to 7 m/s. The structure of the velocity field in the water surface boundary layer was explored at various fetches and depths downstream of the air-water junction plate, using flow visualization techniques based upon dye injection and two-component Laser Doppler velocimeter measurements as described in Caulliez [2]. The water surface displacements were measured simultaneously by means of a thin wire capacitance wave probe. Special measures were taken to minimize perturbations in the air and the water at the entrance of the tanks. In particular, the transition between both flows was made smooth by means of a weakly-inclined rough aluminium plate lengthened by a smooth floating plastic film. The water level at rest being adjusted before and controlled after each experiment, such an arrangement had the great advantage to make the observations insensitive to small water level change that may occur at the entrance of the tank due to evaporation or increase in wind-exerted surface stress. The mean and turbulent properties of the air surface boundary layer over the relevant water surface area at the entrance of the tank was investigated carefully by using Pitot tube and hot X-wire probes.

7.3 Overview of the Water Flow Structure

7.3.1 Visual Observations

A first description of the overall behavior of the water flow was provided by means of the dye visualizations. The typical scenario of the laminar-turbulent transition observed in the wind speed range from 3 to 7 m/s is illustrated in Fig. 7.1 by a series of water flow side views monitored by a photo-camera at increasing fetches. First, just downstream of the air-water junction, the straight marked streamlines parallel to the flat water surface indicate that the surface wind-induced current is laminar (Fig. 7.1**a**). Localized elongated perturbations distinguishable in both plan and side views as Z-like patterns appear from time to time but they evolve to a very small

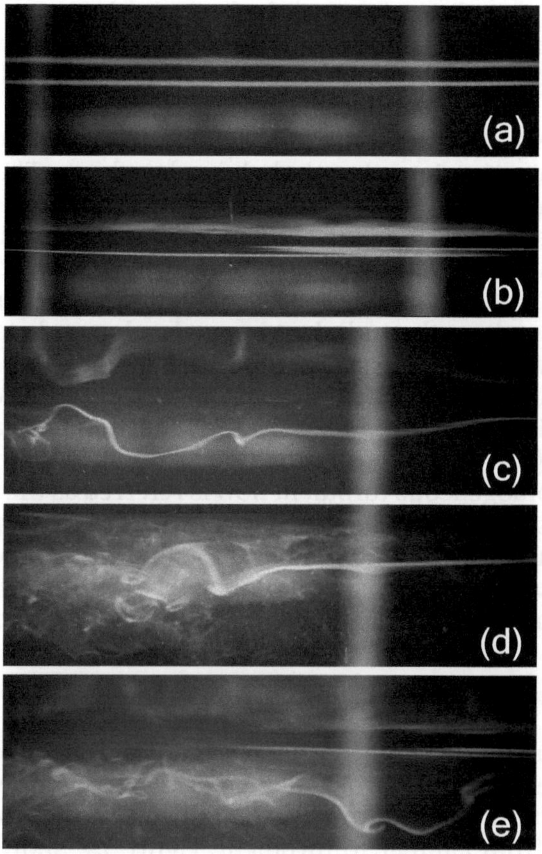

Figure 7.1. Typical evolution of the water flow instabilities traced by colored dye injected 30 cm upstream at a depth of 3 mm and viewed from the side at different fetches in the laminar zone **a**: U = 3.7 m/s, X = 80 cm; **b**: U = 5 m/s, X = 63 cm; **c** U= 5.0 m/s, X = 88 cm; in the transition zone **d**: U = 5 m/s, X = 98 cm; **e**: U = 5 m/s, X = 113 cm. The photograph scaling is 1 cm for 2.8 cm.

extent before vanishing (Fig. 7.1**b**). Then, further downstream, there is a relatively long stage where the perturbations present a consistent tendency to grow, exhibiting typical dye signatures as long wavy instabilities or trickled and skein-like filaments (Fig. 7.1**c**). Finally, at a certain fetch, the flow disturbances start to evolve very rapidly. The dye trajectory first develops instantaneous steep contortions mainly oriented downward. The presence of large oscillations of the residual dye filaments associated with the rapid dye diffusion into a deep layer suggests the formation of turbulent spots (Fig. 7.1**d**). Often after the streamline overturning the resulting turbulent spot disappears, washed up slowly downstream, and the laminar flow in the uppermost surface sublayer is restored until breaking down again, creating an intermittent pattern (Fig. 7.1**e**). Sometimes, the transition to turbulent flow occurs even faster, when the still rectilinear dye filament stops abruptly at a definite fetch and bursts in wreath of dye disseminating into the whole boundary layer. Then, the fetch at which the transition occurs for a particular wind speed widely varies, but the minimal fetch is well-defined and strongly wind-dependent. It is also noticeable that small-scale waves ruffling the free surface become visible at the point of the boundary layer breakdown. They mark the downstream regions where the turbulent spots develop and clearly show how the spots slowly expand spanwise until they merge. Thus, the characteristic V-shaped rough/turbulent streaks appear on the water surface [3]. They are relatively long-lived features for winds higher than 3 m/s. Hereinafter, the region running from the tips of the first V-streaks to the appearance of homogeneous turbulent flow is referred as the *transition zone*, while the use of the term *laminar zone* is restricted to the part of the laminar flow from the air-water junction to the first tips of the V-streaks and the term *turbulent zone* to the part of the turbulent flow where all streaks have already merged and the flow is transversely homogeneous.

7.3.2 Velocity Field Observations

To support quantitatively the visual observations, LDV measurements of the velocity field were performed at increasing fetches downstream of the air-water junction plate for the different stages of the spatial water flow development from laminar to turbulent. Figure 7.2 shows time samples of the longitudinal velocity signals observed at a depth of 2 or 3 mm for seven successive fetches. At the three first fetches located in the laminar zone, time records exhibit only short-lived velocity disturbances of small amplitude and rather long but variable time scale. However, one can easily notice a slow increase of the perturbation amplitude with fetch. Such velocity perturbations might be at the origin of the typical Z-like patterns observed previously by means of the water flow visualizations (Fig. 7.1**b**). At the forth fetch, more pronounced oscillatory fluctuations of typical time scales 2 to 5 s are clearly visible. As they correspond to streamwise flow perturbations of about 10 to 25 cm in wavelength, these disturbances can be very likely associated with the long wavy instabilities of the dye filament detected in flow visualizations just before the laminar boundary layer breakdown (Fig. 7.1**c**). At this stage indeed, these instabilities do not alter dramatically the mean flow field. The laminar to turbulent transition process itself is characterized by intense velocity fluctuations, as seen in the next three time series observed further downstream. At depths comparable to the thickness of the laminar boundary layer, the velocity fluctuations are particularly pronounced at the onset of the transition,

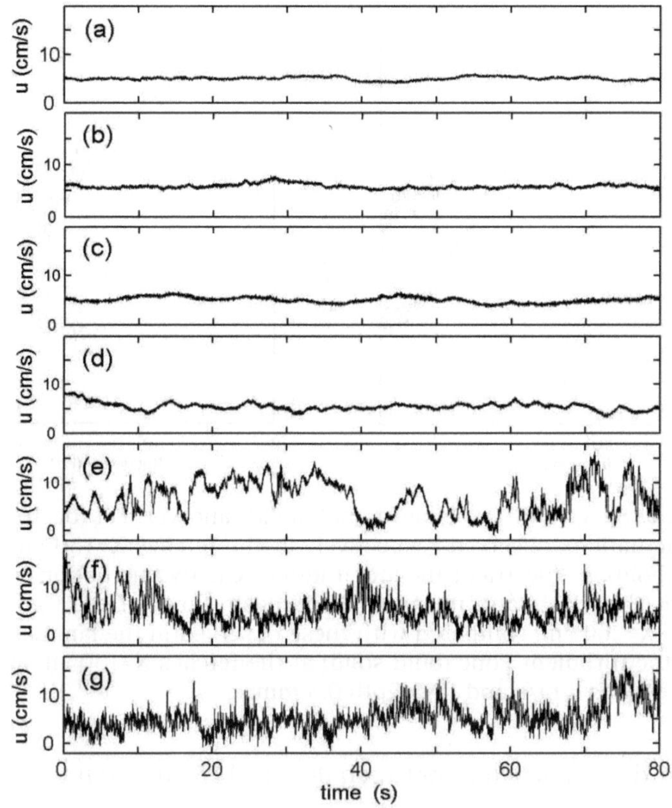

Figure 7.2. Time records of the longitudinal water velocity signals observed at seven fetches located in the laminar (**a**, **b**, **c**, **d**) and transition (**e**, **f**, **g**) zones, for 5 m/s wind speed and a depth of 2 mm (**a**, **b**) or 3 mm (**c**, **d**, **e**, **f**, **g**): **a**: X = 45 cm; **b**: X = 60 cm; **c**: X = 70 cm; **d**: X = 90 cm; **e**: X = 105 cm; **f**: X = 115 cm; **g**: X = 130 cm.

i.e. at 105 cm fetch here, where very rapid jumps up to 10 cm/s in amplitude occur randomly. The duration of the high speed and low speed sequences are extremely variable in time but at these depths, the average duration of the former decreases drastically with fetch in contrast to the latter. In fact, the signal variability observed in the surface sublayer reflects the unsteadiness and the spatial non-homogeneity of the velocity field in the transition zone, which is in direct relation with the intermittent development of localized turbulent spots of low mean velocity inside still-laminar high-velocity flow regions. The increase of the duration of the low-velocity time sequences reflects the crosswise expansion of the turbulent spots with fetch. A rapid increase of the high-frequency velocity fluctuations due to the sudden growth of surface waves is also noticeable

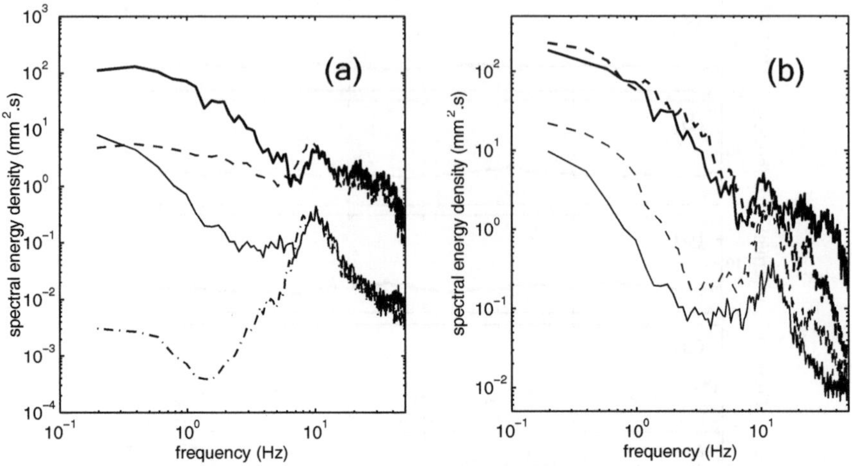

Figure 7.3. a: Spectra of the longitudinal (solid) and vertical (dash) water flow velocity fluctuations observed respectively at the fetches X=130 cm (thin) and X=260 cm (bold). **b**: Spectra of the longitudinal velocity fluctuations observed in the laminar (thin dash) and turbulent (bold dash) flows of the transition zone at the fetch X=200 cm, compared with these observed in the laminar zone (thin solid) and the turbulent zone (bold solid) at the fetches X=130 cm and 260 cm. The wind speed is 4 m/s and the depth 0.5 mm.

in the records, in particular between 90 and 115 cm fetches.

To get an idea of the energy distribution of the velocity perturbations in the streamwise and vertical directions, typical spectra of both velocity fluctuation components observed at two fetches located respectively in the laminar and the turbulent zones are displayed in Fig. 7.3**a**. This figure clearly shows that the energy of streamwise motions far exceeds the energy of vertical ones at any stage of the transition but this trend is more pronounced in the laminar zone. Most of the energy of streamwise fluctuations is also concentrated in the low-frequencies. However, in each zone, the curves of horizontal and vertical spectral energy densities tend to converge at high frequencies in the surface wave domain around 10 Hz. The sharp increase of these high-frequency velocity fluctuations with fetch is due to the rapid growth of surface waves in the transition area [3].

The existence of distinctive stages in the water flow evolution is also borne out by the behavior of the mean velocity vertical profiles (Fig. 7.4**a**). The profile observed at short fetch in the laminar zone shows that the wind-induced drift current develops in a very thin highly-sheared surface layer. The small scatter of the velocity measurements around the mean value, as illustrated in Fig. 7.4**b** for one depth, confirms that this

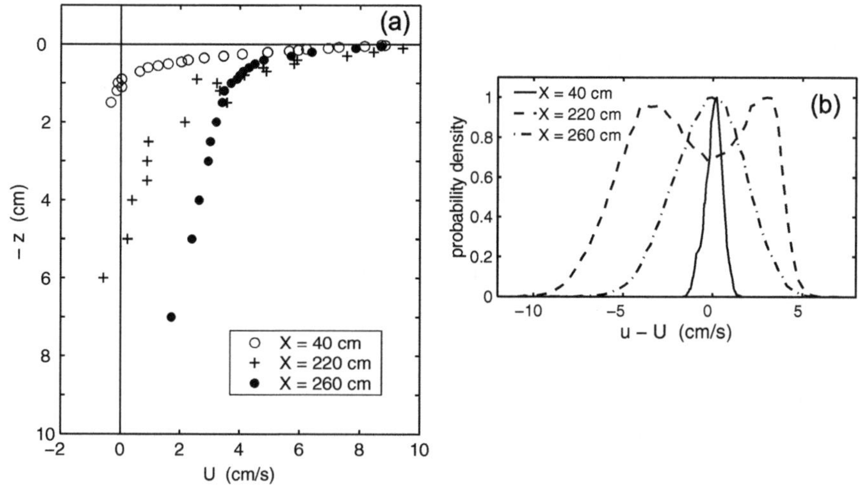

Figure 7.4. Mean vertical distribution **a** and typical histogram **b** of the water velocity observed in the small wind-wave facility at three fetches located respectively in the laminar, transition and turbulent zones, at 4 m/s wind speed. The histograms are evaluated at a depth of 6mm.

early stage of the boundary layer formation is controlled by viscous processes. In the transition zone, a significant increase in the boundary layer thickness is noticeable but the most striking feature of the mean velocity profiles observed at intermediate depths lies in the large data scatter. The characteristic two-peak shape of the related histograms strongly infers that the flow evolves there between two states, namely the laminar state associated with the rather narrow high-velocity peak and a more turbulent state associated with the wide low-velocity peak, the variation in the respective time spent in each state during one particular sequence then explaining the data scatter. The fully turbulent zone is marked by a significant deepening of the boundary layer. The vertical distribution of the mean velocity then tends to resemble a turbulent near-wall velocity profile characterized by a highly-sheared layer close to the surface and a slowly-varying velocity profile at further depths. The wide and nearly Gaussian-shaped histogram of the streamwise velocity signals also suggests that the flow is turbulent and recovers homogeneity at this stage of development.

7.4 The Laminar Zone

To understand the processes involved in the laminar-turbulent transition, first a detailed investigation of the structure of the laminar flow has been

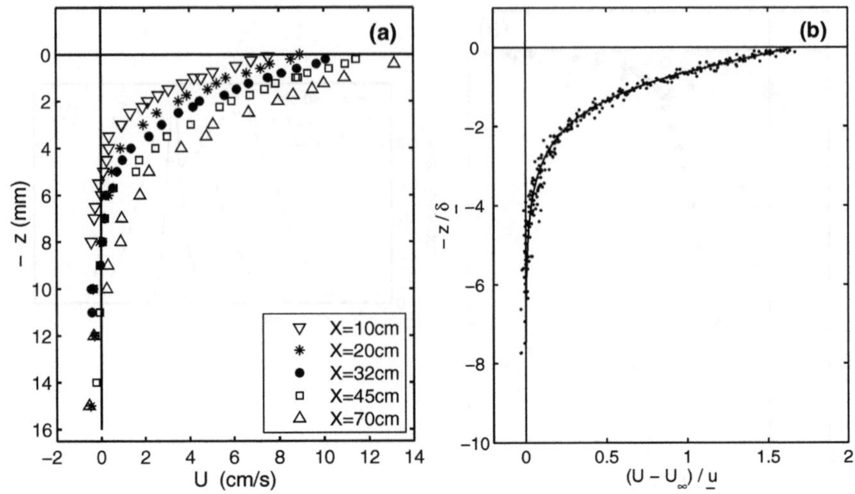

Figure 7.5. a: Vertical profiles of the mean water flow velocity observed at various fetches in the laminar zone at 5 m/s wind speed; **b**: Dimensionless vertical velocity profiles observed in the small-scale facility at various wind speeds and fetches over the laminar zone (the solid line corresponds to the profile described by Equation (7.3).

undertaken. The evolution with fetch of the vertical mean velocity profiles observed in the laminar zone is illustrated in Fig. 7.5**a** for one wind speed. These profiles are characterized by a regular increase of the velocity with fetch coupled with a deepening of the boundary layer. The flow velocity, maximal at the water surface, decreases quasi-linearly in the two first millimeters just underneath the water surface. Below this highly-sheared layer, the velocity gradient decreases gradually down to the depth where the flow velocity reaches zero and reverses. There, in the lowest layers of the water channel, a small velocity return current develops simultaneously with the surface wind-driven flow. The existence of a well-defined linear shear sublayer thus enables us to estimate with a good accuracy the viscous stress exerted by wind at the water surface. This tangential stress was specified in terms of the water friction velocity $u_{\star w}$ using the relationship $\tau_{0w} = \rho_w u_{\star w}^2 = \rho_w \nu_w (dU/dz)_{z=0}$ (Table 7.1). It was found that $u_{\star w}$ no longer varies with fetch, remaining approximately constant over most of the laminar flow development zone except in a narrow zone located immediately downstream of the air-water junction. There, a slight decrease is observed due to the change in the surface conditions from a solid flat wall to a moving liquid surface. Besides, in accordance with the continuity of the viscous stress at the water surface, this evolution in fetch agrees remarkably well with the evolution of the air friction velocity for all

Table 7.1. Characteristic parameters of the water flow surface boundary layer observed in the laminar zone at the various wind speeds and fetches.

U m/s	X m	$u_{\star a}$ cm/s	$u_{\star w}$ cm/s	U_s cm/s	δ_w cm
3	0.30	10.7	0.38	6.0	0.91
3	0.60	9.8	0.32	7.0	1.85
3	1.10	9.5	0.34	9.3	3.0
4	0.13	13.7	0.48	6.1	0.65
4	0.25	13.3	0.46	7.6	0.85
4	0.40	13.1	0.46	9.1	1.0
4	0.62	-	0.48	11.5	1.6
5	0.10	17.2	0.58	7.7	0.55
5	0.21	15.9	0.54	8.9	0.70
5	0.32	16.0	0.55	10.6	0.80
5	0.45	16.3	0.56	12.0	0.85
5	0.70	16.2	0.54	13.8	1.05
6	0.10	19.8	0.66	8.8	0.42
6	0.19	20.0	0.63	11.6	0.57
6	0.29	19.2	0.62	12.9	0.67
6	0.45	-	0.62	14.3	0.90

wind speeds. In view of the similar shape of the velocity profiles observed in the central part of the laminar zone where the tangential stress is constant, the evolution with fetch of both the surface drift current velocity U_s and the flow thickness δ_w was investigated in more details in Fig. 7.6. These quantities were made dimensionless by using viscous scales based on the molecular viscosity ν_w and the water friction velocity $u_{\star w}$. Strictly speaking, the surface boundary layer is defined as the water flow layer driven by the wind stress exerted at the water surface, i.e. the layer from the surface to the depth where the tangential stress crosses zero. However, the depth where the vertical velocity gradient vanishes is rather difficult to estimate because the reverse current observed over most of the deep layer of the water channel exhibits a very flat profile. Therefore, the boundary layer thickness is defined here in a much practical way as the depth where the water flow velocity crosses zero. The surface drift current estimated by extrapolation of the linear velocity profiles to the water surface is nevertheless referred to the constant reverse flow velocity U_∞. Fig. 7.6 shows that the data presented in terms of the non-dimensional quantities U_s^+ and δ_w^+ versus non-dimensional fetch[3] X^+ plotted in logarithmic scale collapse remarkably onto straight lines of slope 1/3. Their evolution is then described by the regression laws:

[3] X^+ corresponds to the Reynold number characteristic of the boundary layer development Re_X when based on the water friction velocity and the fetch X.

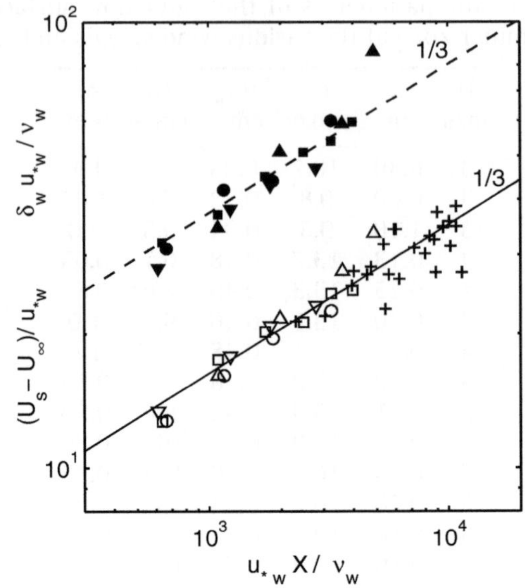

Figure 7.6. Evolution with dimensionless fetch X^+ of the dimensionless mean surface velocity U_s^+ (open symbols and crosses) and water surface boundary layer thickness δ_w^+ (closed symbols) observed in the small-scale wind wave facility for four wind speeds (3 m/s: up triangles; 4 m/s: circles; 5 m/s: squares; 6 m/s: down triangles) and in the large facility at 5 m/s wind speed (+).

$$U_s - U_\infty = 1.66\, u_{\star w}\, (u_{\star w}/\nu_w)^{1/3}\, X^{1/3} \tag{7.1}$$

$$\delta_w = 3.61\, (\nu_w/u_{\star w})^{2/3}\, X^{1/3} \tag{7.2}$$

One can notice that these evolutions are akin to these observed for an accelerated laminar boundary layer driven by a negative pressure gradient (see, *e.g.*, [5]).

The existence of non-dimensional evolution laws for $U_s - U_\infty$ and δ_w confirms *a posteriori* that $u_{\star w}$ and ν_w are indeed the relevant parameters for describing the laminar water flow development. The above findings also strongly suggests the existence of a non-dimensional profile describing the self-similar boundary layer behavior, like the Blasius profile for the laminar boundary layer over a solid flat plate. Using the characteristic flow velocity and length scales $\underline{\mathbf{u}}$ and $\underline{\delta}$ defined as:

$$\underline{\mathbf{u}} = u_{\star w}\, (u_{\star w}/\nu_w)^{1/3}\, X^{1/3}$$
$$\underline{\delta} = (\nu_w/u_{\star w})^{2/3}\, X^{1/3}$$

the evolution of the dimensionless velocity $(U - U_\infty)/\underline{\mathbf{u}}$ versus the dimensionless depth $z/\underline{\delta}$ is displayed in Fig. 7.5**b** for the various wind conditions. It is striking to see that all the experimental points collapse remarkably onto a single curve. Its analytic expression determined by a least

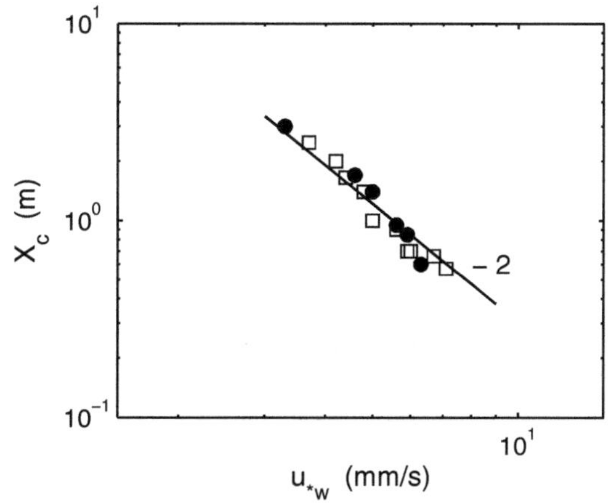

Figure 7.7. Evolution of the critical fetch X_c with water friction velocity $u_{\star w}$ observed for various wind conditions in the small-scale wind wave facility (closed circles) and the large facility (open squares).

square regression method from an exponential power expansion series takes the form:

$$(U - U_\infty)/\underline{\mathbf{u}} = 1.92\, e^{-z/\underline{\delta}} + 0.14\, e^{-2z/\underline{\delta}} - 0.40\, e^{-3z/\underline{\delta}} \qquad (7.3)$$

7.5 The Laminar-Turbulent Transition

As discussed in Sect. 7.3, the wind-induced water boundary layer which develops immediately downstream the leading edge of the water tank undergoes a laminar-turbulent transition. The dye flow visualizations and the time records of the velocity signals enabled us to distinguish two stages in the transition: the first stage where flow perturbations grow gradually embedded in a still-laminar water flow, and the second stage where intense flow instabilities develop suddenly into localized turbulent spots leading to the breakdown of the boundary layer. The natural development of this water flow boundary layer is far from being entirely controlled. Therefore, a detailed quantitative description of the growth of the perturbations which first develop gradually in the laminar zone and then explosively into the turbulent spots from LDV measurements, is strictly out of reach. However, to characterize the transition, the critical transition fetch X_c which corresponds to the longitudinal extent of the laminar zone, was approximately determined by visual water surface observations and dye visualizations in both the small and large wind wave facilities. The evolution of X_c as function of the water friction velocity $u_{\star w}$ observed in the laminar zone is reported in Fig. 7.7. Regarding the

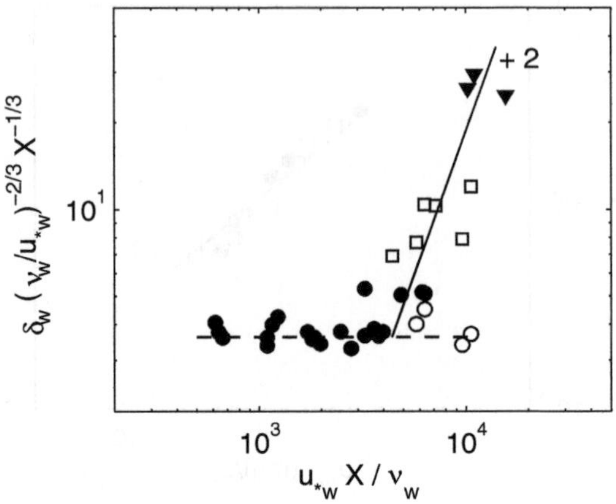

Figure 7.8. Water boundary layer thickness *versus* fetch plotted in dimensionless viscous coordinates for different wind speeds and various fetches located respectively in the laminar zone (closed circles), the laminar (open circles) and turbulent (open squares) regions of the transition zone and the turbulent zone (closed triangles).

large uncertainties of such visual estimations (of order of 10 %), it is somewhat surprising to see that the X_c values observed in both facilities agree very well. Moreover, X_c varies with $u_{\star w}$ roughly as $u_{\star w}^{-2}$.This result can be interpreted in more physical way as follows:

$$\tau_0 \, X_c = const., \tag{7.4}$$

which states that the boundary layer breakdown occurs when the work done by the tangential stress at the water surface τ_0 reaches a certain critical threshold. Then, in contrast to the laminar-turbulent transition observed for a boundary layer over a rigid flat plate, this laminar-turbulent transition is not linked to any critical value of the Reynolds number.

On another hand, to characterize the vertical expansion of the turbulent spots, a rough estimate of the mean boundary layer thickness inside the turbulent regions of the transition zone was made. To that end, the respective mean velocity profiles were derived from the short time-averaged velocity signals by means of a conditional analysis. Figure 7.8 presents in non-dimensional viscous coordinates the evolution with fetch of the boundary layer thickness over the whole boundary layer development area. Compared to the slow 1/3 power law increase observed in the laminar zone, the growth of the boundary layer thickness with fetch observed inside the turbulent spots proves explosive as it follows a 7/3 power law trend. Note that this increase is much higher than those generally observed for a turbulent boundary layer over a rigid flat plate, thus bearing out quantitatively the existence of very intense vertical mixing at this stage of the transition. This result also suggests that the formation of the turbulent spots inside the water flow surface boundary layer is associated with the development of very energetic turbulent motions. The high intensity of the streamwise velocity fluctuations inside the turbulent spots compared with those observed at the same fetch inside the still-laminar flow regions of the transition zone is illustrated in Fig. 7.3**b** where the respective velocity spectra observed in these regions are displayed, as well as the 'reference' spectra given in Fig. 7.3**a** for the same depths in the laminar and turbulent zones. This figure clearly shows that the level of low-frequency perturbations inside the turbulent spots of the transition zone rises abruptly in comparison with those observed inside the laminar flow regions. It is also noteworthy that this quantity reaches or even exceeds those observed downstream in the turbulent zone, corroborating again that turbulent motions inside the turbulent spots develop very actively.

7.6 Concluding Remarks

The investigation of the wind-driven water surface boundary-layer development made by visualizations and LDV measurements enables us to identify the main features of the laminar-turbulent transition in such a flow, and then to establish a number of specific quantitative relations governing the flow behavior. Summarizing, first it was shown that the thin water sublayer induced by wind at the leading edge of the water sheet is indeed laminar. The drift current surface velocity and the boundary layer depth increase gradually with fetch, following 1/3 power laws characteristic of the accelerated flat-plate laminar boundary layers. Further downstream, the formation and the development of longitudinal vortices within the laminar boundary layer are suggested by the presence of low frequency velocity perturbations of streamwise amplitude much greater than the vertical one. The laminar-turbulent transition manifests itself by a sudden development of turbulent motions of large amplitude and broad-band frequencies but it remains localized both in space and time. The critical fetch where the first turbulent spots appear is not related to any critical Reynolds number, but was found to depend on friction velocity as $u_\star^{-2}$. Then, these turbulent spots evolve with fetch inside longitudinal elongated patches characterized by a slow crosswise expansion, while the boundary layer thickness inside these patches increases with fetch explosively.

These observations then suggest a plausible scenario of the transition to turbulence of wind-driven boundary layer essentially characterized by two distinct stages. By considering the accelerated non-dimensional velocity profile described by Equation (7.3), Shrira et al. [6] has shown on the basis of a linear stability analysis that unstable modes do not exist and so, the classical Tollmien-Schlichting-like perturbations of rigid wall boundary layers do not grow. Hence, the streamwise elongated structures observed at this first stage of development can emerge only from natural noise disturbances always present inside the water flow. Then, owing to the development of nonlinear processes, they grow slowly without significant alteration of the mean flow field. Finally, the development of secondary instabilities appears to be responsible for their breakdown at a critical fetch. This picture resembles the scenario of the rigid plate laminar-turbulent transition caused by a by-pass mechanism as reported by Wu et al. [10] or Andersson et al. [1] and investigated by Tsai [9] for temporally-developing wind-driven subsurface water flow. In contrast, the second stage, where the turbulent spots created by the collapse of the shear boundary layer expand and merge, differs dramatically from the rigid plate scenario, in particular by the abruptness and the intensity with which the vertical mixing phenomena occur. The observed differences can be attributed to the intermittent breakdown of the mean flow field in localized areas caused by the flow deceleration following the breakdown of the elongated coherent structures. This induces a cascade of inflexional instability processes within the whole boundary layer, that leads to the formation of turbulent spots and the related sharp enhancement of the turbulent diffusion in the vertical direction.

References

[1] Andersson P., Brandt L., Bottaro A., Henningson D.S. (2001) On the breakdown of boundary layer streaks. *J Fluid Mech* 428:29–60

[2] Caulliez G. (1987) Measuring the wind-induced water surface flow by laser Doppler velocimetry. *Exp Fluids* 5:145–153

[3] Caulliez G., Ricci N., Dupont R. (1998) The generation of the first visible wind waves. *Phys Fluids* 10:4:757–759

[4] Pierson W.L., Banner M.L. (2003) Aqueous surface layer flows induced by microscale breaking wind waves. *J Fluid Mech* 479:1–38

[5] Schmid P.J., Henningson D.S. (2001) Stability and Transition in Shear Flows. *Appl Math Sci* 142, Springer-Verlag New York

[6] Shrira V.I., Ivonin D.V., Caulliez G. (2002) Laminar-turbulent transition in the wind driven free-surface boundary layer. In: *Progress in Nonlinear Science, Proceed Int Conf Nizhny Novgorod* vol.II:270–275

[7] Siddiqui M.H.K., Loewen M.R., Richardson C., Asher W.E., Jessup A.T. (2001) Simultaneous particle image velocimetry and infrared imagery of microscale breaking waves. *Phys Fluids* 13:7:1891–1903

[8] Veron F., Melville W.K. (2001) Experiments on the stability and transition of wind-driven water surfaces. *J Fluid Mech* 446:25–65

[9] Tsai W. (2001) On the formation of the streaks on wind-driven water surfaces. *Geophys Res Lett* 28, 20:3959–3962

[10] Wu X., Jacobs R.G., Hunt J.C.R., Durbin P (1999) Simulation of boundary layer transition induced by periodically passing wakes. *J Fluid Mech* 398:109—153

8

Defining the Enhancement of Air-Water Interfacial Oxygen Exchange Rate due to Wind-Forced Microscale Waves

William L. Peirson[1], James W. Walker[1], Chani Welch[2] and Michael L. Banner[3]

[1] Water Research Laboratory, School of Civil and Environmental Engineering University of New South Wales, King St, Manly Vale, 2093, Sydney, Australia, W.Peirson@unsw.edu.au
[2] SMEC International Pty Ltd., Australia, chani.welch@smecbd.com
[3] School of Mathematics, University of New South Wales Sydney, NSW, 2052, Australia, M.Banner@unsw.edu.au

Abstract Over the last few years, compelling evidence has emerged that the exchange of low-solubility gases across air-water interfaces is strongly enhanced by microscale breaking (e.g. Jähne and Haußecker [12], Zappa et al. [28]). Jähne and Haußecker [12] observe that low-solubility gas flux rates are enhanced by up to a factor of 5 in the presence of small scale waves. Investigations using surface infrared imagery [10, 22, 27, 28] have demonstrated a strong correlation between total flux and a proportional area of surface with a high infra-red radiation emission associated with the passage of microscale breaking waves. The mechanisms causing this significant enhancement in exchange rate remain unclear. Zappa et al. [28] proposed that thinning of the aqueous diffusion sublayer by subsurface turbulence in the vicinity of the high infra-red emission region was primarily responsible for this enhancement. Alternate to this is a relationship between the air-water surface exchange rate and the passage rate of wind-forced microscale breaking waves proposed by Peirson and Banner [21]. They have suggested that subduction of the aqueous diffusion sublayer by the microscale wave spilling regions coupled with a weak surface divergence on the upwind faces of the waves primarily determines the microscale-breaking associated flux rate. We have completed a sequence of precise oxygen re-aeration measurements with the specific objective of testing the findings of Peirson and Banner [21]. Specifically, we have compared the flux rates of wind-forced, flat water surfaces in the absence of waves with those in the presence of wind-forced, steep, unbroken waves and wind-forced, microscale breaking waves. With the introduction of steep, unbroken micro-scale waves the surface exchange rate is enhanced by a factor of approximately 2.5. The transition from incipient breaking of the waves to the microscale breaking state induces a significant increase in the associated wind stress [1]. The observed rapid increase in flux rate is approximately proportional to the increase in the wind stress. For the microscale-breaking state, the observed flux rates show good agreement with the predictions of Peirson and Banner [21].

C.S. Garbe, R.A. Handler, B. Jähne (eds.): Transport at the Air Sea Interface
pp. 119-131, 2007, © Springer-Verlag Berlin, Heidelberg 2007

8.1 Introduction

Existing parameterizations of air-sea exchange of low solubility gases, including major greenhouse gases such as carbon dioxide and oxygen, are based primarily on the intensity of wind forcing or changes in surface colour associated with bubble entrainment (so-called *whitecap area*) (e.g. Monahan [19]). These proposed relationships show only limited agreement with available field and laboratory data, with the scatter in gas exchange rate in laboratory data greater than one order of magnitude. There is now very strong evidence that the microscale breaking wave field is the primary determining factor in gas exchange once wind speeds exceed 4 ms^{-1} and remains the dominant factor at high wind speeds where bubble entrainment becomes widespread (e.g. Zappa et al. [28]).

There have been two broad approaches to developing parameterizations of air-water exchange of low-solubility gases. The first approach has been to characterise transfer velocity in terms of a surface renewal process whereby turbulence adjacent to the interface forces exchange between the aqueous diffusion layer and the bulk. For example, Komori et al. [15] carefully monitored turbulent activity adjacent to a wind-forced air-water interface and related an observed characteristic turbulent time scale to the observed gas flux.

The attractiveness of this approach is that it has the potential to produce a unified parameterization of low-solubility gas flux for open waters independent of whether the surface disturbances enhancing exchange arise from sub-surface generated turbulence or wind-generated waves. The recent study by Brocchini and Peregine [5] has shown possibility for the development of more generalised descriptions of the relationship between turbulence and surface waves. However, we note that present evidence is that the dominant contribution to wind-forced wave development is pressure asymmetry on the waves themselves (e.g. Banner [1]).

An alternative approach to quantify surface renewal has recently been examined by McKenna and McGillis [18]. They examined oxygen flux in the presence of clean and contaminated surfaces exposed to standing waves and subsurface grid-generated turbulence. Using particle image velocimetry, they measured the root mean of the surface divergence magnitude and found that it provided good collapse for their measured surface flux rates.

The second approach to developing parameterizations of low-solubility gas flux has been process-based. Deacon [8] developed a boundary layer turbulence model to quantify the increase in gas flux across the interface due to turbulence generated immediately adjacent to the interface. However, Deacon [9] subsequently found that this characterisation significantly underestimated the observed gas flux levels. He anticipated that this observed disparity as due to microscale breaking at the air-sea interface [3].

A sequence of theoretical investigations by Longuet-Higgins and his co-workers (see especially, Longuet-Higgins [16] and Longuet-Higgins [17]) have been directed at understanding the inception of small-scale wave breaking and its potential links with the formation of parasitic capillary waves on the lee-ward faces of the small gravity waves. In particular, he noted the possibility of the formation of surface stagnation at the troughs of the parasitic capillaries with the consequent inception of flow separation. Ostensibly ignoring Longuet-Higgins's work, Szeri [23] completed a theoretical study of the enhancement of gas flux due to parasitic capillary waves and found that substantial enhancement could be anticipated due thinning of the aqueous diffusion sublayer at the parasitic capillary wave troughs. To formally link Szeri's findings to wind-forced wave surfaces, the dynamical linkages between parasitic capillaries and the underlying gravity wave will be necessary (e.g. Longuet-Higgins [17]).

Other investigators have followed Deacon's idea and draw direct linkages between microscale breaking and gas flux. Csanady [7] constructed a surface vortex model to simulate the effects of microscale breaking on surface exchange. He found that strong, localised surface divergence on the upwind face of the waves significantly reduced the thickness of the aqueous diffusion sublayer locally thereby enhancing the transfer rate. The development and application of infrared imagery to the ocean surface has shown that the upwind faces and crest are regions of significant heat flux [10] in general agreement with Csanady's findings. In spite of the significance differences in Schmidt number between heat and low solubility gases, strong correlations have been found between a surface heat signature and gas transfer velocity [27].

Based on observations of the aqueous flows immediately adjacent to the interface, Peirson and Banner [21] proposed a simple model description of the impact of microscale breaking on gas transfer. Their observations showed relatively weak divergence ($<10\mathrm{s}^{-1}$) on the upwind faces of micro-scale breaking wind waves but intense convergence at the toe of the spilling regions of these waves. Assuming complete subduction at the toe of the spilling region, Peirson and Banner produced a simple model of exchange that incorporated their observations of the surface drift current and the probability of microscale breaking.

In this contribution, we describe a process-based investigation of the air-water gas transfer of a low solubility gas for water surfaces exposed to wind. By carefully controlling the wave properties, we have investigated the rates of gas flux that occur in the presence of different characteristic microscale wave states.

8.2 Methodology

A significant difficulty facing air-water flux experimentalists is the fetch dependence of the surface wave state of wind-forced water surfaces and the associated transfer velocities (e.g. Wanninkhof and Bliven [24]). Provided the wind-forcing is sufficient ($\gtrsim$4m/s), microscale breaking waves form at a fetch of approximately 2m whose frequency and breaking probability decrease with increasing fetch [2].

Fetch-dependency can be alleviated using a circular wind tank (e.g. Jähne et al. [14]). However, the geometry of such facilities generates lateral overturning of both the air and water cavities that is not representative of mean wind-wave interaction on open waters. Wave interaction at the outer tank wall also promotes stem wave formation that is not common on open air-water interfaces.

During this investigation, we have adopted an alternative approach. By carefully triggering of high frequency waves using a paddle at the windward end of a linear tank (see Figure 8.1) and coupling these monochromatic wave fields with appropriate levels of wind forcing, we have been able to create wind wave fields with minimal variation in surface wave state with fetch. By varying the degree of wind forcing and surface wave state, we have been able to investigate changes in flux rate due to changes in characteristic surface condition.

The primary difficulty associated with undertaking these experiments is maintaining an approximately consistent set of wave conditions along a fetch of several metres. This is for two reasons: first, monochromatic waves are inherently unstable [4] and form groups as the sidebands develop; secondly, viscosity and breaking dissipate short water waves and consequently the wind energy density input to the waves must match the dissipation rate. Our investigations were undertaken in a small glass-walled tank (0.2m wide, 0.5m deep by 9m length) installed in a purpose-built, temperature-regulated room at the Water Research Laboratory. The water depth in the tank remained approximately 250mm and the air temperature remained 21$\pm$1°C throughout testing. Appropriate flow straightening was installed at the upwind end of the tank to ensure a uniform and symmetric distribution of air flow entering the tank. A small cantilever paddle was installed at the upwind end of the tank of minimal height to avoid disruption to the air flow but avoiding the splashing of generated wave crests over the top of the paddle. The side walls of the tank were insulated with 50mm polystyrene blocks.

To ensure that the tank water remained clean and free from biological or surface active contaminants, the tank was subdivided into two compartments. The upwind compartment was used for the experiments and carefully separated from the downwind chamber by a sealed dividing wall. When not in active use, a pump system gently circulates flow from a depth of 1m in the downwind chamber to a depth of 250mm in the upwind

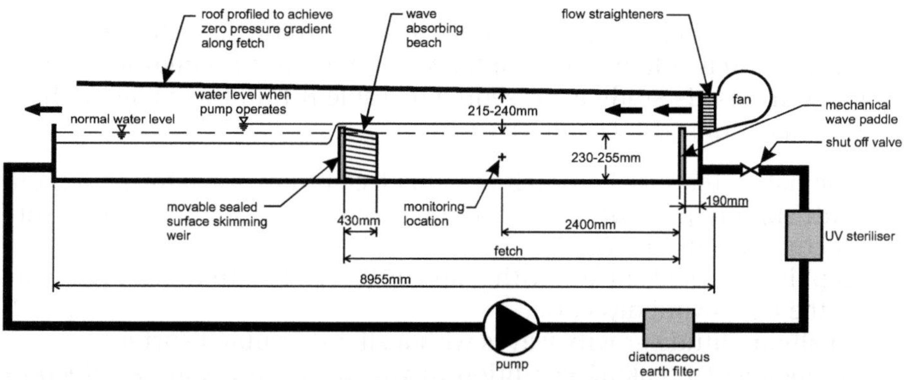

Figure 8.1. Shematic of wind-wave flume. The surface skimming weir ensured a slick free clean water surface for all experiments. Waves were generated using a combination of the mechanical paddle and wind forcing.

chamber. With assistance from a light wind, any surface active material residing on the surface of the upwind chamber is driven over the dividing wall to reside on the surface of the downwind chamber. A high-resolution imaging system similar to that developed by Peirson [20] has shown that the surface remains free of surface contamination. The pump system also incorporates a 5μm particulate filter and an ultraviolet source to inhibit slime growth. Periodically, the entire system is flushed.

Two series of experiments have been completed so far. The first series were undertaken with a total fetch of 4m. We selected identical wave conditions to those examined by Banner [1] and found that we were able to maintain waves in an approximately consistent microscale breaking state along our entire fetch. However, when we reduced the paddle amplitude, lower steepness waves proceeded to grow to a microscale breaking state with fetch. To maintain consistent wave conditions with fetch, we had to decrease the wind speed. To examine the impact of low steepness waves on the oxygen flux rate, we undertook a reference test in the absence of waves.

Capacitance wave probes were positioned in the wind-wave flume at locations: near the wave paddle, in the vicinity of the measurement section and near the dissipative beach. Wave data was logged to a computer and wave energies at each of the three locations calculated. This data was used to verify that mean wave energy did not vary by more than ±10% about the mean along the fetch.

Unsatisfied with undertaking testing with a varying wind speed, we found that by reducing the fetch slightly (to 3m) we were able to undertake a series of tests with a fixed reference wind speed and paddle frequency with consistent wave conditions along the fetch varying from low steepness to a continuous microscale breaking condition. However, with the paddle switched off, the fetch became covered with capillary ripples of approximately 30mm wavelength. A summary of the conditions investigated during this study is presented in Table 8.1. In this table, we have characterised the surface into five states:

1. The flat water surface which remains free of any wave formation but with slightly perceptible high-frequency motion due to the pressure fluctuations in the wind.
2. Capillary rippled surface with 30mm waves of length which propagate in the downwind direction.
3. Small-amplitude gravity waves with a slightly rippled surface.
4. An incipient breaking condition in which the waves are as steep as possible without lapsing into a microscale breaking state. Such waves have a smooth steady crest accompanied by well-developed parasitic capillaries leeward of the wave crests.
5. The microscale breaking condition characterised by parasitic capillary waves visible ahead of a distinct jump in the water surface slope and pulsations in the crest due to the unsteady motions associate with breaking.

For our investigations, we selected oxygen as our tracer gas. Bulk dissolved oxygen levels in the wind-wave flume were measured using an Orion model 835A dissolved oxygen meter and flow cell arrangement with data logged to a computer at 5s intervals. Calibration of the dissolved oxygen meter was undertaken using the standard 2 point method. Measurement of the overall transfer rate of oxygen from the air to the water was undertaken by monitoring oxygen levels after initial de-aeration and use of the conventional expression (e.g. Woodrow and Duke [26]):

$$\ln \frac{C_I - C(t)}{C_I - C(0)} = -\frac{k_v}{h} t \tag{8.1}$$

where C_I is the interfacial concentration derived from Henry's Law, $C(t)$ is the bulk concentration at time t, $C(0)$ is the initial bulk concentration, k_v is the gas transfer rate (ms^{-1}), and, h is the depth. The intakes of pump-through dissolved oxygen probes were at mid-depth within the tank to monitor changes in bulk oxygen concentration with time and thereby derive the transfer rate. The results of our measurements are summarised in Table 8.1.

8.3 Results and Discussion

In the absence of waves, flux rates should be well represented by the Deacon [8] expression. However, repeat measurements for this case showed a systematically higher (~+35%) transfer rate than predicted. An ancillary

Table 8.1. Summary of measurements obtained during this investigation.

Designation	U (m/s)	fp (Hz)	Surface characteristic state	u_* (m/s) [10%]	ak	T_{water} (°C)	k_{v,O_2} (m/hr) [5%]	Prediction Method	Predicted k_v (m/hr)	difference
f00u21	2.10	0.00	Flat water	0.121	0.00	19.7	0.032	Deacon [8]	0.021	-35.70%
f34u21	2.10	3.40	low steepness waves	0.130	0.25	19.0	0.054			
f34u30	3.00	3.40	incipient breaking	0.218	0.27	18.8	0.086			
f34u57	5.70	3.40	microscale breaking	0.418	0.32	18.0	0.324	Peirson and Banner [21]	0.351	8.50%
f29u21	2.10	2.90	low steepness waves	0.103	0.17	18.7	0.061			
f29u36	3.60	2.90	incipient breaking	0.263	0.24	18.7	0.115			
f29u57	5.70	2.90	microscale breaking	0.467	0.27	18.4	0.396	Peirson and Banner [21]	0.325	-18.00%
f00u32	3.20	0.00	capillary ripples	0.260	unknown	18.4	0.068			
f21u32	3.20	2.10	low steepness waves	0.270	0.10	18.8	0.112			
f21u32	3.20	2.40	incipient breaking		0.24	18.8	0.126			
f21u32	3.20	3.10	microscale breaking	0.340	0.28	18.4	0.259	Peirson and Banner [21]	0.336	29.50%

observation was that after application of wind and waves, the water temperature decreased to the stable value as indicated in Table 8.1, whilst air temperature in the entire facility remained the same. This is indicative of the evaporative heat flux taking place at the surface eventually achieving balance with input heat fluxes through the surface, walls and base of the tank. Consequently, the stable temperature of the water must be dependent on the rate of overall heat transfer taking place within the facility.

Other investigators (Woodrow and Duke [25], Figure 16) have observed directly plumes of oxygen-rich water leaving wind-forced free surfaces. The evaporatively-induced temperature instability at the interface could be a source of surface convergence, generating vertical transport and transfer rates systematically higher than Deacon's predictions. The difference between the measured flat water flux rate and the estimated value is comparable with the enhancement that occurs with the application of small amplitude waves.

At present, we have no direct particle image velocimetry observations with which to compare our data in the form of McKenna and McGillis [18]. However, for wind-forced, microscale breaking waves, Peirson and Banner [21] showed that the mean surface divergence on the up-wind faces of wind-forced microscale breaking waves is relatively weak and the mean surface velocity field can be well approximated by a 5th order Stokes estimate with a superimposed, weakly modulated surface drift. Computations of the mean divergence using a 5th order Stokes description of the surface velocity for the wave forms of the present study shows differences of only about 10% from the simple expression:

$$\langle |\Gamma| \rangle = 4akf \tag{8.2}$$

where $|\Gamma|$ is the magnitude of the local surface divergence, the angle brackets denote spatial averaging, ak is the mean wave steepness and f is the wave frequency. Figure 8.2 presents our data plotted as a function of square root of the divergence magnitude evaluated using Equation (8.2). The collapse of the non-breaking wave data is modest with levels of transfer velocity observed for non-breaking waves are similar to those observed by McKenna and McGillis [18]. Figure 8.2 highlights the significant differences in transfer velocity between non-breaking waves and microscale breaking wave conditions.

Whilst the transfer rates at the incipient breaking limit are higher than for those of lower mean wave steepness, there is no observable jump in transfer that suggests strong flux enhancement due to parasitic capillary presence.

The most significant increases in transfer velocity can be observed to be associated with the transition from incipient to microscale breaking. Consistent with the observations of Banner [1], there is a also substantial jump in the momentum flux as exhibited by the increases in the friction

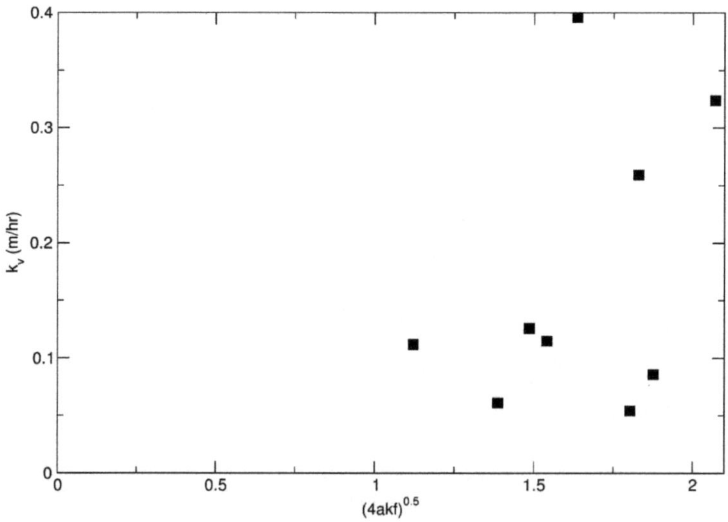

Figure 8.2. Oxygen transfer velocities as a function of the estimated square root of the divergence magnitude obtained during the present study. Note the significantly higher transfer velocities (>0.25m/hr) associated with microscale breaking in comparison with those measured for non-breaking waves (<0.15m/hr).

velocity $u_{\star a}$. This observation highlights the importance of the microscale breaking process to the air-water gas exchange process.

Shown in Table 8.1 are the predictions of Peirson and Banner [21] for these conditions. The predictions are reasonable in views of the the substantial jump in transfer rate associated with the transition to microscale breaking and the simplicity of the theoretical approach,. The predicted values assume 100% breaking occurrence and minimal impact of the wind-drift current. The formation of wave groups along the wave tank may have reduced the breaking probabilities at the downwind end of the test chamber to about 80%. These two factors allow scope to reduce the predicted values by about 20% but no scope to increase them.

The measured values have been converted to equivalent CO_2 flux rates at 20°C (with a maximum 5% error due to Schmidt number exponent uncertainties). The present results are shown in the context of laboratory data assembled from other studies by Komori et al. [15] and reproduced here in Figure 8.3. In the regime of moderate wind forcing ($u_{\star a} < 0.5$m/s), the present data span the range of observed flux rates and suggest regimes of surface wave impact on low solubility gas transfer. The results of our investigations to date indicate that a process-related categorisation of low solubility air-water gas exchange processes may be possible. As shown in Figure 8.3, very low exchange rates occur in the absence of any wave activity. The presence of small scale wave activity in the absence of any micro-scale breaking can enhance the exchange rate to the level indicated.

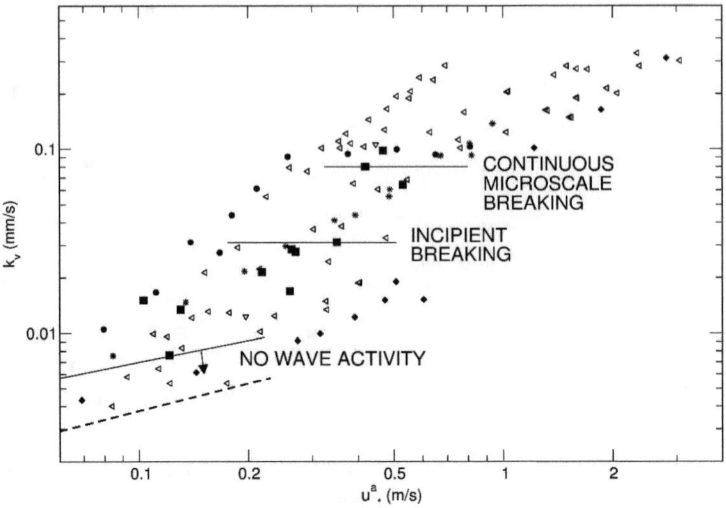

Figure 8.3. Laboratory measured transfer velocities as a function of friction velocity in the air obtained during the present study (solid squares ■) normalised for CO_2 at 20°C and shown in comparison with data assembled by Komori et al. [15] (their CO_2 data shown as solid circles •); Jähne [11] CO_2 data, left pointing triangles ◁; Broecker et al. [6] CO_2 data, downward pointing triangles ▽; Jähne et al. [13] CO_2 data, solid diamonds ◆; Wanninkhof and Bliven [24] SF_6 data, asterices $*$. The dashed line shows the Deacon [8] relationship and solid lines and annotation delineate regimes of wave-related transfer velocity indicated by this investigation.

Increasing the probability of microscale wave breaking to 100% results in a maximum transfer rate in the vicinity of 0.09mm/s under moderate wind forcing. At high wind forcing ($u_{\star a} > 0.5$m/s), exchange can be further enhanced possibly due to spray or bubble generation but a process-based investigation of exchange at high wind forcing remains to be undertaken.

8.4 Conclusions and Recommendations

Careful observations of re-oxygenation have been undertaken in a linear wind-wave interaction tank with minimal surface contamination. A unique characteristic of these experiments is that the wind and paddle-generated wave conditions have been selected to minimise the fetch-dependence of the wind-wave field.

Our observation of transfer rate in the absence of surface waves are systematically a factor of 3 times higher than the predictions of Deacon [8]. A concomitant cooling of the water with the application of light winds indicates significant heat flux, presumably due to evaporation. Density instability induced by evaporative cooling could promote the formation of localised surface divergences and direct transport away from the interface consistent with the observations of Woodrow and Duke [26]. The potential contribution of direct transport away from the surface war-rants further investigation.

Transfer velocity increases systematically with the steepness of small unbroken waves up to the point of incipient breaking. Our observed increase in transfer velocity appears to be consistent with the levels observed by McKenna and McGillis [18]. Further investigation of the role of surface divergence in the enhancement of transfer velocity due to unbroken waves is presently been undertaken. An increase in total stress of approximately a factor of two is required to support the transition from a continuous incipient breaking to a continuous microscale breaking wave state, consistent with the observations of Banner [1]. However, the associated enhancement of the transfer velocity associated with the transition from incipient to active microscale breaking is approximately a factor of 2 to 3. The measured transfer rates in the presence of microscale breaking are consistent with the predictions of Peirson and Banner [21]. Investigations to determine those processes causing further enhancement of transfer velocity under high wind forcing is recommended.

References

[1] M. L. Banner. The influence of wave breaking on the surface pressure distribution in wind-wave interactions. *J.Fluid Mech.*, 211:463–495, 1990.

[2] M. L. Banner and W. L. Peirson. Tangential stress beneath wind-driven air-water interfaces. *J.Fluid Mech.*, 364:115–145, 1998.

[3] M. L. Banner and O. M. Phillips. On the incipient breaking of small-scale waves. *J.Fluid Mech.*, 65:647–656, 1974.

[4] T. B. Benjamin and J. E. Feir. The disintegration of wave trains on deep water. *J.Fluid Mech.*, 27:417–430, 1967.

[5] M. Brocchini and D. H. Peregine. The dynamics of strong turbulence at free surfaces. part 1. description. *J.Fluid Mech.*, 449:225–254, 2001.

[6] H. C. Broecker, W. Siems, and J. Petermann. The influence of wind on co_2-exchange in an wind-wave tunnel, including effect of monolayers. *Journal of Marine Research*, 36:595–610, 1978.

[7] G. T. Csanady. The role of breaking wavelets in air-sea gas transfer. *Journal of Geophysical Research*, 95(C1):749–759, 1990.

[8] E. L. Deacon. Gas transfer to and across an air-water interface. *Tellus*, 29:363–374, 1977.

[9] E. L. Deacon. Sea-air gas transfer: The wind speed dependence. *Boundary-Layer Meteorology*, 21:31–37, 1981.

[10] A. T. Jessup, C. J. Zappa, and H. H. Yeh. Defining and quantifying microscale wave breaking with infrared imagery. *Journal of Geophysical Research*, 102(C10):23145–23153, 1997.
[11] B. Jähne. *Parametrisierung des Gasaustausches mit Hilfe von Laborexperimenten.* PhD thesis, Institut für Umweltphysik, University of Heidelberg, 1980.
[12] B. Jähne and H. Haußecker. Air-water gas exchange. *Annual Reviews Fluid Mechanics*, 30:443–468, 1998.
[13] B. Jähne, K. O. Münnich, and U. Siegenthaler. Measurements of gas exchange and momentum transfer in a circular wind-water tunnel. *Tellus*, 31:321–329, 1979.
[14] B. Jähne, K. O. Münnich, R. Bösinger, A. Dutzi, W. Huber, and P. Libner. On the parameters influencing air-water gas exchange. *Journal of Geophysical Research*, 92(C2):1937–1949, 1987.
[15] S. Komori, R. Nagaosa, and Y. Murakami. Mass transfer across a sheared air-water interface. *J.Fluid Mech.*, 249:161–183, 1993.
[16] M. S. Longuet-Higgins. Capillary rollers and bores. *J.Fluid Mech.*, 240: 659–679, 1992.
[17] M. S. Longuet-Higgins. Parasitic capillary waves: a direct calculation. *J.Fluid Mech.*, 301:79–107, 1995.
[18] S. P. McKenna and W. R. McGillis. The role of free-surface turbulence and surfactants in air-water gas transfer. *Int.J.Heat Mass Transfer*, 47(3):539–553, 2004.
[19] E. C. Monahan. The physical and practical implications of a co2 gas transfer coefficient that varies as the cube of the wind speed. *Geophysical Monograph 127, American Geophysical Union*, pages 193–197, 2002.
[20] W. L. Peirson. Measurement of surface velocities and shears at a wavy air-water interface using particle image velocimetry. *Expt.in Fluids*, 23:427–437, 1997.
[21] W. L. Peirson and M. L. Banner. Aqueous surface flows induced by microscale breaking wind waves. *J.Fluid Mech.*, 479:1–38, 2003.
[22] M. H. K. Siddiqui, M. R. Loewen, C. Richardson, W. E. Asher, and A. T. Jessup. Simultaneous particle image velocimetry and infrared imagery of microscale breaking waves. *Phys.Fluids*, 13:1891–1903, 2001.
[23] A. J. Szeri. Capillary waves and air-sea transfer. *J.Fluid Mech.*, 332: 341–358, 1997.
[24] R. H. Wanninkhof and L. F. Bliven. Relationship between gas exchange, wind speed and radar backscatter in a large wind wave tank. *J.Geophys.Res.*, 96(C2):2785–2796, 1991.
[25] P. T. Woodrow and S. R. Duke. Laser-induced fluorescence studies of oxygen transfer across unsheared flat and wavy air-water interfaces. *Eng.Chem.Res.*, 40(8):1985–1995, 2001.

[26] P. T. Jr. Woodrow and S. R. Duke. Lif measurements of oxygen concentration gradients along flat and wavy air-water interfaces. *Geophysical Monograph 127, American Geophysical Union*, pages 83–88, 2002.

[27] C. J. Zappa, A. T. Jessup, and W. E. Asher. Microscale wave breaking and air-water gas transfer. *J.Geophys.Res.*, 106(C5):9385–9391, May15 2001.

[28] C. J. Zappa, W. E. Asher, A. T. Jessup, J. Klinke, and S. R. Long. Effect of microscale wave breaking on air-water gas transfer. *Geophysical Monograph 127, American Geophysical Union*, pages 23–29, 2002.

9

Momentum Flux and Energy Dissipation Associated with Breaking Waves

Johannes Gemmrich

University of Victoria,
Physics & Astronomy,
PO Box 3055,
Victoria, BC,
V8W 3P6, Canada
gemmrich@uvic.ca

Abstract Breaking waves dissipate energy in the oceanic surface layer (top few meters) and also support the air-sea momentum flux. Spectrally resolved energy dissipation and momentum flux are extracted from open ocean observations of the breaking crest length distribution $\Lambda(c)$. This concept, first introduced by Duncan and Phillips more than 2 decades ago, includes an unknown proportionality factor b. Independent estimates of direct turbulence measurements are used to evaluate this proportionality factor.

9.1 Introduction

Surface waves are sometimes called the gearbox between atmosphere and ocean. Wave breaking plays an important role in many air-sea exchange processes and, in the analogy of the gearbox, breaking waves indicate the highest gear. At moderate to high wind speed the momentum transfer from wind to ocean currents passes through the wave field via wave breaking. The breaking of surface waves is responsible for the dissipation of wave energy, thus limiting wave growth. Furthermore, wave breaking is a source of enhanced turbulence kinetic energy (TKE) levels in the near-surface layer and therefore plays an important role in upper ocean processes. Breaking waves not only transfer energy, momentum, heat and gases from the atmosphere to the ocean surface layer but also enhance aerosol generation and latent heat fluxes due to sea spray. Breaking waves also disperse pollutants and generate underwater sound. Comprehensive overviews of the role of wave-induced turbulence in upper-ocean dynamics and air-sea exchange processes are given by Thorpe [17], Melville [12] and Duncan [3].

It is now accepted that dissipation in the near-surface layer, say the top few meters, of a wind driven ocean departs significantly from the

C.S. Garbe, R.A. Handler, B. Jähne (eds.): Transport at the Air Sea Interface
pp. 133-144, 2007, © Springer-Verlag Berlin, Heidelberg 2007

classic constant stress layer description in terms of magnitude as well as its depth dependence [18]. Direct measurement of the fine-scale velocity field in the ocean and especially in the near-surface layer is extremely challenging. Therefore, oceanic turbulence is commonly characterized by the dissipation rate of turbulence kinetic energy ε, which may be inferred from the turbulence velocity shear $\partial u/\partial z$ or rate of strain $\partial u/\partial x$. Alternatively, ε may be obtained from the wavenumber velocity spectra $S(k)$. Thus, two fundamentally different approaches exist in oceanic turbulence measurements:

i) observation of the velocity shear or the rate of velocity strain (the basis of measurements with microstructure profilers), and
ii) observation of the velocity field in space or time (obtained by single point velocity meters or high resolution acoustic Doppler profilers).

Kolmogorov's inertial subrange hypothesis states that the spectral level of this subrange is only a function of the energy dissipation occurring at much smaller scales,

$$S(k) = A\varepsilon^{2/3}k^{-5/3}, \tag{9.1}$$

where k is the wave number and A a universal constant. Thus, if a wave number velocity spectrum resolves the inertial subrange, estimation of energy dissipation is relatively straightforward. In the case of single point observations, the frequency spectra have to be converted into wave number spectra, invoking Taylor's hypothesis of frozen turbulence, which introduces additional uncertainties.

Two decades ago, Phillips [15] developed the equilibrium range theory for wind-generated gravity waves. This theory assumes that there exists a spectral range where wind input, nonlinear wave-wave interactions and energy dissipation are all equally important. He also suggested that wave breaking may be characterized by the length of the breaking crest (in the along-crest direction) and its propagation speed, and introduced the breaking crest length distribution $\Lambda(c)$. This distribution is defined in a way that $\Lambda(c)dc$ describes the average total length of breaking wave crest, per unit area, that propagate at speeds in the range c to $c + dc$. The fraction of surface turned over by breaking waves, per unit time, is

$$R = \int c\Lambda(c)dc. \tag{9.2}$$

Laboratory experiments with towed hydrofoils initialized following similarity arguments that relate the breaking wave phase speed to the rate of energy loss [2]. The breaker zone on the forward face of the crest of a stationary breaker, such as generated by a hydrofoil, covers a fixed fraction of the wave amplitude. Assuming the breaking waves are self-similar, the cross-sectional area of the breaking zone, per unit crest length, is proportional to the square of the wavelength $(c^2/g)^2$ where g is the gravitational acceleration. The weight of this breaking zone exerts a tangential force proportional to c^4/g which acts against the orbital motion. On the forward face of the wave the orbital motion is directed upslope and the relevant speed is proportional to c. Thus, the rate of energy loss is

$$\varepsilon(c) = b\rho g^{-1} c^5 \Lambda(c), \tag{9.3}$$

where ρ is the water density and b is a non-dimensional factor, assumed to be constant. The momentum flux from waves to underlying currents follows as

$$m(c) = b\rho g^{-1} c^4 \Lambda(c). \tag{9.4}$$

The total average energy dissipation and momentum flux associated with breaking waves are

$$\dot{E} = b\rho g^{-1} \int c^5 \Lambda(c) dc \tag{9.5}$$

and

$$\dot{M} = b\rho g^{-1} \int c^4 \Lambda(c) dc. \tag{9.6}$$

This concept is very attractive since it opens the possibility of remote measurements of wave dynamics. However, so far only very limited data of $\Lambda(c)$ distributions exist [7, 13, 16] and estimates of the proportionality factor b span at least the range 7×10^{-4} to 8×10^{-3} [13, 16]. Furthermore, it is not established that b indeed is a universal constant. More observations of the breaking crest distribution are required and calibration of inferred energy dissipation and/or momentum fluxes are necessary before remote sensing may be applied successfully to infer wave dynamics. Here we describe such a first attempt.

9.2 Observations

Observations of the surface wave field as well as the subsurface turbulence were taken as part of the FAIRS (Fluxes, Air-sea Interaction and Remote Sensing) experiment aboard the research platform FLIP in the open ocean 150 km offshore of the central Californian coast. Turbulence observations were made from a surface-following float tethered to *R/P FLIP* [9]. The main components of the turbulence package were high resolution Doppler sonars (Dopbeam, Sontek) and custom-made acoustic resonators for measuring bubble size distributions [4]. The Dopbeams acquire velocity profiles at 0.15 m to 0.87 m range from the sonar head, at a radial resolution of 6×10^{-3} m. The two Dopbeams utilized for turbulence measurements were mounted at 0.8 m depth, one pointing vertically down, the other in the nominal cross-wind direction. The raw velocity sampling rate was 20 Hz; dissipation estimates are obtained from mean velocity wavenumber spectra at a 1Hz sampling rate from Equation (9.1).

Video recordings of the ocean surface were made simultaneously with some of the turbulence measurements. Two analog black/white video cameras mounted on *R/P FLIP* yielded recordings with overlapping fields of view of 15.4×20.5 m (Camera 1) and 9×12 m (Camera 2). The video recordings were digitized at 640×480 pixels, ensuring that even the smallest visible whitecaps were resolved. Differential images were generated by subtracting successive video frames. These images highlight the front of propagating breaking crests and filter out all stationary signals including foam and, to some extent, sun glitter. An image processing scheme generates an ellipse which covers the entire breaking crest front, while minimizing the area of the ellipse. The major axis of this ellipse defines half the length of the breaking crest $L_{br}/2$. Displacement of the ellipse's centroid yields the raw propagation speed of the breaking crest $\tilde{c}_{br}$. Even for the lowest speeds considered here this scheme yields a centroid displacement of about 7 pixels, which is well resolved. Subtracting the potential advection by the orbital motion u_{orb} of underlying longer waves yields the true breaker speed $c_{br} = \tilde{c}_{br} - u_{orb}$. The equivalent linear phase speed c of the wave associated with the breaking crest is somewhat larger [13]; here we take $c = c_{br}/0.9$ (Banner, personal communication 2006).

The FAIRS experiment included wind conditions ranging from almost calm up to $15\,\mathrm{m\,s^{-1}}$. The resulting wave field ranged from purely swell conditions with significant wave height $H_s < 1$ m to developed seas with $H_s > 4$ m. Here we discuss four fetch-unlimited data sets of 2 - 5 hour duration, recorded under various wind forcing and wave field conditions.

Data set I follows a period of increasing wind speed. For several days prior to this data set winds were very light ($< 4\ \mathrm{m\,s^{-1}}$) and the wave field was dominated by swell with $H_s < 1$ m. For the period of about 11 hours prior to the start of this data set the wind speed u_{10} increased steadily and peaked at $u_{10} = 12.8\,\mathrm{m\,s^{-1}}$. Then, throughout the data set, the wind speed stayed nearly constant at $12\,\mathrm{m\,s^{-1}}$. At the beginning of the data set the significant wave height had increased to $H_s = 2.8$ m and continued to rise to 3.1m. The dominant wave period was $f_p = 0.13$ Hz and the wave age $c_p/u_\star \approx 26$, where c_p is the phase speed of the dominant waves and $u_\star$ is the friction velocity in air. This data set I is more than twice as long as the other data sets and also occurred at a time of rapid wave development. Thus, in the following this data set is divided into two parts (Ia, Ib) of equal numbers of breaking events. Both data segments represent a *developing sea*.

Data set II occurred after three days of sustained wind speed $u_{10} > 10\,\mathrm{m\,s^{-1}}$ with well-developed wind waves at wave age $c_p/u_\star \approx 33$, significant wave height and dominant frequency $f_p = 0.11$ Hz. The wind speed was $u_{10} \approx 11.5\ \mathrm{m\,s^{-1}}$. This data set represents a *developed sea*.

The third data set represents a fully developed sea. The dominant frequency remained unchanged at $f_p = 0.1$ Hz, the wind speed had increased

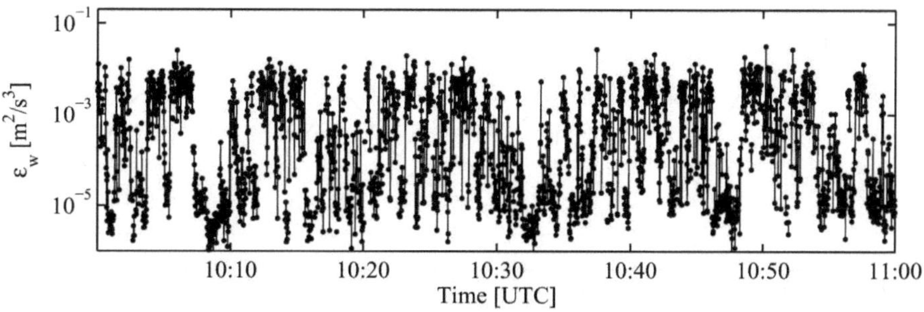

Figure 9.1. Example of observed dissipation time series during deployment I

to $u_{10} \approx 12.5\,\mathrm{m\,s^{-1}}$ and the significant wave height reached $H_s = 3.2\,\mathrm{m}$ at a wave age $c_p/u_\star \approx 33$.

Data set IV occurred at the end of a rapid increase of wind speed from $< 5\,\mathrm{m\,s^{-1}}$ to $13\,\mathrm{m\,s^{-1}}$. The significant wave height increased from $< 2\,\mathrm{m}$ to $> 4\,\mathrm{m}$. The data set itself covers the period of slow increase in wind speed, from $11.8\,\mathrm{m\,s^{-1}}$ to $13\,\mathrm{m\,s^{-1}}$, but a significant increase in wave height from $H_s = 2.5\,\mathrm{m}$ to $H_s = 3.9\,\mathrm{m}$ and reduction in dominant wave frequency from $f_p = 0.16\,\mathrm{Hz}$ to $f_p = 0.1\,\mathrm{Hz}$. The wave age was $c_p/u_\star \approx 29$. This data set represents a growing sea superimposed onto significant swell and will be labeled *mixed sea*.

9.3 Energy Dissipation and Momentum Fluxes

9.3.1 Turbulence Observations

Wave breaking is a very intermittent phenomenon and the resulting turbulence fields are very patchy. Several observations of turbulence in a wind-driven sea reveal average dissipation levels that are 10 to 100 times larger than expected in an equivalent flow along a rigid boundary $\varepsilon_{wl} = u_\star^3(\kappa z)^{-1}$ (For references see Gemmrich and Farmer [8] and Gemmrich and Farmer [9]).

$$10 \le \frac{\bar{\varepsilon}}{\varepsilon_{wl}} \le 100. \tag{9.7}$$

Our turbulence measurements with the Dopbeams revealed that instantaneous dissipation levels can be much higher than seen in the mean profiles. Dissipation levels more than four orders of magnitude larger than ε_{wl} were observed beneath an active breaking wave. These high values persist only for a few seconds (Fig. 9.1) but are likely to play an important role in the breakup of entrained air bubbles [5, 9]. Beneath individual breaking waves, turbulence decayed as

$$\varepsilon \propto t^m, m = -4.3. \tag{9.8}$$

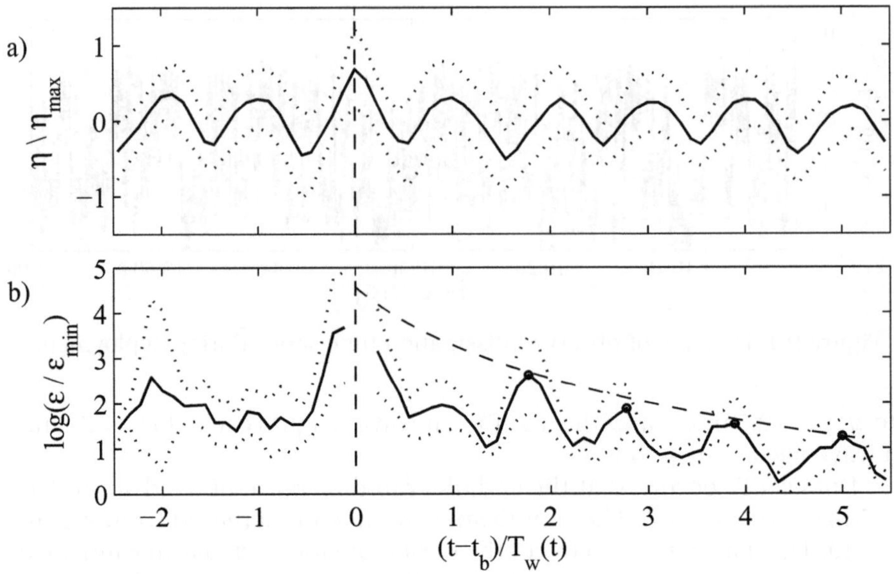

Figure 9.2. Mean normalized surface elevation (a) and normalized energy dissipation (b) associated with breaking waves. Time is normalized by the wave period of the breaking wave and centered at the passage of the breaking crest. The minimum dissipation observed within the records ε_{min} is commonly less than ε_{wl}. After five wave periods dissipation levels are roughly consistent with wall layer scaling. The curved dashed line depicts the turbulence decay rate $m = -4.3$. Dotted lines represent one standard deviation

Closer to the surface, within the layer of direct turbulence injection, the decay rate is even faster. Temperature and air fraction measurements at 0.18 m depth yielded $m = -7.6$ [6]. Approximately five wave periods after the onset of breaking the turbulence levels have decayed to the background level which is consistent with wall-layer flows (Fig. 9.2). The observed decay rates are faster than expected for isotropic turbulence, where $m = -2.2$. Our dissipation estimates are sampled at 1 Hz, much faster than previously reported dissipation profiles. Histograms of dissipation values obtained at this high sampling rate reveal the coexistence of two distinct contributions, a wide distribution centered on constant stress layer turbulence levels ($\log(\varepsilon/\varepsilon_{wl}) \approx 0$) and a smaller and narrower distribution representing breaking waves and centered on ($\log(\varepsilon/\varepsilon_{wl}) \geq 2$). The broader distribution of lower enhancement rates is associated with periods between breaking events and is broadly consistent with a wall-layer flow. A more detailed discussion of the turbulence observations is given in [9].

9.3.2 Breaking Crest Length Observations

For each breaking event the whitecap propagation speed c_{br}, the length of the major object axis L_{br} and the event duration t_{br} are known. At an arbitrary instant during the total observation period T, the expected breaking crest length of an individual event is $L_{br}t_{br}/T$. The expected breaking crest length of an event in the speed range c to $c + dc$ is the summation over all events which are within this speed range, $\sum L_{br}t_{br}/(TA_v)$. Here the transformation of the breaking crest propagation speed c_{br} into the linear phase speed of the breaking wave, $c = c_{br}/0.9$, is applied. Thus, the average length of breaking crest per unit area per unit speed interval is

$$\Lambda(c)dc = \sum L_{br}t_{br}/(TA_v), \tag{9.9}$$

where A_v is the area of the video footprint.

Our data sets cover only a small range of wind speeds, but a significant variation in wave development.

For all five data segments, $\Lambda(c)$ peaks at intermediate wave scales with phase speeds of $3 - 4\,\mathrm{m\,s^{-1}}$, corresponding to $c/c_p \approx 0.3$. In the peak region, $\Lambda(c)$ values for the different data sets vary by roughly a factor two, whereas at the smallest and the largest wave scales the different data sets spread more than one order of magnitude. Melville and Matusov [13] found a scaling factor $(10/u_{10})^3$, where u_{10} is the 10 m height wind speed value in m $\mathrm{s^{-1}}$, to collapse their data sets recorded at wind speeds between 7.2 m $\mathrm{s^{-1}}$ and 13.6 m $\mathrm{s^{-1}}$. For our four data sets this scaling factor varies by less than 15% and therefore does not significantly reduce the spreading between $\Lambda(c)(10/u_{10})^{-3}$ values of the different data sets. Hence, this scaling factor, which is purely based on the wind speed, is not universally applicable. Our data indicate that wave development is a more relevant factor for determining the peak levels of $\Lambda(c)$.

Momentum flux and energy dissipation due to breaking are given by the fourth and fifth moment of the breaking crest length distribution $\Lambda(c)$. These higher moments are weighted towards larger wave scales (larger c). However, a priori the proportionality factor b (see Equations (9.3) and (9.4)) is unknown and only the relative spectral distributions to momentum flux and energy dissipation are available, assuming b is scale-independent. Momentum fluxes supported by larger wave scales fluctuate considerably between the four data sets (Fig. 9.3). The momentum fluxes due to breaking waves are strongest at wave scales corresponding to phase speeds of about $6 - 9\,\mathrm{m\,s^{-1}}$. At wave scales with $c < 5\,\mathrm{m\,s^{-1}}$ momentum fluxes supported by white capping waves fall off by roughly three orders of magnitude. Energy dissipation due to whitecaps is even more dominated by larger wave scales. The spectral distributions of the energy dissipation peak at $c = 8 - 10\ \mathrm{m\,s^{-1}}$. Only in the case of the developing sea are dominant breakers involved in energy dissipation; in old seas the largest breaker scales are smaller than the dominant waves.

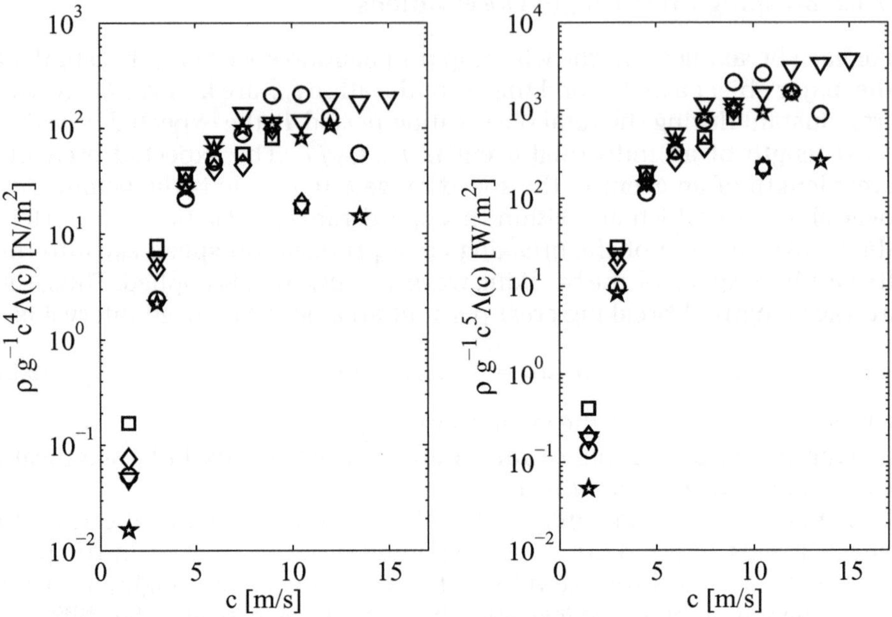

Figure 9.3. Momentum flux (left) and energy dissipation (right) associated with breaking waves, scaled by the unknown factor b [Equations (9.4) and (9.3)] , as function of linear wave phase speed. The symbols (○, ▽, □ , ◇, ★) correspond to data sets (Ia, Ib, II, III, IV)

9.3.3 The Proportionality Factor *b*

The concept of relating breaking wave dynamics to remotely observed whitecap properties promises new observational insight in wave breaking processes. However, any quantitative, and maybe even qualitative, assessments of energy dissipation and momentum fluxes based on this concept depend on the proportionality factor b. So far, only very limited data exist [13, 16] and estimates of b are inconclusive; in fact it is not even established that b is constant, or universal.

In principle, b could be evaluated from observations of the breaking crest length distribution $\Lambda(c)$ combined with independent estimates of the i) wave-induced momentum flux, or ii) measurements of the wave-induced energy dissipation. Assuming b indeed is constant, i.e. independent of the scale of the breaking wave, it may be estimated as

$$b = \frac{\tau_w}{\rho g^{-1} \int c^4 \Lambda(c) dc} \tag{9.10}$$

where τ_w is the total air-sea momentum flux supported by the form drag of the waves, and the integral has to include all scales which are involved in breaking. In developed wave fields, $\tau_w \geq 0.85\tau$, where $\tau = \rho_{air}\overline{u'w'}$

is the vertical momentum flux in the atmospheric boundary layer [1]. Our video data do not resolve micro breakers as well as some weaker whitecaps with insufficient visual contrast. Therefore, the integral in Equation (9.10) is incomplete. This fact can be remedied by reducing the air-sea momentum flux in Equation (9.10) to the fraction supported by waves that are also resolved in our whitecap observations $\tau_w = \gamma\tau$, with $\gamma \ll 0.85$. The exact value of γ is not known, but model results suggest the value for our observations could be as low as $\gamma \approx 0.2$ [14].

The energy input from the wind into the wave field is

$$E_{in} = \rho\, c_{\text{eff}}\, u_\star^2 \tag{9.11}$$

where c_{eff} is the effective phase speed of waves acquiring energy from the wind [10]. In developed seas the energy input roughly equals the energy dissipation. Therefore, b may be estimated from the energy balance in a similar manner as could be done from the momentum flux balance.

$$b = \frac{c_{\text{eff}}\,\tau}{\rho g^{-1} \int c^5 \Lambda(c) dc} \tag{9.12}$$

If the concept is consistent, Equations (9.10) and (9.12) will give the same value for b. Results of these approaches are presented in Gemmrich et al. [11].

Alternatively, b may be estimated from evaluating the breaking crest length approach against the direct turbulence observations, and b may be estimated from following balance:

$$\int \varepsilon(c)\, dc = \iint \varepsilon_{obs}(t,z)\, dt\, dz \tag{9.13}$$

The Dopbeam observations yielded information of the temporal evolution of the breaking wave induced dissipation ε_{obs} as well as its vertical structure, although limited to three equally spaced locations within the 0.7m velocity profile [9]. Combination of the observational results with modeled dissipation profiles provide an estimate of b according to Equation (9.13):

$$b = \frac{\iint \varepsilon_{obs}(t,z)\, dt\, dz}{\rho g^{-1} \int c^5 \Lambda(c) dc} \tag{9.14}$$

To this extent, a dissipation profile $\varepsilon \propto z^n$ has been fitted to our observation, yielding n=-3, where we used $z_0 = 0.2\,\text{m}$ as the depth of the TKE injection layer [8] (Fig. 9.4). The observed dissipation time series (Fig. 9.1) provide the temporal evolution of wave-induced turbulence. Based on this observed temporal and modeled spatial functional dependence of the dissipation field, Equation (9.14) yields $b \approx 10^{-5}$. The direct turbulence observations provide only a crude estimate of the integrated energy dissipation, and the obtained estimate of b should be considered as order of magnitude estimate only. In particular the choice of z_0 and the poor

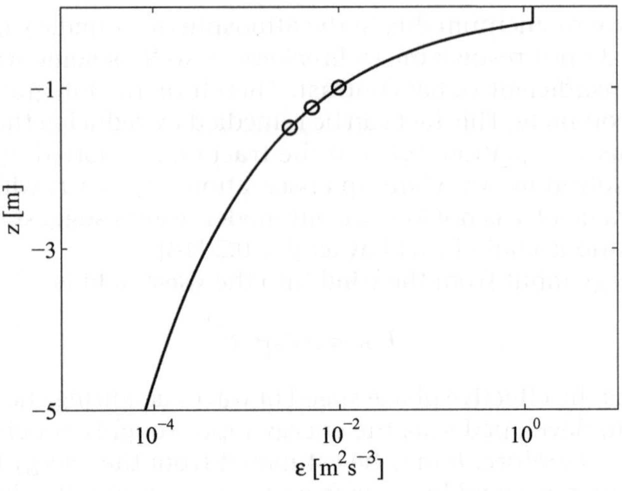

Figure 9.4. Modeled depth dependence of energy dissipation beneath breaking waves. Circles depict depth of observations

depth resolution dominate the uncertainty of b, and the uncertainty in the integrated fifth moment of the breaking crest length distribution, estimated to be < 15%, is less crucial. However, the estimated b-value is also in good agreement with our results obtained from evaluating Equations (9.10) and (9.12) (reported in Gemmrich et al. [11]).

9.4 Conclusions

Previously, near-surface dissipation levels in a wind-driven sea, estimated from observed velocity wavenumber spectra, established the picture of patchy turbulence with short-lived turbulence levels up to five orders of magnitude higher than expected in an equivalent flow along a solid wall [8]. Simultaneous to these turbulence observations, the white capping activity has been monitored with video recordings. From these data the distribution of the breaking crest length $\Lambda(c)$ has been extracted. Phillip's equilibrium range theory predicts that wave-induced momentum flux and energy dissipation are directly proportional to the fourth and fifth moment of this distribution. However, the proportionality factor b is not known a priori and only very limited estimates are available so far. Scaling the observed dissipation levels with a modeled depth dependence, we could evaluate the predicted dissipation against the energy dissipation in the water column. This evaluation yields an estimate for the proportionality factor $b = \mathcal{O}(10^{-5})$. In fully developed wave fields the energy dissipation is limited to wave scales much smaller than the dominant waves. However, in developing seas all wave scales, including the dominant waves are part of the breaking spectrum; in fact energy dissipation is dominated by intermediate to large scale waves. This is of relevance for refining the energy dissipation function S_{dis} in spectral wave models.

Acknowledgement. I gratefully acknowledge the funding by the Canadian Foundation for Atmospheric and Climate Sciences (CFCAS). Many discussions with Michael Banner and Chris Garrett guided the white cap analysis. Part of the analysis was done during a visit at the Institute for Baltic Sea Research, Warnemünde, and I thank Hans Burchard and Lars Umlauf for the hospitality and the many insightful discussions.

References

[1] Banner M.L., and Pierson W.L. (1998) Tangential stress beneath wind-driven air-water interfaces. J Fluid Mec 364: 115–145.

[2] Duncan J.D., (1981) An experimental investigation of breaking waves produced by a towed hydrofoil. Proc R Soc Lond A 377: 331–348.

[3] Duncan, J.D., (2001) Spilling breakers. Ann Rev Fluid Mech 33: 517–547.

[4] Farmer D.M., Vagle S., Booth A.D., (1998) A free-flooding acoustical resonator for measurement of bubble size distribution. J Atm Ocean Tech 15: 1132–1146.

[5] Garrett C., Li M. and Farmer D.M. (2000) The connection between bubble size spectra and energy dissipation rates in the upper ocean. *J Phys Oceanogr* 30: 2163–2171.

[6] Gemmrich J.R. (2000) Temperature anomalies beneath breaking waves and the decay of wave-induced turbulence. *J Geophys Res* 105: 8727–8736.

[7] Gemmrich J.R. (2005) On the occurrence of wave breaking. *'Aha Huliko'a Hawaiian winter workshop*, January 2005, University Hawaii. 123–130.

[8] Gemmrich J.R. and Farmer D.M. (1999) Near surface turbulence and thermal structure in a wind driven sea. *J Phys Oceanogr* 29: 480–499.

[9] Gemmrich J.R. and Farmer D.M. (2004) Near surface turbulence in the presence of breaking waves. *J Phys Oceanogr* 34: 1067–1086.

[10] Gemmrich J.R., Mudge T.D. and Polonichko V.D. (1994) On the energy input from wind to surface waves. *J Phys Oceanogr* 24: 2413–2417.

[11] Gemmrich J.R., Banner M. and Garrett C. (2007) Spectrally resolved energy dissipation and momentum flux associated with wave breaking. *J Phys Oceanogr.* Submitted.

[12] Melville, W.K., (1996) The role of surface-wave breaking in air-sea interaction. *Annual Rev Fluid Mech* 26: 279–321.

[13] Melville W.K. and Matusov P., (2002) Distribution of breaking waves at the ocean surface. *Nature* 417: 58–62.

[14] Kudryavtsev V.N., Makin V.K. and Chapron B., (1999) Coupled sea surface-atmosphere model 2. Spectrum of short wind waves. *J Geophys Res* 104: 7625–7639.

[15] Phillips, O.M., (1985) Spectral and statistical properties of the equilibrium range in wind-generated gravity waves. *J Fluid Mech* 156: 505-531.

[16] Phillips, O.M., Posner F.L., Hansen J.P., (2001) High range resolution radar measurements of the speed distribution of breaking events in wind-generated ocean waves: Surface impulse and wave energy dissipation rates. *J Phys Oceanogr* 31: 450–460.

[17] Thorpe, S.A., (1995) Dynamical processes of transfer at the sea surface. *Progress Oceanogr* 35: 315–352.

[18] Thorpe, S.A., (2005) The Turbulent Ocean. Cambridge University Press.

10

The Interaction of Ocean Surface Processes, Waves, and Turbulence in the Adjacent Boundary Layers

Alastair D. Jenkins[1,2]

[1] Geophysical Institute
University of Bergen
alastair.jenkins@gfi.uib.no
[2] Bjerknes Centre for Climate Research
Allégaten 70, 5007 Bergen, Norway
alastair.jenkins@bjerknes.uib.no

Abstract In order for air–sea exchange processes to be estimated in a dynamically consistent manner in the coupled atmosphere-ocean boundary-layer system, it is necessary to account for the dynamics of surface waves and other movements of the air-water interface. This is also necessary for the interpretation of turbulent flux observations made within the boundary layers, particularly those made from non-stationary measurement platforms. In recent years, considerable progress has been made in observational technology for the direct determination of the vertical flux of momentum, heat, and mass by eddy covariance techniques.

We present an approach to the study of atmosphere–ocean boundary-layer fluxes which employs a general time-dependent coordinate formulation. To provide a uniform treatment both above and below the water surface it is advantageous to use the instantaneous sea surface as a coordinate surface. Reynolds covariances for turbulent flux are then replaced by more complex expressions, and we explore the implications for the design of flux measurement systems and modelling of the coupled interfacial/boundary-layer system. We show that for the flux of trace substances, and other scalar quantities such as heat, measurements of averaged fluxes from moving measurement platforms are subject to biases which are proportional to the square of the amplitude of the platform displacement or of the wave slope, and we indicate how such biases may be corrected or allowed for.

10.1 Introduction

The coupled boundary-layer system in the vicinity of the sea surface is the site of numerous complex physical processes and their interactions, contributing to the exchange of mass, momentum, and energy, between atmosphere and the ocean. Reliable assessments of local and global bud-

C.S. Garbe, R.A. Handler, B. Jähne (eds.): Transport at the Air Sea Interface
pp. 145-158, 2007,

gets of gas species and other atmospheric and oceanic constituents are made difficult by motions of the air–sea interface, as is the construction of coupled atmosphere–ocean model systems. It is necessary to put in place a dynamically consistent framework for exchange processes, which must, directly or indirectly, account for the effect of surface waves. Waves are known to affect the air–sea momentum flux via the effective aerodynamic roughness of the sea surface [9, 13, 14, 15, 52, 53, 56], the drift current near the sea surface [10, 22, 33, 39], and the rate coefficients for gas transfer between the atmosphere and the ocean [19, 21]. Theoretical and numerical modelling studies of wave effects on air–sea momentum flux include those of Jacobs [23], Janssen [25, 26], and Jenkins [30, 31]. Such studies provide appropriate boundary conditions for modelling and predicting ocean drift currents, including those within the upper few centimetres and metres of the water column [28, 29, 36, 43, 55, 63].

For gas transfer, it is not yet straightforward to undertake systematic theoretical studies of the effect of waves, primarily because measurement difficulties still lead to considerable uncertainties in estimates of the variation of gas transfer rate with wind speed [11, 24, 42, 50, 58, 59, 60, 62]. Since pressure fluctuations do not transport mass (as they do for momentum), and diffusion across the interface is inhibited by the presence of laminar boundary layers, the direct dynamical effect of wave motions on gas (and sensible heat) flux will be less than on momentum flux (cf. [37, 38]), though it will be enhanced by the breakup of the water surface and laminar boundary layers via breaking waves [7, 12, 32, 35, 48, 54] and bubbles [16, 49, 57, 61, 64]. However, there is scope for studies of the kinematics and dynamics of wave effects on fluid flow in the near-surface boundary layers, as regards the estimation of fluxes from, in particular, non-stationary instrument platforms.

In recent years, considerable progress has been made in observational technology for the direct determination of the vertical flux of momentum, heat, and mass by eddy covariance techniques, thus releasing the flux estimates from the assumptions which were previously necessary and based on empirical boundary-layer studies or the theory of the Kolmogorov inertial subrange. However the processing of eddy covariance flux measurements over the ocean may be rather difficult, as corrections must be made for variations in the attitude, velocities, and positions of the sensors, as well as for possible flow distortion and time-response effects (see [1, 15, 20, 44]). In addition, even if the instantaneous motions are completely corrected for, a systematic bias may arise due to averaging over the non-stationary path of the measurement platform. Also, the amplitude of the waves may be sufficiently large that for at least part of the wave cycle the instrument platform lies below the height of adjacent wave crests, and in such a situation the concept of mean concentration, vertical flux, and so on, at a fixed level, becomes difficult to define, since the given level will be partly in the air and partly underwater.

We give a discussion of some of the difficulties described above, and how one may approach them by employing different types of reference coordinate system, in the following sections of this paper.

10.2 Coordinate Systems

10.2.1 Alternatives

There are advantages to using curvilinear coordinate systems which follow the water surface. We may resolve vertical variations at small distances from the surface, which is useful for computing heat, gas, and particle exchange through the water surface, and ice formation. Time-independent curvilinear coordinates (e.g. [43]) can be used with surface waves of fixed form. The coordinates of the Lagrangian formulation of hydrodynamics (e.g. [8, 27, 28, 63], in which the coordinates are fixed with respect to the fluid particles, provide straightforward boundary conditions at the air–water interface, and are useful if the mean velocities are not too large: however, the coordinate transformation between the Lagrangian particle coordinates and the spatial ('Eulerian') coordinates may distort unacceptably for long times (t).

The generalised Lagrangian mean (GLM) formulation [2, 3] provides an elegant method of accounting for mean and wave-induced fluxes and flows: it provides an approximately Eulerian representation for the mean variables with a superimposed approximately Lagrangian representation of the wave-induced oscillations. It is, however, unsuitable in the case of strong mean flows (e.g. wind over waves) where there are 'critical levels' where the mean flow speed coincides with the speed of the wave crests, as in that case the oscillatory part of the coordinate transformation becomes singular. In such a case, more general coordinate systems should be used.

If the coordinate transformation consists of vertical displacements only, we obtain a 'sigma-coordinate' representation [47]. However, this may be insufficiently general for some purposes: an approach using general curvilinear coordinates was employed by Jenkins [30], and will be described below in more detail.

10.2.2 General Coordinate System Approach

The approach using general curvilinear coordinate systems provides a representation which encompasses other coordinate system representations (Eulerian/Cartesian, Lagrangian, generalised Lagrangian mean formulation, etc.) as special cases. One simplification which we do employ is that the directions of vector and tensor components remain those of the physical Cartesian coordinate system, the underlying physical space being Euclidean, so that the complication of introducing covariant and contravariant vector/tensor components is not necessary. The use of a general coordinate system is also useful for interpreting measurements from moving instruments, such as time-dependent or time-averaged gas concentration measurements from sensors at moving locations.

In the original Cartesian coordinate system $\mathbf{x} = (x_1, x_2, x_3)$, we have the following equations for momentum, mass, and tracer concentration, respectively:

$$\rho^{\mathbf{x}}\left[u^{\mathbf{x}}_{j,t} + u^{\mathbf{x}}_l u^{\mathbf{x}}_{j,l} + \Phi^{\mathbf{x}}_{,j} + 2(\boldsymbol{\Omega}\times\mathbf{u}^{\mathbf{x}})_j\right] - \tau^{\mathbf{x}}_{jl,l} = 0, \tag{10.1}$$

$$\rho^{\mathbf{x}}_{,t} + u^{\mathbf{x}}_l \rho^{\mathbf{x}}_{,l} + \rho^{\mathbf{x}} u^{\mathbf{x}}_{l,l} = 0, \tag{10.2}$$

$$C^{\mathbf{x}}_{,t} + u^{\mathbf{x}}_l C^{\mathbf{x}}_{,l} + C^{\mathbf{x}} u^{\mathbf{x}}_{l,l} + F^{\mathbf{x}}_{l,l} = 0, \tag{10.3}$$

where ρ is the fluid density, C is tracer concentration, F_l is tracer flux, $\mathbf{u} = (u_1, u_2, u_3)$ is the velocity, $\boldsymbol{\Omega}$ is the rotational angular velocity vector, Φ is a force (e.g. gravitational) potential, and the tensor τ_{jl} incorporates both pressure and shear stress. Repeated indices are summed from 1 to 3.

In the curvilinear coordinate system, with coordinates $\mathbf{y} = (y_1, y_2, y_3)$, the Jacobian coordinate transformation determinant $J = \det[x^{\mathbf{y}}_{j,l}]$ has cofactors K_{jl}, and Equations (10.1–10.3) may be written as

$$P_{j,t} - T_{jl,l} = S_j, \tag{10.4}$$

$$(\rho^{\mathbf{y}} J)_{,t} + [K_{ml}\rho^{\mathbf{y}}(u^{\mathbf{y}}_m - x^{\mathbf{y}}_{m,t})]_{,l} = 0, \tag{10.5}$$

$$(C^{\mathbf{y}} J)_{,t} + \{K_{ml}[C^{\mathbf{y}}(u^{\mathbf{y}}_m - x^{\mathbf{y}}_{m,t}) + F^{\mathbf{y}}_m]\}_{,l} = 0, \tag{10.6}$$

where $P_j = \rho^{\mathbf{y}} J u^{\mathbf{y}}_j$ is the 'concentration of x_j-momentum in $\mathbf{y}$-space', $T_{jl} = \left[\tau^{\mathbf{y}}_{jm} - \rho^{\mathbf{y}} u^{\mathbf{y}}_j \left(u^{\mathbf{y}}_m - x^{\mathbf{y}}_{m,t}\right)\right] K_{ml}$ is minus the flux of x_j-momentum across y_l-surfaces, and $S_j = -\rho^{\mathbf{y}}\Phi^{\mathbf{y}}_{,l} K_{jl} - 2\rho^{\mathbf{y}} J(\boldsymbol{\Omega}\times\mathbf{u}^{\mathbf{y}})_j$ is a source function which incorporates potential and Coriolis forces.

For more detailed information on how to derive Equations (10.1–10.3) from Equations (10.4–10.6), the reader is referred to [30, 34]. The principle is to convert each conservation-law equation in three dimensions into an equation for four-dimensional divergence, before performing the coordinate transformation. Andrews and McIntyre [2] show how to manipulate the Jacobian coordinate transformation determinant and its cofactors.

10.2.3 Vertical Fluxes

Vertical fluxes of momentum, heat, mass, and so on, in the atmospheric and/or ocean boundary layers may be treated by considering time or ensemble averages of Equations (10.1–10.3) and Equations (10.4–10.6). The measurement and computation of Reynolds stress and the flux of heat and gas species in the terrestrial atmospheric boundary is considered in detail by Lee et al. [40], Foken et al. [18], Finnigan [17], Leuning [41], Malhi et al. [45], and Massman [46]. Lee et al. [40] demonstrate the advantages of using orthogonal curvilinear coordinate systems for airflow over hilly terrain: however, it is inconvenient to restrict oneself to coordinate systems with specified local geometry if a surface-following coordinate system over a moving wavy surface is desired. The simplification which we employ in this paper is the employment of fixed directions for the coordinate directions of velocity, momentum, and stress components, and such a restriction is justified by the law of conservation of momentum.

We split the dependent variables into steady and fluctuating parts. For the coordinate transformation we have

$$\mathbf{x}^{\mathbf{y}} = \mathbf{y} + \boldsymbol{\xi}. \tag{10.7}$$

For other dependent variables ϕ, we write

$$\phi^{\mathbf{y}} = \overline{\phi}^{\mathbf{y}} + \phi^{\mathbf{y}\prime}, \tag{10.8}$$

where the averaging is performed at points fixed with respect to the curvilinear coordinate system $\mathbf{y}$, which may thus be moving with respect to the Cartesian coordinate system $\mathbf{x}$.

For simplicity, we now restrict the motions to the vertical plane (x_1, x_3), and assume that mean quantities are independent of time and of the horizontal coordinates:

$$\overline{(\cdot)}_{,t} = \overline{(\cdot)}_{,1} = \overline{(\cdot)}_{,2} = 0. \tag{10.9}$$

Then from Equation (10.7) we obtain the following two-dimensional matrices (with respect to the x_1 and x_2 coordinate directions) for the coordinate transformation derivatives and the cofactors of the Jacobian coordinate transformation determinant, respectively:

$$\frac{\partial \mathbf{x}}{\partial \mathbf{y}} = \begin{bmatrix} 1+\xi_{1,1} & \xi_{1,3} \\ \xi_{3,1} & 1+\xi_{3,3} \end{bmatrix}, \qquad \mathsf{K} = \begin{bmatrix} 1+\xi_{3,3} & -\xi_{3,1} \\ -\xi_{1,3} & 1+\xi_{1,1} \end{bmatrix}. \tag{10.10}$$

Equation (10.6) for the tracer concentration can then be averaged and integrated with respect to y_3, so that

$$\begin{aligned} &\overline{K_{m3}[C^{\mathbf{y}}(u^{\mathbf{y}}_m - x^{\mathbf{y}}_{m,t}) + F^{\mathbf{y}}_m]} \\ &\quad = \overline{K_{13}[C(u_1 - x_{1,t}) + F_1]} + \overline{K_{33}[C(u_3 - x_{3,t})]} \\ &\quad = \text{constant}, \end{aligned} \tag{10.11}$$

where we have dropped the $(\cdot)^{\mathbf{y}}$-superscripts for convenience.

Then if we substitute $u^{\mathbf{y}}_1 = \overline{u_1} + u_1{}'$, $u^{\mathbf{y}}_3 = u_3$, $C^{\mathbf{y}} = \overline{C} + C'$, $F^{\mathbf{y}}_1 = F_1{}'$, and $F^{\mathbf{y}}_3 = \overline{F_3} + F_3{}'$, i.e. we assume that we may neglect $\overline{F_1}$ and $\overline{u_3}$, and neglect averaged products of more than two fluctuating quantities, we obtain

$$\begin{aligned} \overline{F_3} + \overline{F_3{}'\xi_{3,3}} - \overline{F_1{}'\xi_{3,1}} + \overline{C}[\overline{(u_3{}' - \xi_{3,t})\xi_{3,3}} - \overline{(u_1{}' - \xi_{1,t})\xi_{3,1}}] \\ + \overline{C'(u_3{}' - \xi_{3,t} - \overline{u_1}\xi_{3,1})} = \text{constant}. \end{aligned} \tag{10.12}$$

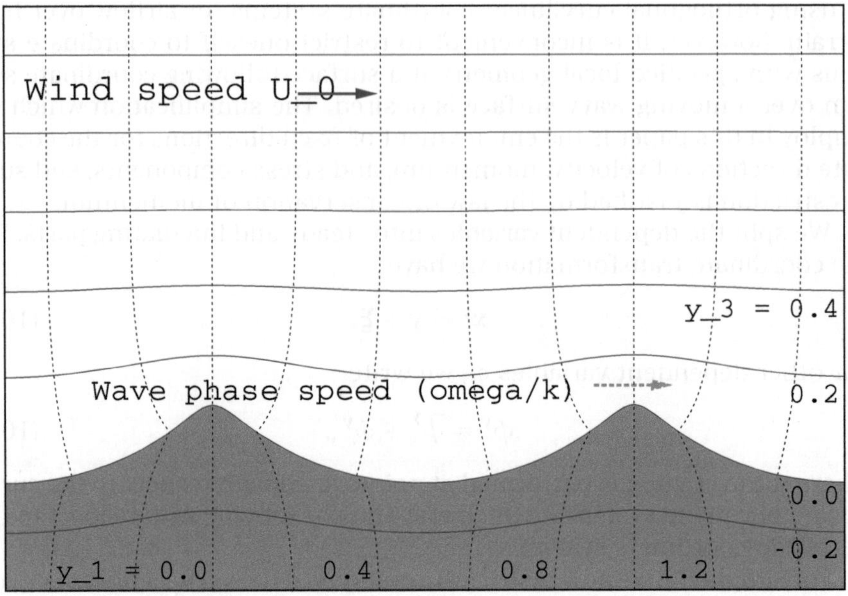

Figure 10.1. Coordinate system used in the simple example to illustrate tracer flux.

Simple Example

We apply Equation (10.12) in a simple example, using a coordinate system shown schematically in Fig. 10.1. The coordinate system is given by

$$\begin{bmatrix} x_1 \\ x_3 \end{bmatrix} = \begin{bmatrix} y_1 - ae^{-k|y_3|}\sin(ky_1 - \omega t) \\ y_3 + ae^{-k|y_3|}\cos(ky_1 - \omega t) \end{bmatrix} \tag{10.13}$$

We assume that the airflow over the wavy water surface is potential flow, with uniform horizontal velocity U_0 at large heights, and will thus follow the surfaces of constant y_3. We assume also that the tracer concentration is equal to C_0 in the water $y_3 \leq 0$, and above the water surface it is subject to a constant diffusion coefficient D, and has reached an equilibrium concentration profile with a vertical gradient α at large heights. In a frame of reference (moving with the wave phase speed ω/k) in which the wave crests are stationary, the velocity potential and stream function for the airflow and the tracer concentration all obey Laplace equations, and we obtain the following expressions for the mean and fluctuating parts of the dependent variables above the water surface in the **y**-coordinate system:

$$\overline{u_1} = U_0 + \mathcal{O}[(ak)^2]$$
$$u_1' = a(kU_0 - \omega)\mathbf{e}^{-ky_3}\cos(ky_1 - \omega t) + \mathcal{O}[(ak)^3]$$
$$u_3' = -a(kU_0 - \omega)\mathbf{e}^{-ky_3}\sin(ky_1 - \omega t) + \mathcal{O}[(ak)^3]; \quad (10.14)$$

$$C = \overline{C} = C_0 + \alpha y_3; \qquad C' = 0; \quad (10.15)$$

$$F_1' = \alpha Dak\mathbf{e}^{-ky_3}\sin(ky_1 - \omega t) + \mathcal{O}[(ak)^3];$$
$$F_3' = -\alpha Dak\mathbf{e}^{-ky_3}\cos(ky_1 - \omega t) + \mathcal{O}[(ak)^3];$$
$$\overline{F_3} = -\alpha D[1 + (ak)^2\mathbf{e}^{-2ky_3}] + \mathcal{O}[(ak)^3]. \quad (10.16)$$

We note that $\overline{F_3}^{\mathbf{y}}$ differs from the value it would have in the absence of surface wave motions by an amount proportional to the square of the wave slope ak. This is compensated for in Equation (10.12) by the two terms $\overline{F_3'\xi_{3,3}}$ and $-\overline{F_1'\xi_{3,1}}$, which to $\mathcal{O}[(ak)^2]$ each have a value of $^1/_2\alpha D(ak)^2\mathbf{e}^{-2ky_3}$. To the same approximation, the terms $\overline{(u_3' - \xi_{3,t})\xi_{3,3}}$ and $-\overline{(u_1' - \xi_{1,t})\xi_{3,1}}$ vanish since their individual factors are in quadrature, and the final term $\overline{C'(u_3' - \xi_{3,t} - \overline{u_1}\xi_{3,1})}$ vanishes since $C' = 0$ by Equation (10.15).

We see, therefore, that the mean vertical tracer flux determined by averaging at a fixed point in the curvilinear coordinate system differs from that which we would have obtained by averaging at a fixed point in (**x**-) space, so in order to obtain a 'correct' value for the mean flux we must take account of fluctuating terms in both horizontal and vertical directions and also the coordinate transformation. Note that in a more general case the tracer concentration will also fluctuate at fixed **y**-coordinate positions, so additional terms in Equation (10.12) will have to be taken into account.

Momentum flux

The above discussion of tracer flux may also, of course, be applied to other scalar quantities such as heat, and, with appropriate additional terms, to reactive substances and to the concentration of particles subjected to additional forces such as gravity. In the case of momentum, the situation is made more complex by its vector character, the momentum flux being a tensor quantity, and due to the fact that momentum may also be transferred via pressure forces. A treatment of the effect of a wavy surface on the various contributions to the vertical flux of horizontal momentum was given by Jenkins [30], who also accounted for a continuous spectrum of waves, instead of a single wave Fourier component, by integrating the $\mathcal{O}[(ak)^2]$ terms appropriately over the wave spectrum (see also Janssen [25]). Figure 10.2 shows the results of a computation of the various contributions to vertical momentum flux, for a specific wind-wave spectrum, using a turbulence closure scheme of eddy-viscosity type and allowing the airflow profile to adjust in a self-consistent manner.

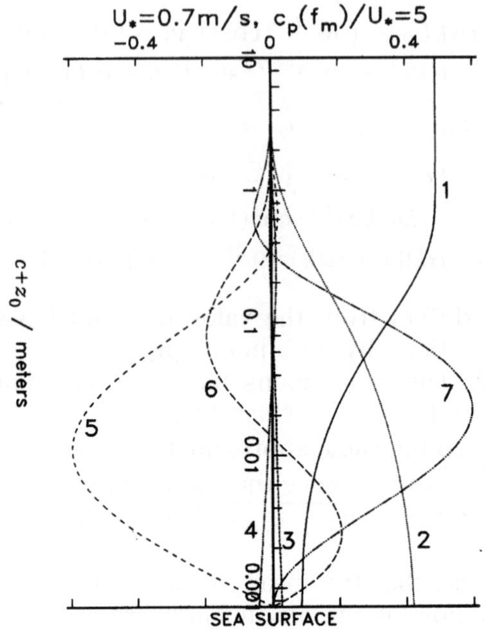

Figure 10.2. Computed vertical profile of the various contributions to the downward momentum flux over wind waves, calculated by the quasi-linear eddy-viscosity-based model of Jenkins [30]. 1, $\overline{\tau_{13}}/\rho$; 2, $\overline{p'\xi_{3,1}}/\rho$; 3, $-\overline{\sigma'_{11}\xi_{3,1}}/\rho$; 4, $\overline{\tau'_{13}\xi_{1,1}}/\rho$; 5, $\overline{u_1'\xi_{1,t}}$; 6, $-\overline{u_1'u_3'}$; 7, $\overline{\overline{u_1}u_1'\xi_{3,1}}$. Notation: $c = y_3$ is the vertical curvilinear coordinate, with $c = 0$ being the water surface; ξ_1 and ξ_3 are the horizontal and vertical coordinate displacements; $\overline{u_1}$ is the mean horizontal velocity; u_1 and u_3 are the horizontal and vertical velocity components; $\sigma_{jl} = \tau_{jl} + p\delta_{jl}$ is the traceless stress tensor; over-bars and primes denote mean and fluctuating values with respect to the curvilinear coordinate system.

10.3 Discussion

Since gas flux through the sea surface is controlled by processes which have a much smaller vertical scale than the height of surface waves, there are advantages in analysing measurements and applying a model framework in time-dependent curvilinear coordinates. We see from the analysis of a simple case of potential flow over waves with tracer flux governed by a hypothetical constant diffusion coefficient, that spatial and temporal averaging is possible with respect to such coordinates, but that it is important to be aware of inherent second-order bias effects. It should be possible to extend the technique to continuous wave spectra, as has been done in the case of momentum flux [30]. A more realistic airflow profile should be estimated numerically, using, for example, a suitable turbulence closure scheme, or a more realistic analytical turbulent boundary-layer flow profile such as those [5, 6] derived from rapid distortion theory [4]. In such a way we should be able to estimate the spatial and temporal derivatives of the coordinates and hydrodynamic variables which arise in Equations (10.4–10.6) and Equations (10.11–10.12).

Awareness of the dynamical consequences of using such time-dependent coordinates is valuable in interpreting the results of observations from moving (and stationary) instruments located near the water surface, since coordinate-system-dependent biases of averaged measurements may occur. In our simple example, we allowed the measurement platform to move with the airflow streamlines, but in general this will not be the case. A coordinate system should be chosen which, if possible, follows both the air-water interface and the position of the measurement platform: such a coordinate system will in practice need to be computed numerically. Where wave-induced flow is important, the measurement system should also, if possible, monitor the phase and surface displacement of the wave field.

Nevertheless, the main result of this paper should still apply: even after completely correcting for platform motion, we obtain averaged fluxes which differ by a quantity proportional to the square of the wave slope, when we compare observations from a fixed point in space with those from a platform with wave-induced motions. Note that our result, that the difference is proportional to $\mathbf{e}^{-2k|y_3|}$, only applies if the platform motion decreases as fast as $\mathbf{e}^{-k|y_3|}$. If the measurement platform movement decreases more slowly with increasing distance from the interface, the difference between 'fixed' and 'moving' averaged measurements will also decrease more slowly. Pollard [51] found that if an ideal vector-averaging current meter was suspended from a surface-following buoy in deep water, it would measure an average current which differed from the mean current averaged at an equivalent fixed point in space by an amount proportional to $\mathbf{e}^{-k|y_3|}$, even though the difference between the Eulerian and Lagrangian mean current is proportional to $\mathbf{e}^{-2k|y_3|}$.

In our analysis we have only taken into account fluctuations in airflow velocity, concentration, and flux, which are associated with oscillatory movements of the surface-following coordinate system. Hence we are in principle assuming that the flow is laminar. Of course in practice the flow will be turbulent, and the flux components will be given by covariances of turbulent fluctuations of velocity and concentration. Nevertheless, the laminar flow approximation presented in this paper enables us to demonstrate salient features of the representation and the measurement of fluxes in the vicinity of moving interfaces and from moving measurement platforms, and will provide a basis which may be extended to take account of more complex turbulent dynamics.

It may be argued that it will be difficult to apply the method under realistic field conditions, since it is in practice impossible to obtain a complete three-dimensional time-dependent picture of the fluid velocity and tracer concentration field. The author admits that this is often the case, but, nevertheless, it may be possible to make simplifying assumptions, in a manner analogous to the employment of Taylor's frozen turbulence hypothesis to infer spatial structure from temporal fluctuations. From a simplified view of the flow field it may be possible to estimate the potential bias arising from averaging observations from moving platforms: the 'simplest' coordinate transformation, to a Cartesian framework in which the instrument platform coordinates are time-independent, may prove useful in this case.

Acknowledgement. I am grateful to Will Drennan for a useful discussion and for the provision of observational data which was useful in the formulation of this work. I also thank Bill Peirson and Guillemette Caulliez for helpful discussions. The work was supported by the Research Council of Norway, projects 155923/700 and 175763/V30. This is publication No. A 157 from the Bjerknes Centre for Climate Research.

References

[1] F. Anctil, M. A. Donelan, W. M. Drennan, and H. C. Graber. Eddy-correlation measurements of air–sea fluxes from a discus buoy. *J Atmos Oceanic Technol*, 11:1144–1150, 1994.

[2] D. G. Andrews and M. E. McIntyre. An exact theory of nonlinear waves on a Lagrangian-mean flow. *J Fluid Mech*, 89:609–646, 1978.

[3] D. G. Andrews and M. E. McIntyre. On wave-action and its relatives. *J Fluid Mech*, 89:647–664, 1978.

[4] G. K. Batchelor and I. Proudman. The effect of rapid distortion of a fluid in turbulent motion. *Q J Mech Appl Math*, 7:83–103, 1954.

[5] S. E. Belcher and J. C.R. Hunt. Turbulent flow over hills and waves. *Ann Rev Fluid Mech*, 30:507–538, 1998.

[6] S. E. Belcher, J. A. Harris, and R. L. Street. Linear dynamics of wind waves in coupled turbulent air-water flow. Part 1. Theory. *J Fluid Mech*, 271:119–151, 1994.

[7] P. Bonmarin. Geometric properties of deep-water breaking waves. *J Fluid Mech*, 209:405–433, 1989.

[8] M.-S. Chang. Mass transport in deep-water long-crested random gravity waves. *J Geophys Res*, 74:1515–1536, 1969.

[9] H. Charnock. Wind stress on a water surface. *Q J R Meteorol Soc*, 81: 639–640, 1955.

[10] A. D. D. Craik. The drift velocity of water waves. *J Fluid Mech*, 116: 187–205, 1982.

[11] E. L. Deacon. Sea-air gas transfer: The wind speed dependence. *Bound-Layer Meteorol*, 21:31–37, 1980.

[12] D. G. Dommermuth, D. K. P. Yue, W. M. Lin, R. J. Rapp, E. S. Chan, and W. K. Melville. Deep-water plunging breakers: A comparison between potential theory and experiments. *J Fluid Mech*, 189:423–442, 1988.

[13] M. A. Donelan, W. M. Drennan, and K. B. Katsaros. The air-sea momentum flux in mixed wind sea and swell conditions. *J Phys Oceanogr*, 27:2087–2099, 1997.

[14] W. M. Drennan, H. C. Graber, and M. A. Donelan. Evidence for the effects of swell and unsteady winds on marine wind stress. *J Phys Oceanogr*, 29:1853–1864, 1999.

[15] W. M. Drennan, H. C. Graber, D. Hauser, and C. Quentin. On the wave age dependence of wind stress over pure wind seas. *J Geophys Res*, 108(C3), 2003. 8062, doi:10.1029/2000JC000715.

[16] D. M. Farmer, C. L. McNeil, and B. D. Johnson. Evidence for the importance of bubbles in increasing air–sea gas flux. *Nature*, 361:620–623, 1993.

[17] J. Finnigan. Advection and modeling. In X. Lee, W. Massman, and B. Law, editors, *Handbook of Micrometeorology: A Guide for Surface Flux Measurement and Analysis*, chapter 10, pages 209–244. Kluwer, 2004.

[18] T. Foken, M. Göckede, M. Mauder, L. Mahrt, B. Amiro, and W. Munger. Post-field data quality control. In Xuhui Lee, William Massman, and Beverly Law, editors, *Handbook of Micrometeorology: A Guide for Surface Flux Measurement and Analysis*, chapter 9, pages 181–208. Kluwer, 2004.

[19] N. M. Frew, E. J. Bock, U. Schimpf, T. Hara, H. Haußecker, J. B. Edson, W. R. McGillis, R. K. Nelson, S. P. McKenna, B. M. Uz, and B. Jähne. Air–sea gas transfer: Its dependence on wind stress, small-scale roughness, and surface films. *J Geophys Res*, 109, 2004. C08S17, doi:10.1029/2003JC002131.

[20] H. C. Graber, E. A. Terray, M. A. Donelan, W. M. Drennan, J. C. van Leer, and D. B. Peters. ASIS – a new air–sea interaction spar buoy: Design and performance at sea. *J Atmos Oceanic Technol*, 17:708–720, 2000.

[21] T. Hara, E. J. Bock, N. M. Frew, and W. R. McGillis. Relationship between air–sea gas transfer velocity and surface roughness. In B. Jähne and E. C. Monahan, editors, *Selected Papers from the Third International Symposium on Air–Water Gas Transfer, July 24–27, 1995, Heidelberg University*. AEON Verlag & Studio, 1995.

[22] N. E. Huang. On surface drift currents in the ocean. *J Fluid Mech*, 91: 191–208, 1979.

[23] S. J. Jacobs. Effective roughness length for turbulent flow over a wavy surface. *J Phys Oceanogr*, 19:998–1010, 1989.

[24] B. Jähne, W. Huber, A. Dutzi, T. Wais, and J. Ilmberger. Wind/wave tunnel experiment on the Schmidt number – and wave field dependence of air/water gas exchange. In W. Brutsaert and G. H. Jirka, editors, *Gas Transfer at Water Surfaces*, pages 303–309. Reidel, 1984.

[25] P. A. E. M. Janssen. Wave-induced stress and the drag of air flow over sea waves. *J Phys Oceanogr*, 19:745–754, 1989.
[26] P. A. E. M. Janssen. Quasi-linear theory of wind-wave generation applied to wave forecasting. *J Phys Oceanogr*, 21:1631–1642, 1991.
[27] A. D. Jenkins. A theory for steady and variable wind and wave induced currents. *J Phys Oceanogr*, 16:1370–1377, 1986.
[28] A. D. Jenkins. Wind and wave induced currents in a rotating sea with depth-varying eddy viscosity. *J Phys Oceanogr*, 17:938–951, 1987.
[29] A. D. Jenkins. The use of a wave prediction model for driving a near-surface current model. *Dt Hydrogr Z*, 42:133–149, 1989.
[30] A. D. Jenkins. A quasi-linear eddy-viscosity model for the flux of energy and momentum to wind waves, using conservation-law equations in a curvilinear coordinate system. *J Phys Oceanogr*, 22:843–858, 1992.
[31] A. D. Jenkins. A simplified quasilinear model for wave generation and air-sea momentum flux. *J Phys Oceanogr*, 23:2001–2018, 1993.
[32] A. D. Jenkins. A stationary potential-flow approximation for a breaking-wave crest. *J Fluid Mech*, 280:335–347, 1994.
[33] A. D. Jenkins. Comparison of current measurements over the Norwegian continental shelf, using near-surface moored current meters and surface drifters. In *Proceedings of Conference, "Current Measurements Offshore", London, 17 May 1984*, London, 1984. Soc Underwater Technol.
[34] A. D. Jenkins. Conservation form of the momentum equation in a general curvilinear coordinate system. *Ocean Modelling (newsletter)*, 84:6–8, 1989. Available from the Robert Hooke Institute, Dept of Atmospheric, Oceanic and Planetary Physics, Clarendon Laboratory, Parks Road, Oxford, OX1 3PU, U.K. May be downloaded from `http://www.gfi.uib.no/~jenkins/papers/JenkinsAD_OM-1989-6.ps.gz`.
[35] A. D. Jenkins. Geometrical and kinematic properties of breaking waves in the framework of a stationary flow approximation. In M. Olagnon and G. Athanassoulis, editors, *Rogue Waves 2000: Proceedings of Workshop, Brest, 29–30 November 2000*, pages 221–226. Ifremer, 2001.
[36] A. D. Jenkins and F. Ardhuin. Interaction of ocean waves and currents: How different approaches may be reconciled. In *Proc 14th Int Offshore & Polar Engng Conf, Toulon, France, May 23–28, 2004*, volume 3, pages 105–111. Int Soc Offshore & Polar Engrs, 2004.
[37] A. D. Jenkins and B. Ward. A simple model for the short-time evolution of near-surface current and temperature profiles. *Deep-Sea Res II*, 52:1202–1214, 2005.
[38] A. D. Jenkins and B. Ward. Reply to: Comment on the paper: A simple model for the short-time evolution of near-surface current and temperature profiles. *Deep-Sea Res II*, 52:1218–1219, 2005.

[39] A. D. Jenkins, R. B. Olsen, and S. Christianidis. Intercomparison trials: Near-surface current measurements over the Norwegian continental shelf. In *Proc IEEE Third Working Conf on Current Measuremt, Airlie, Virginia, Jan 1986*, pages 20–25. Inst. Electrical & Electronic Engrs, 1986.

[40] X. Lee, J. Finnigan, and K. T. P. U. Coordinate systems and flux bias error. In X. Lee, W. Massman, and B. Law, editors, *Handbook of Micrometeorology: A Guide for Surface Flux Measurement and Analysis*, chapter 3, pages 33–66. Kluwer, 2004.

[41] R. Leuning. Measurements of trace gas fluxes in the atmosphere using eddy covariance: WPL corrections revisited. In X. Lee, W. Massman, and B. Law, editors, *Handbook of Micrometeorology: A Guide for Surface Flux Measurement and Analysis*, chapter 6, pages 119–132. Kluwer, 2004.

[42] P. S. Liss and L. Merlivat. Air–sea gas exchange: Introduction and synthesis. In P. Buat-Ménard, editor, *The Role of Air–Sea Exchange in Geochemical Cycling*, pages 113–127. Reidel, 1986.

[43] M. S. Longuet-Higgins. Mass transport in water waves. *Phil Trans R Soc Lond Ser A*, 245:535–581, 1953.

[44] L. Mahrt, D. Vickers, W. M. Drennan, H. C. Graber, and T. L. Crawford. Displacement measurement errors from moving platforms. *J Atmos Oceanic Technol*, 22:860–868, 2005.

[45] Y. Malhi, K. McNaughton, and C. von Randow. Low frequency atmospheric transport and surface flux measurements. In X. Lee, W. Massman, and B. Law, editors, *Handbook of Micrometeorology: A Guide for Surface Flux Measurement and Analysis*, chapter 5, pages 101–118. Kluwer, 2004.

[46] W. Massman. Concerning the measurement of atmospheric trace gas fluxes with open- and closed-path eddy covariance system: The WPL terms and spectral attenuation. In X. Lee, W. Massman, and B. Law, editors, *Handbook of Micrometeorology: A Guide for Surface Flux Measurement and Analysis*, chapter 7, pages 133–160. Kluwer, 2004.

[47] G. Mellor. The three-dimensional current and surface wave equations. *J Phys Oceanogr*, 33:1978–1989, 2003.

[48] W. K. Melville, F. Veron, and C. J. White. The velocity field under breaking waves: Coherent structures and turbulence. *J Fluid Mech*, 454:203–233, 2002.

[49] L. Memery and L. Merlivat. Modelling of gas flux through bubbles at the air–water interface. *Tellus*, 37B:272–285, 1985.

[50] L. Merlivat and L. Memery. Gas exchange across an air–water interface: Experimental results and modeling of bubble contribution to transfer. *J Geophys Res*, 88:707–724, 1983.

[51] R. T. Pollard. Interpretation of near-surface current meter observations. *Deep-Sea Res*, 20:261–268, 1973.

[52] A. Sjöblom and A. Smedman. Comparison between eddy-correlation and inertial dissipation methods in the marine atmospheric surface layer. *Bound-Layer Meteorol*, 110:141–164, 2004.
[53] A. Sjöblom and A. Smedman. The turbulent kinetic energy budget in the marine atmospheric surface layer. *J Geophys Res*, 107(C10):3142, 2002. doi: 10.1029/2001JC001016.
[54] P. P. Sullivan, J. C. McWilliams, and W. K. Melville. The oceanic boundary layer driven by wave breaking with stochastic variability. Part 1. Direct numerical simulations. *J Fluid Mech*, 507:143–174, 2004.
[55] C. L. Tang, W. Perrie, A. D. Jenkins, B. M. DeTracey, Y. Hu, B. Toulany, and P. C. Smith. Observation and modelling of surface currents on the Grand Banks—a study of the wave effects on surface currents. J Geophys Res - Oceans (submitted), 2006.
[56] P. K. Taylor and M. J. Yelland. The dependence of the sea surface roughness on the height and steepness of the waves. *J Phys Oceanogr*, 31:572–590, 2001.
[57] S. A. Thorpe. On the determination of K_V in the near-surface ocean from acoustic measurements of bubbles. *J Phys Oceanogr*, 14:855–863, 1984.
[58] R. Wanninkhof. Relationship between wind speed and gas exchange over the ocean. *J Geophys Res*, 97:7373–7382, 1992.
[59] R. Wanninkhof and L. Bliven. Relationship between gas exchange, wind speed and radar backscatter in a large wind-wave tank. *J Geophys Res*, 96:2785–2796, 1991.
[60] R. Wanninkhof and W. R. McGillis. A cubic relationship between air-sea CO_2 exchange and wind speed. *Geophys Res Letters*, 26:1889–1892, 1999.
[61] R. Wanninkhof, W. Asher, and E. Monahan. The influence of bubbles on air-water gas exchange: Results from gas transfer experiments during WABEX-93. In B. Jähne and E. C. Monahan, editors, *Selected Papers from the Third International Symposium on Air-Water Gas Transfer, July 24–27, 1995, Heidelberg University*, pages 239–253. AEON Verlag & Studio, 1995.
[62] A. Watson, R. Upstill-Goddard, and P. S. Liss. Air–sea exchange in rough and stormy seas. *Nature*, 349:145–147, 1991.
[63] J. E. Weber. Steady wind- and wave-induced currents in the open ocean. *J Phys Oceanogr*, 13:524–530, 1983.
[64] D. K. Woolf and S. A. Thorpe. Bubbles and the air–sea exchange of gases in near-saturation conditions. *J Marine Res*, 49:435–466, 1991.

11

A Numerical Study on the Characteristic Flow Structures of a Micro-Breaking Wind Wave

Wu-ting Tsai[1,2] and Li-ping Hung[2]

[1] Institute of Hydrological Sciences
National Central University, Taoyuan, Taiwan
[2] Department of Civil Engineering
National Chiao Tung University, Hsinchu, Taiwan

Abstract We present results from numerical simulation of an aqueous turbulent boundary layer underneath a dynamic air–water interface and driven by wind stress and pressure. The simulation results reveal distinct surface and flow structures of micro-breaking wind waves, including a bore-like crest preceded by parasitic capillary waves riding along the forward face and elongated streamwise velocity streaks in the backward face, and confirm the observations in the laboratory and field experiments. The results also highlight the potential impacts caused by the short-wavelength capillaries on the gravity dominant free-surface flows, and consequently reveal the necessity in incorporating such microscale processes in the parameterizations of fluxes across the air–sea interface.

11.1 Introduction

Breaking of very short gravity waves without air entrainment, which is termed 'micro-breaking' [1], occurs ubiquitously over the oceans. It is found that the occurrence of micro-breaking is far more widespread than that of whitecapes caused by spilling breakers in open oceans [2]. This also suggests significant impacts of microscale wave breaking on the fluxes of gas and heat across the air–sea interface [4, 19]. In contrast to the visible manifestation of whitecapping, micro-breaking is more difficult to detect due to the lack of air entrainment. A popular approach in recent years to identify and quantify microscale wave breaking is to infer the event from infrared imagery of the water surface [7]. The accuracy of the technique thus relies on relating quantitatively the small scale surface signatures to the underlying turbulence processes.

The surface deformation of a micro-breaking wavelet is characterized by a bore-like crest accompanied by parasitic capillary waves distributed along the forward face [5, 7]. The typical wavelength is within $\mathcal{O}(0.1 \sim 1)$ m and the amplitude is a few centimeters. In addition to the parasitic capillary waves trapped on the forward face of the dominant wave, Ebuchi et

C.S. Garbe, R.A. Handler, B. Jähne (eds.): Transport at the Air Sea Interface
pp. 159-168, 2007, © Springer-Verlag Berlin, Heidelberg 2007

al. [5] also observed streamwise streaky structures on the windward face. The underlying flow structures were investigated in detail in laboratory flumes by Okuda [10], Siddiqui et al. [12], Peirson and Banner [11] among others. In these measurements, flow separation with strong vorticity near the crest was identified.

We postulate that the observed distinct surface and flow structures within the surface layer of a micro-breaking wind wave are induced by the interaction between the surface gravity waves and the sheared turbulent flow underneath. We therefore pose a numerical simulation by considering a three-dimensional turbulent shear flow beneath surface waves, and driven by pressure forcing and shear stress.

11.2 Numerical Simulation

The underlying numerics of the computational model are based on that of Tsai et al. [14], which consider incompressible and non-hydrostatic flow, and employs mixed pseudo-spectral and finite-difference method. The present numerical model [15], however, imposes the fully-nonlinear boundary conditions of mass and momentum conservations on the exact water surface; thus is capable of resolving surface dynamics ranging from gravity to capillary waves. This requires tracking the free-surface Lagrangian boundary and solving the governing equations in a time-dependent domain. Consequently, the computational demand of the presented "free-surface" simulation greatly exceeds those models based on boundary approximation or parameterization [13, 16, 17]. Details of the mathematical formulation and the numerical implementation of the simulation are reported in Tsai and Hung [15].

To demonstrate the effectiveness of the numerical model, the evolution of a two-dimensional gravity–capillary wave is simulated, and the result is compared with the weakly nonlinear solution of Fedorov and Melville [6]. The computation is conducted using a resolution of 128×128 grids, and is initiated with the surface profile and velocity field of a third-order weakly nonlinear Stokes gravity wave without any prescribed parasitic capillary waves. The pressure forcing is imposed spontaneously on the water surface as in the weakly nonlinear solution [6]. The simulated results of the surface elevations η and slopes η_x are shown in Figure 11.1 at the time interval one and half linear period from the start of the simulation. Shortly after the start of the simulation, short-wavelength capillaries form from the crest and propagate downstream along the forward face of the dominant gravity wave. For the present fully nonlinear computation, there exists no steady state for the posed initial condition and the surface forcing. However, at the time when the fully nonlinear wave evolves to a steepness close to that of the steady weakly nonlinear wave, as depicted in Figure 11.1, the two results share qualitatively identical surface pro-

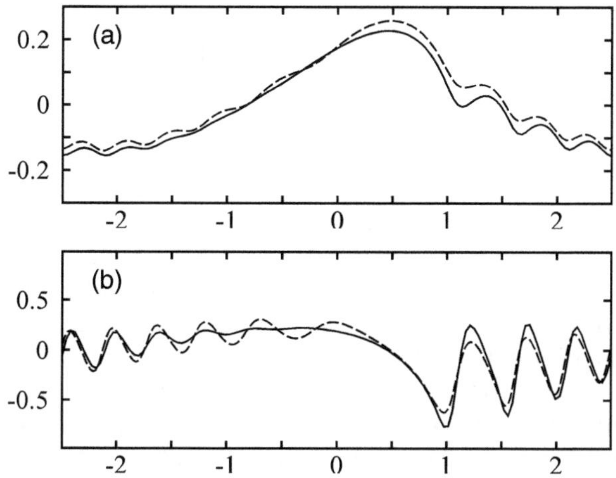

Figure 11.1. Surface elevations η (a) and slopes η_x (b) of a two-dimensional gravity–capillary wave at the time interval 1.5 linear periods from the start of the simulation computed by the present fully nonlinear numerical model (solid lines). The dashed lines are the weakly nonlinear solutions of Fedorov and Melville [6]. To better reveal the surface variation, the frame of reference is moved with the linear phase speed.

files with the same number of parasitic capillaries riding on the dominant gravity wave.

For the posed three-dimensional model flow, an initial gravity wave with a wavelength of 7.5 cm and steepness of 0.25 is considered. The distribution of the surface-normal pressure forcing is also specified as in Fedorov and Melville [6]; and the surface-tangential stress following the measurement of Banner and Peirson [3]. The normal-stress amplitude is $\|\tau_n^s\| = 0.963$ dyn/cm^2. To initiate the turbulence simulation, the flow field is posed by superposition of the irrotational velocity field of a two-dimensional nonlinear Stokes wave, a two-dimensional mean shear profile and a three-dimensional fluctuating velocity field. The initial mean shear profile is chosen such that the mean surface velocity $U_0 = 10$ cm/s, the mean tangential shear stress $\tau_0 = 2$ dyn/cm^2. The initial fluctuating velocity field is homogeneous in both streamwise and spanwise directions. Thus the only flow structure of the initial velocity field is the two-dimensional wavy motion in the along-wind vertical plane associated with the gravity wave. The initial surface elevation and the distribution of the velocity (spanwise component) on the water surface are shown in Figure 11.2**a**. Both the width and the depth of the computational domain are chosen to be the same as the wavelength. The simulations is carried out with 128^3 grid points.

11.3 Surface Structures

Despite the idealizations inherent in conducting the simulation, the computed flow exhibits the major surface features, qualitatively identical to that observed in the experiment of Ebuchi et al. [5] and others. These surface fine structures are illustrated by depicting in Figure 11.2**b** the three-dimensional prospective surface profiles with the turbulent cross-wind velocity distribution superimposed. The simulation results reveal distinct surface structures, including parasitic capillary waves riding along the forward face of the dominant gravity wave and elongated velocity streaks on the backward face. The picture is clearly resembled to that observed by Ebuchi et al. [5] (Figure 11 in their paper).

To further demonstrate the characteristic structures on the water surface, instantaneous contour distributions of the free-surface deformation (η), the streamwise and spanwise surface slopes (η_x, η_y), and the turbulent fluctuating velocities (u', v', w') are shown in Figure 11.3.

Initially, the water surface is prescribed by only a gravity wave deformation and a random roughness without any other fine structures. As the flow evolves, the distribution patterns of streamwise surface slopes (Figure 11.3**b**) clearly demonstrate the generation of parasitic capillary waves on the forward face of the predominant gravity wave as observed in wind–wave flumes [5, 7, 11, 12] (see also Figure 11.2**b**). The train of capillary ripples extend from the crest towards the trough of the gravity wave. The maximum slopes (negative), which can reach as low as -0.9, occur at the first trough on the forward face immediately next to the crest. The steeper downward surface and the dimpled trough form a bore-like crest. The crest/trough lines of the parasitic ripples are more or less parallel to that of the dominant gravity wave but exhibit spanwise undulation.

Accompanying the formation of the parasitic capillaries, elongated streaky structures also become apparent on the fluctuating velocity distributions (Figure 11.3**a**,**b**). On the backward face near the crest of the dominant gravity wave, both the along and cross-wind fluctuating velocities exhibit streaky structures. The major streaks are arranged with somewhat equal transverse spacing. The average interval is about 0.625 cm, which is close to the observation of Ebuchi et al. [5].

In the region where parasitic capillary waves exist (forward face from the crest), the streamwise velocity streaks are annihilated. Instead, the fluctuating velocity distributions exhibit two-dimensional oscillatory patterns. For the streamwise and vertical components, the patterns resemble a chess board. The positive and negative extreme velocity regions are arranged alternately in both streamwise and spanwise directions of the chess-board patterns; with the streamwise wavelength identical to that of the parasitic capillaries, and the spanwise spacing close to that of the velocity streaks. These surface structures imply strong interactions between the streaky velocity field associated with the turbulent shear flow and the

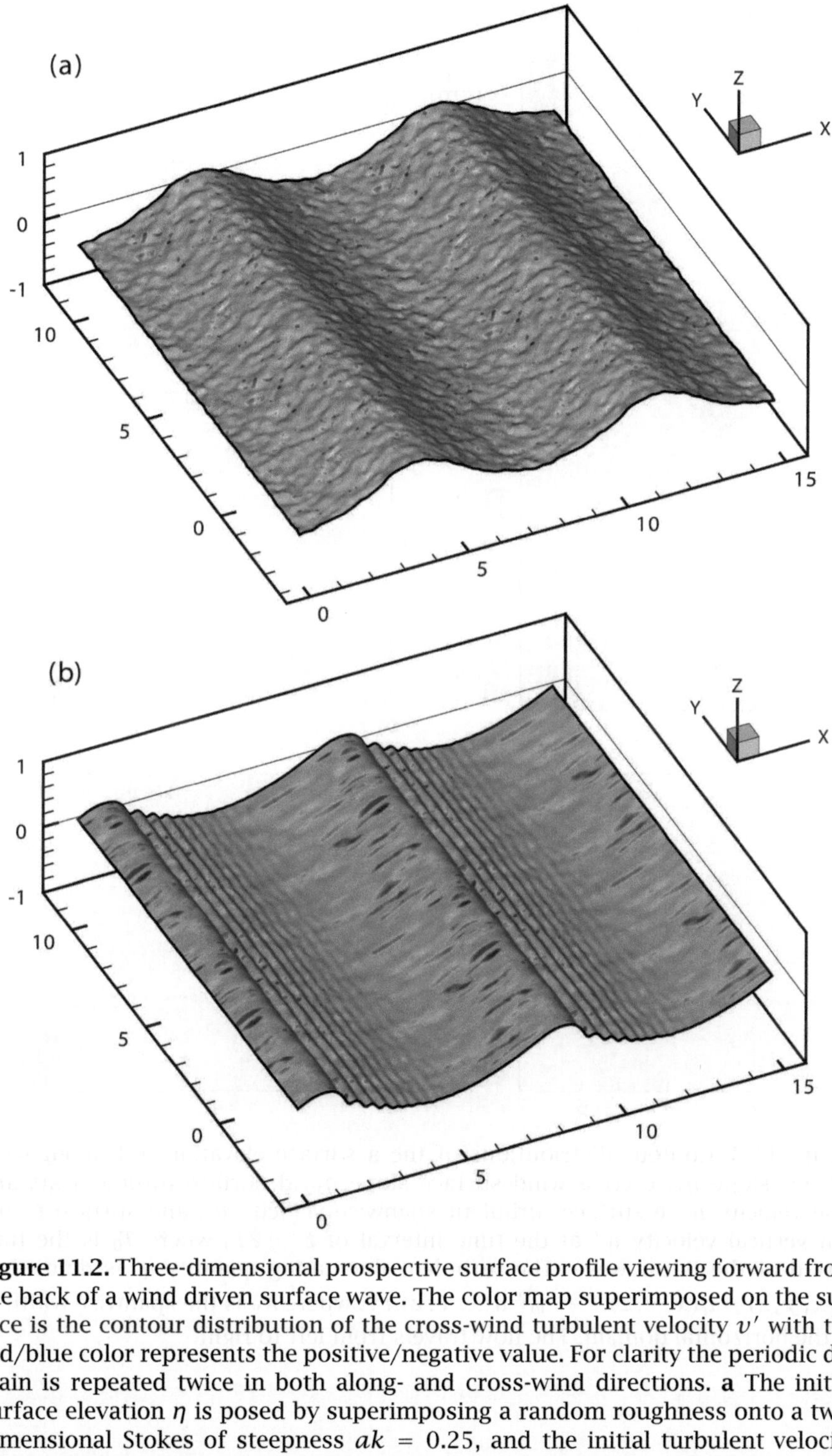

Figure 11.2. Three-dimensional prospective surface profile viewing forward from the back of a wind driven surface wave. The color map superimposed on the surface is the contour distribution of the cross-wind turbulent velocity v' with the red/blue color represents the positive/negative value. For clarity the periodic domain is repeated twice in both along- and cross-wind directions. **a** The initial surface elevation η is posed by superimposing a random roughness onto a two-dimensional Stokes of steepness $ak = 0.25$, and the initial turbulent velocity distribution is homogeneous in both streamwise and spanwise directions. **b** Surface elevation and velocity distribution showing the parasitic capillary waves on the forward face and the streaky velocity structures on the backward face.

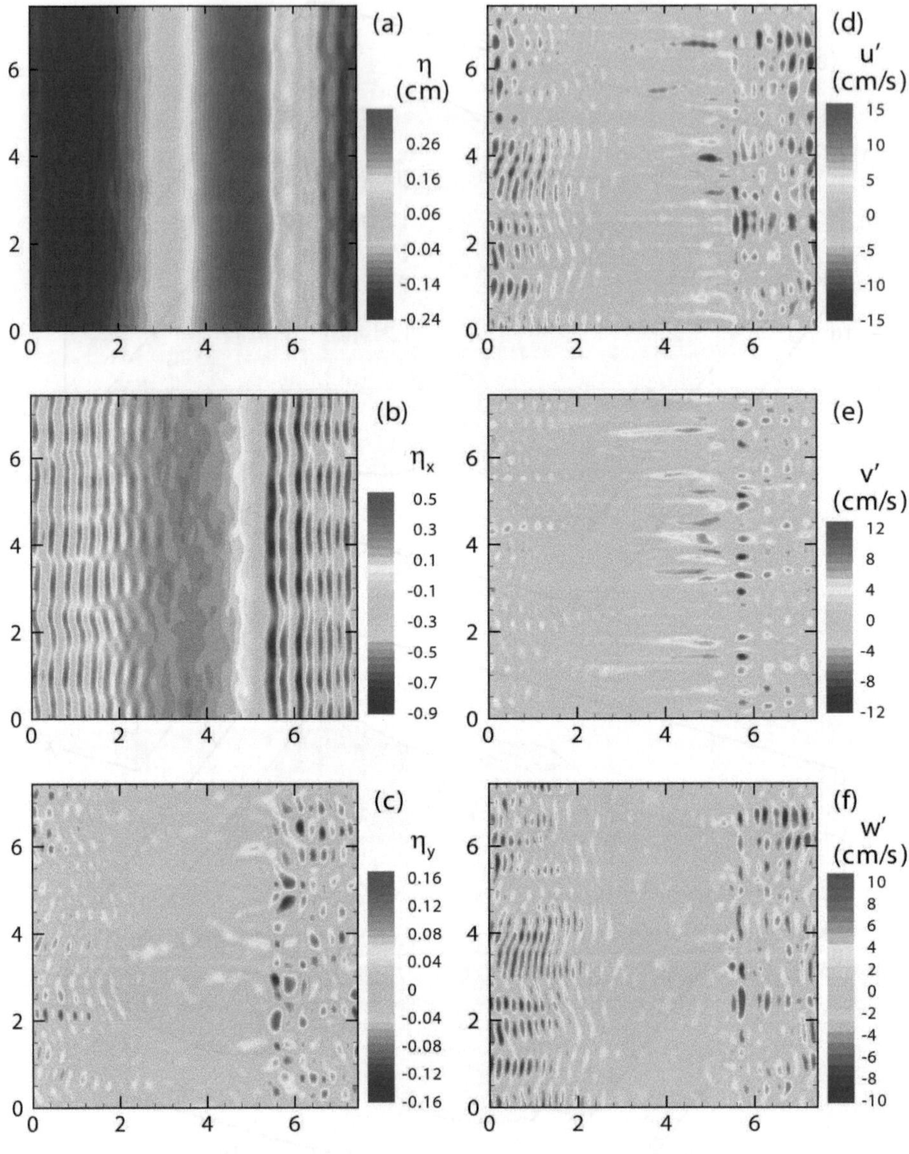

Figure 11.3. Contour distributions of the **a** surface elevation η, **b** along-wind surface slope η_x, **c** cross-wind surface slope η_y, **d** surface turbulent streamwise velocity u', **e** surface turbulent spanwise velocity v', and surface turbulent vertical velocity w' at the time interval of $t = 2T_0$ where T_0 is the time period of the linear wave. The turbulent fluctuating quantity q' is defined as $q'(x,y,z) = q(x,y,z) - \frac{1}{\lambda_y}\int_0^{\lambda_y} q(x,\tilde{y},z)d\tilde{y}$, where λ_y is the spanwise width of of the horizontal domain. The flow travels from left to right.

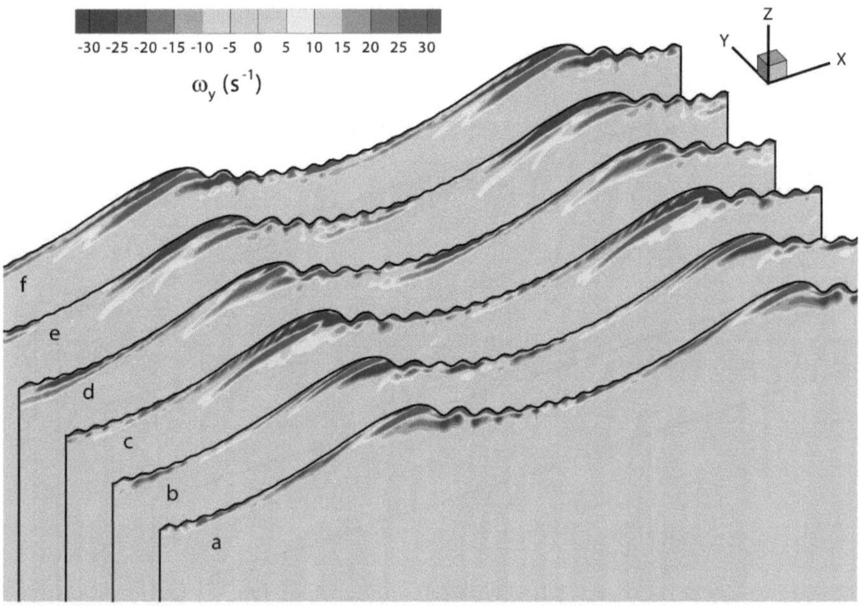

Figure 11.4. Contour distributions of the spanwise turbulent vorticity (ω_y) on various vertical planes ($x - z$ planes) parallel to the wind direction. The flow travels from left to right. The red/blue color represents the clockwise/counter-clockwise vorticity. The alphabets labels the planes that are referred to in the text. For clarity the periodic domain is repeated twice in the streamwise direction.

orbital velocities induced by the parasitic capillary waves. Further studies are needed to unravel the formation mechanisms of the parasitic capillary waves and the streamwise surface streaks, as well as the interaction between the two.

11.4 Underlying Vortical Structures

To illustrate the corresponding vortical structures beneath the characteristic surface features of Figure 11.2, contour distributions of the spanwise turbulent vorticity (ω'_y) on representative vertical planes parallel to the wind direction are shown in Figure 11.4, and the streamwise turbulent vorticity (ω'_x) on the cross-wind planes are shown in Figure 11.5. The regions of high vorticities are confined to near the water surface. Away from the surface the flows are virtually irrotational.

For the distributions of spanwise vorticity, as shown in Figure 11.4, some of the vortical structures have been identified in previous experimental studies, although the results exhibit cross-wind variability [7]. The typical vorticity distributions confirm the laboratory observations

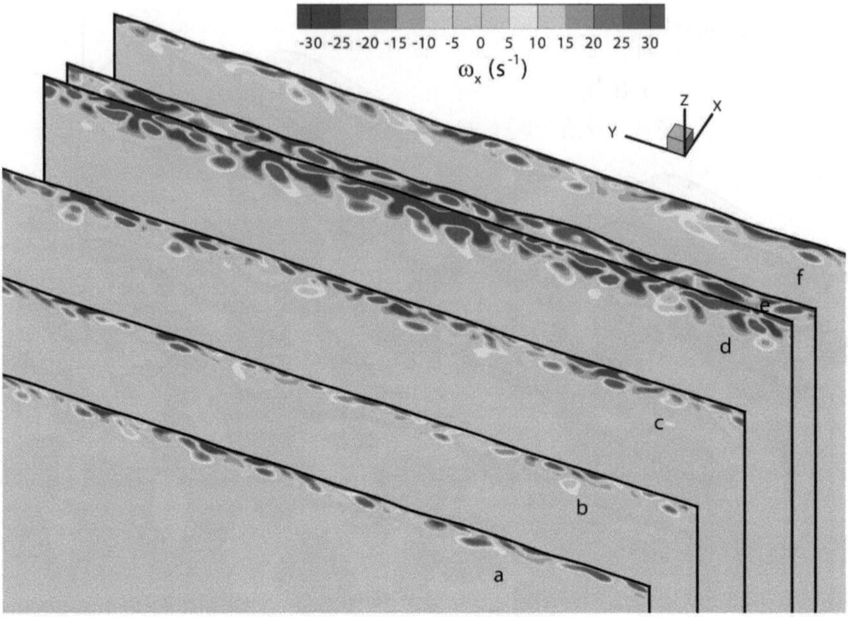

Figure 11.5. Contour distributions of the streamwise turbulent vorticity (ω_x) on various vertical planes ($y-z$ planes) perpendicular to the wind direction. The flow travels from left to right. The red/blue color represents the clockwise/counter-clockwise vorticity. The alphabets labels the planes that are referred to in the text.

[5, 10, 11, 12] that a high vorticity region exists near the crest of the dominant gravity wave (e.g. planes a and c in Figure 11.4). The distributions reveal that such a vortical roller originates from the accompanying parasitic capillaries [9], which induce vortex shedding by the curvature effect. Yoshikawa et al. [18] and Komori et al. [8] reported evidences of downward bursts underneath micro breakers in their laboratory experiments of wind waves. Such events can also be educed from the downward-drafted and backward-extended tails of the eddies beneath the crests. The present numerical model should provide the detailed simulations necessary to unravel the generation mechanisms and evolution processes of these vortical structures.

The characteristic spanwise vorticity distributions, which have been identified in the previous experiments, however, are not the only structures observed in the present simulations. For instance, there also exist vortical structures with reverse rotations, as shown in planes c and f of Figure 11.4. The complete picture of the coherent vortical structures beneath a microscale breaking wave is yet to be studied. These coherent vortices, which can be considered as the surface renewal eddies, however, imply the potential effects of micro-breaking waves on the transfer at the air-water interface.

Similar to the distributions of spanwise vorticity, dominant streamwise vortices arise in the region underneath the crest, as depicted in planes d and e of Figure 11.5. The vortical structure exhibits anorganized pattern with alternating vortices resembling that of Langmuir circulatory cells. In addition, the spacing between each pair of the counter-rotating vortices is roughly equal to that between the elongated surface streaks as shown in Figures 11.2 and 11.3. Further study is required to reveal the formation mechanisms of such surface and underlying structures in a micro-breaking wave, and also the impact of these structures on the transport at the air–water interface.

11.5 Concluding Remark

In this study, we consider the development of a three-dimensional micro-breaking wave by simulating a canonical problem with a two-dimensional mean initial field subject to two-dimensional surface forcing of shear stress and pressure. Despite the three dimensionalities of the surface signatures and the underlying vortical structures, the inherent mean flow is still two dimensional. On natural water surfaces, micro-breaking wind waves commonly appear with more irregular patterns and exhibit high intermittency. In addition, the simulated flow field, which is maintained by applying a given exerting stress and pressure on the water surface, may not be identical to the actual flow of water driven by the boundary forcing of a developing wind field. The focuses of the present simulation, however, are on the inception of the coherent surface signatures and the underlying vortical structures, as well as the formation mechanisms of these structures. For such purposes, the present simulation of the idealized canonical problem should be sufficient. In fact, as we have discussed in sections 11.3 and 11.4, the present numerical simulation results reproduce all the surface features observed in the experiments. The results also highlight the potential impacts caused by the micro-breaking waves, and consequently reveal the necessity in incorporating such microscale processes in the parameterizations of fluxes across the air–sea interface.

References

[1] Banner M.L., Phillips O.M. (1974) On the incipient breaking of small scale waves. *J Fluid Mech* 65:647–656

[2] Banner M.L., Peregrine D.H. (1993) Wave breaking in deep water. *Annu Rev Fluid Mech* 25:373–397

[3] Banner M.L., Peirson W.L. (1998) Tangential stress beneath wind-driven air–water interfaces. *J Fluid Mech* 364:115–145
[4] Csanady G.T. (1990) The role of breaking wavelets in air–sea gas transfer. *J Geophys Res* 95:749–759
[5] Ebuchi N., Kawamura H., Toba Y. (1987) Fine structure of laboratory wind–wave surfaces studied using an optical method. *Bound-Layer Meteor* 39:133–151
[6] Fedorov A.V., Melville W.K. (1998) Nonlinear gravity–capillary waves with forcing and dissipation. *J Fluid Mech* 354:1–42
[7] Jessup A.T., Zappa C.J., Yeh H. (1997) Defining and quantifying microscale wave breaking with infrared imagery. *J Geophys Res* 102:23,145–23,153
[8] Komori S., Nagaosa R., Murakami Y. (1993) Turbulence structure and mass transfer across a sheared air–water interface in wind-driven turbulence. *J Fluid Mech* 249:161–183
[9] Longuet-Higgins M. (1992) Capillary rollers and bores. *J Fluid Mech* 240:659–679
[10] Okuda K. (1982) Internal flow structure of short wind waves Part 1. On the internal vorticity structure. *J Oceanogr Soc Japan* 38:28–42
[11] Peirson W.L., Banner M.L. (2003) Aqueous surface layer flows induced by microscale breaking wind waves. *J Fluid Mech* 479:1–38
[12] Siddiqui M.H.K., Loewen M.R., Richardson C., Asher W.E., Jessup A.T. (2001) Simultaneous particle image velocimetry and infrared imagery of microscale breaking waves. *Phys Fluids* 13:1891–1903
[13] Sullivan P.P., McWilliams J.C., Melville W.K. (2004) The oceanic boundary layer driven by wave breaking with stochastic variability. Part.1 Direct numerical simulations. *J Fluid Mech* 507:143–174
[14] Tsai W.-T., Chen S.-M., Moeng C.-H. (2005) A numerical study on the evolution and structure of a stress-driven, free-surface turbulent shear flow. *J Fluid Mech* 545:163–192
[15] Tsai W.-T., Hung L.-P. (2006) Three-dimensional modeling of small-scale processes in the upper boundary layer bounded by a dynamic ocean surface. To appear in *J Geophys Res*
[16] Tsai W.-T. (1998) A numerical study of the evolution and structure of a turbulent shear layer under a free surface. *J Fluid Mech* 354:239–276
[17] Tsai W.-T., Chen S.-M., Lin M.-Y., Hung L.-P. (2003) Molecular sublayers beneath the air-sea interface relative to momentum, heat and gas transports. *Geophys Res Lett* 30, 1968, doi:10.1029/2003GL018164
[18] Yoshikawa I., Kawamura H., Okuda K., Toba Y. (1988) Turbulent structure in water under laboratory wind waves. *J Oceangr Soc Japan* 44:143–156
[19] Zappa C.J., Asher W.E., Jessup A.T. (2001) Microscale wave breaking and air–water gas transfer. *J Geophys Res* 106:9385–9391

12

The Effect of Raindrops on Interfacial Turbulence and Air-Water Gas Transfer

Satoru Komori, Naohisa Takagaki, Rina Saiki, Naoya Suzuki and Kenji Tanno

Department of Mechanical Engineering and Science and Advanced Research Institute of Fluid Science and Engineering
Kyoto University, Kyoto 606-8501, Japan
komori@mech.kyoto-u.ac.jp

Abstract The effects of impinging raindrops on both turbulence below the air-water interface and CO_2 transfer across the air-water interface are discussed using laboratory measurements by Takagaki and Komori [1]. The measurements of CO_2 absorption rate and turbulence quantities in an open-channel flow show that impinging raindrops enhance both turbulent mixing near the free surface on the liquid side and CO_2 transfer across the air-water interface, and that the mass transfer velocity due to impinging raindrops is well correlated with the mean vertical momentum flux of raindrops. The reason why the mass transfer velocity is well correlated by the mean vertical momentum flux is explained by showing the instantaneous velocity vectors induced by a falling single droplet. Further, in order to clarify the effects of rainfall on the global and local CO_2 transfer across the air-sea interface, the mean annual net air-sea CO_2 flux was estimated using both the daily precipitation data set and the empirical correlation [1] between the mass transfer velocity and mean vertical momentum flux. The rainfall effects are also compared with wind shear effects. The results show that rainfall effects are significant for the local CO_2 budget between atmosphere and ocean in equatorial and mid-latitude regions, but are not so important for global budget, compared to the wind shear effect.

12.1 Introduction

Numerical predictions of the CO_2 exchange rate across the air-sea interface are so sensitive to the air-sea CO_2 transfer velocity that uncertainty in the CO_2 air-sea exchange rate may lead to uncertainty in future predictions of the global carbon budget. It is, therefore, of great importance to investigate all the fluid-mechanical factors that control the CO_2 transfer across the air-sea interface in precisely estimating the CO_2 exchange rate across the air-sea interface. The effects of wind shear on the air-sea CO_2 transfer have been investigated as the most important factor in previous

C.S. Garbe, R.A. Handler, B. Jähne (eds.): Transport at the Air Sea Interface
pp. 169-179, 2007,

studies [2, 3, 4, 5, 6] and some empirical equations based on the relation between the mass transfer velocity and wind speed have been proposed. However, there exist other control factors that have not been well investigated.

One such control factor is rainfall, which is expected to promote CO_2 transfer across the air-sea interface [7, 8]. Ho et al. [8] showed that the mean kinetic energy flux of raindrops, KEF, is a main parameter for determining the effects of rainfall on the CO_2 transfer across the air-water interface. Here, KEF is defined as

$$KEF = 0.5\rho R v_p^2, \qquad (12.1)$$

where ρ is the density, R the rain rate and v_p the impinging velocity of raindrops. On the other hand, Takagaki and Komori [1] measured the mass transfer velocity by changing the impinging velocity of raindrops, v_p, and they concluded that the mean kinetic energy flux KEF is not always the predominant parameter and the mean vertical momentum flux of raindrops, MF, is the most suitable parameter. Here, MF is defined as

$$MF = \rho R v_p. \qquad (12.2)$$

However, it has not been clarified why MF is a more suitable rain parameter than KEF.

Impinging raindrops on the air-water interface enhance the turbulence below the interface and the enhancement of turbulence results in promotion of the CO_2 transfer. However, previous studies have not investigated whether the rainfall effect on local or global air-sea CO_2 exchange is really significant, compared to the wind shear effect. In addition to this mass transfer due to the impingement of raindrops on the air-sea interface, raindrops themselves absorb CO_2 while falling. Direct numerical simulations by Sugioka and Komori [9] have suggested that the CO_2 concentration inside a raindrop reaches equilibrium before the raindrop impinges on the air-sea interface. It is, therefore, of interest to estimate the total rainfall effects including both CO_2 transfer across the air-sea interface and CO_2 absorption into raindrops through the raindrop surface by using the empirical relation between k_{LR} and MF, and precipitation data.

The purpose of this paper is to investigate the precise relation between rainfall and air-water CO_2 transfer and to estimate the rainfall effect on the air-sea CO_2 transfer.

12.2 Experiments

Figure 12.1 shows the open channel and water-velocity measuring system used here. The open flume was 7.6m long, 0.5m wide and 0.2m deep.

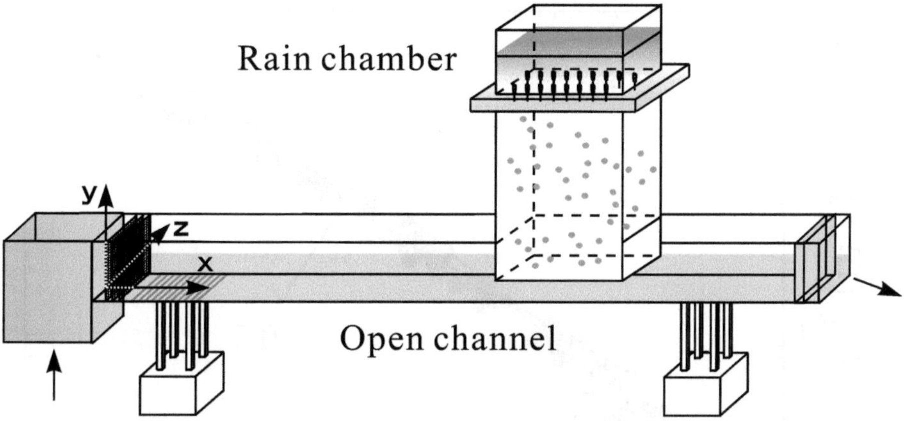

Figure 12.1. Experimental apparatus and measuring system [1].

The flume was filled with tap water. The flow depth in the open channel ranged from 0.06 to 0.17m, and the open-channel flow was fully developed. The cross-sectional mean velocity, U_{ave}, ranged from 0.056 to 0.164m/s, and the Reynolds number based on the hydraulic radius and U_{ave} ranged from 22700 to 31700. Turbulence quantities were measured using a laser Doppler velocimeter (LDV).

In order to conduct CO_2 absorption experiments, a closed rain chamber was set on the free surface in an open-channel flow. The rain chamber was 1.00m long and 0.40m wide, and the chamber height was changed from 0.38 to 1.20m. Uniform raindrops were generated in the closed rain chamber by using many uniform needles mounted at the rooftop of the chamber, i.e., at the bottom of a head water tank. By changing needle diameter and tap water level in the head tank, the droplet diameter, d_r, was varied from 2.1 to 5.7mm. The height of the rain chamber was changed to several values, 0.38, 0.49, 1.09 and 1.20m, to obtain a spread of a factor of 2 in raindrop velocity. The impinging raindrop velocity was measured by a high-speed video system; it ranged from 2.37 to 4.69m/s. The rain rate R ranged from 1 to 435mm/h.

Pure CO_2 was injected into the rain chamber at atmospheric pressure, and the CO_2 absorption rate was measured using a soap-film meter. The absorption rates due to both original turbulent motion in an open-channel flow and falling of raindrops before impinging on the free surface were subtracted from the absorption rate measured by the soap-film meter [1]. From the measured CO_2 flux per unit area at the air-water interface, the mass transfer velocity (mass transfer coefficient on the liquid side) due to impinging raindrops, k_{LR}, was estimated and k_{LR} was normalized to the mass transfer velocity for fresh water at 20°C. The details of the experiments are described in Takagaki and Komori [1].

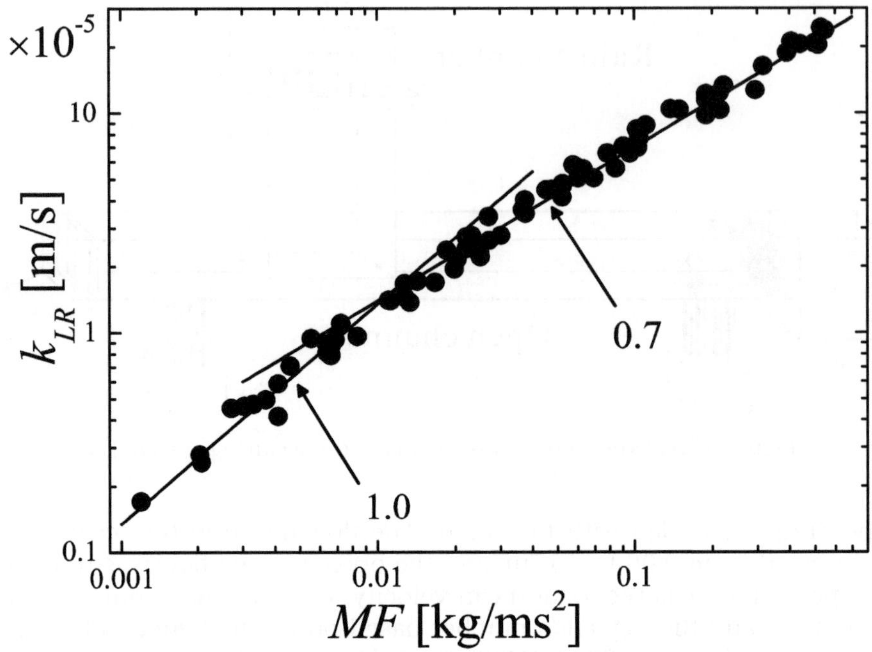

Figure 12.2. Relation between mass transfer velocity k_{LR} due to impinging raindrops and mean vertical momentum flux MF [1].

12.3 Results and Discussion

12.3.1 Mass Transfer Velocity

Figure 12.2 shows the relation between the mass transfer velocity k_{LR} due to impinging raindrops and the mean vertical momentum flux MF defined by Equation (12.2).

It is found that the mass transfer velocity k_{LR} is well correlated with MF. Takagaki and Komori [1] have confirmed that this relation holds for various rain rates, raindrop diameters, impinging raindrop velocities and horizontal distances between raindrops. Furthermore, Takagaki and Komori [1] showed that k_{LR} normalized by the Schmidt number for 3.5wt% salt water is the same between two open channel flows filled with fresh water and 3.5wt% salt water.

On the other hand, Ho et al. [8] concluded that the mean kinetic energy flux of raindrops KEF defined by Equation (12.1) is a main parameter for determining the rainfall effect on gas transfer. However, Ho et al. [8] measured k_{LR} without significantly changing the impinging velocity. In fact, we could not find a good correlation between k_{LR} and KEF for different impinging velocities [1]. This suggests that MF is more appropriate as a rainfall parameter for representing air-water CO_2 transfer due to imping-

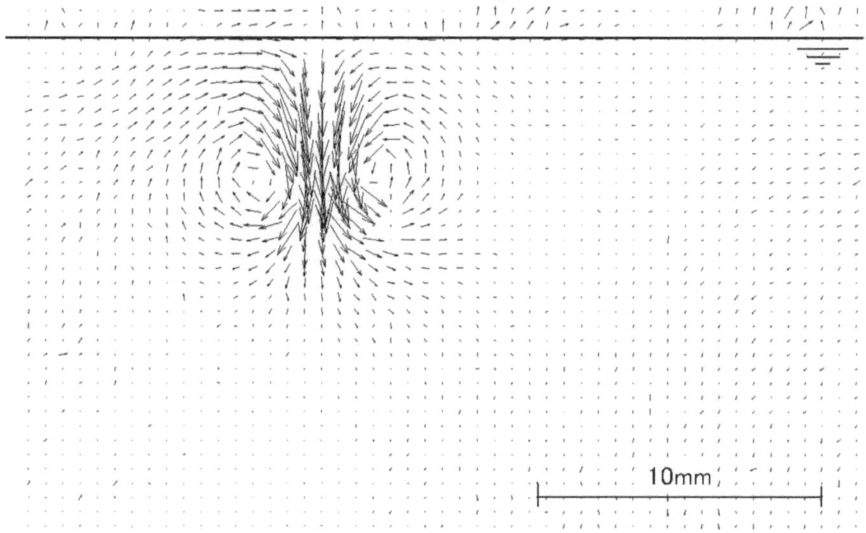

Figure 12.3. Instantaneous velocity vectors at 400ms after the impingement of a single raindrop with the diameter of d_r=2.2mm and impinging velocity of v_p=3.5m/s.

ing raindrops than *KEF*. In order to find the reason why *MF* is a significant rainfall parameter, we measured the instantaneous velocity vectors around an impinging raindrop with the diameter of d_r=2.2mm and impinging velocity of v_p=3.5m/s by means of a Particle Image Velocimeter (PIV). Figure 12.3 shows the instantaneous velocity vectors at 400ms after the impingement of a single raindrop. It is found that surface renewal eddies are generated behind the impinging raindrop. Considering that the surface renewal motions responsible for the mass transfer are generated by exchanging momentum with falling raindrops, we can easily understand that the vertical momentum flux should be relevant to the mass transfer.

12.3.2 Estimation of Air-Sea CO_2 Flux

The CO_2 flux across the air-sea interface was estimated by

$$F_R = k_{LR} S_s \Delta p\mathrm{CO}_2, \tag{12.3}$$

where k_{LR} is the mass transfer velocity due to an impinging raindrop, S_s is the solubility of CO_2 in sea water and $\Delta p\mathrm{CO}_2$ is the partial pressure difference between atmosphere and ocean. k_{LR}(m/s) is given by the following best-fit curves [1] shown in Figure 12.2:

$$\begin{aligned} k_{LR} &= 0.00135MF && \text{for} \quad 0 < MF < 0.0111, \\ k_{LR} &= 0.00035MF^{0.7} && \text{for} \quad 0.0111 \leq MF, \end{aligned} \tag{12.4}$$

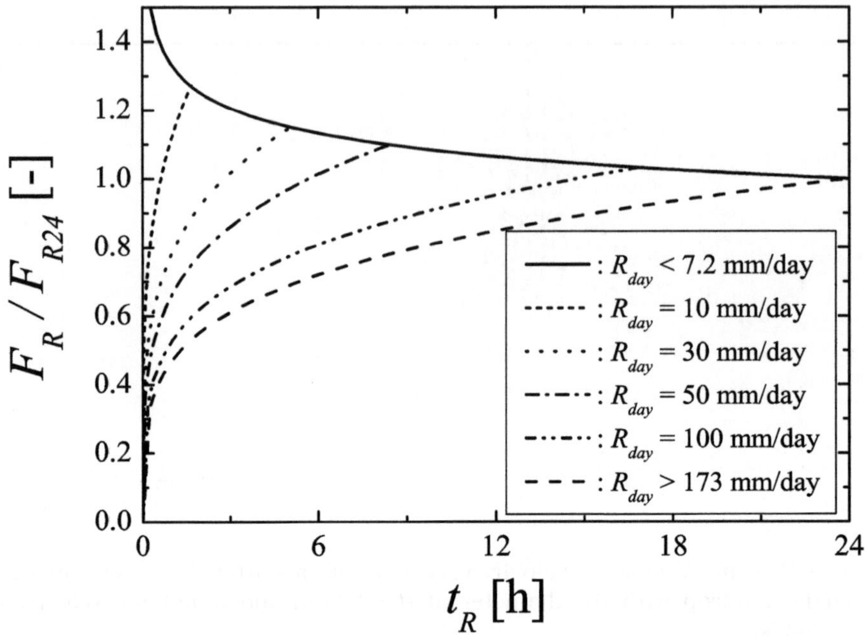

Figure 12.4. Relation between rainfall time t_R and CO_2 flux normalized by the full day rainfall flux for t_R=24hr.

where MF(kg/ms^2) is given by Equation (12.2). The mean vertical momentum flux MF cannot be explicitly determined for unsteady natural rainfall. In order to overcome this problem we derived a relationship between MF and the rain rate in a natural environment, R_n(mm/h), by using the raindrop size distribution [10] and the terminal velocity of the droplets [11]:

$$MF = 1.29 \times 10^{-3} R_n^{1.09}. \tag{12.5}$$

The same method was used by Ho et al. [11] to calculate KEF. Time records of global rain rate R_n are required to estimate the air-sea CO_2 flux from this MF. However, such global data of temporal R_n are not available except for some local observing stations with buoys. Therefore, we were forced to use the GPCP (Global Precipitation Climatology Project) One-Degree Daily Precipitation Data Set (World Climate Research Program [13], Huffman et al. [14]) and we estimated R_n from the daily rain rate, R_{day}, by using the following method. When rain is assumed to continue for t_R hours, R_n is given by R_{day}/t_R and the daily CO_2 flux due to the impinging raindrops can be calculated by integrating F_R in Equation (12.3) over the time period t_R. Figure 12.4 shows the CO_2 flux F_R against t_R. Here, F_R is normalized by the flux F_{R24} for t_R=24hr and the best-fit curve for each R_{day} is overlapped with the bold solid line from each intersection point. From the flux curves for each R_{day}, we chose the value of t_R to give the

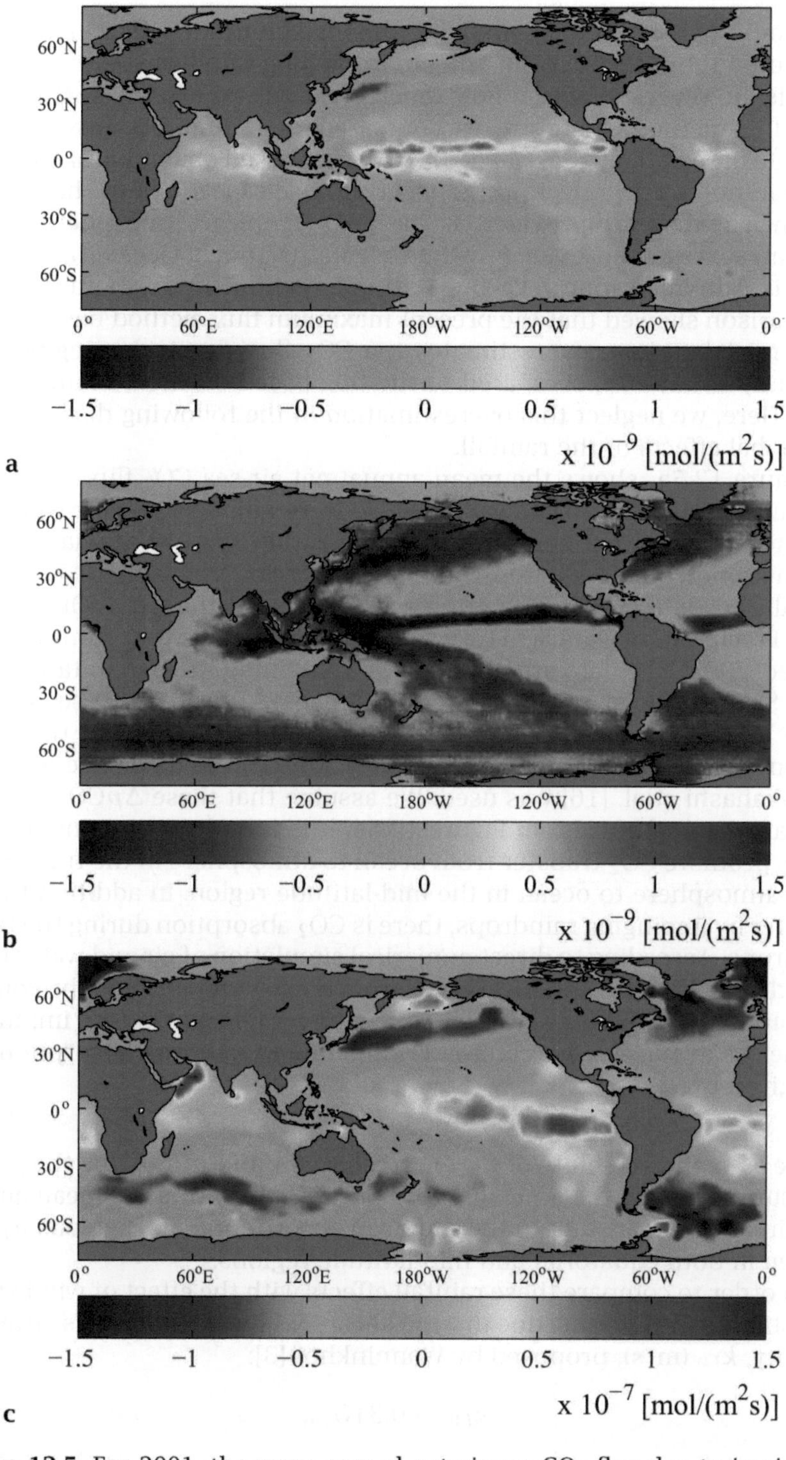

Figure 12.5. For 2001, the mean annual net air-sea CO_2 flux due to impinging raindrops, raindrop absorption and wind shear in **a, b** and **c** respectively.

maximum CO_2 flux F_R. By applying this maximum CO_2 flux method, we computed the global CO_2 flux due to impinging raindrops.

It is, however, uncertain how much error will be caused by this maximum flux method. Therefore, we compared the rainfall CO_2 flux based on the data base of daily rain rate R_{day} with that based on temporal data base of 10 minutes rain rate R_{10min}, observed at 28 locations in the tropical region. The data are provided as the TAO (Tropical Atmosphere Ocean) buoy array data set provided by NOAA/PMEL (National Oceanic and Atmospheric Administration / Pacific Marine Environmental Laboratory). The comparison showed that the present maximum flux method based on the daily rain data base overestimates the CO_2 flux due to impinging raindrops by about 10%, compared to the flux based on the 10 minute data base. Here, we neglect this overestimation in the following discussion of the global effects of the rainfall.

Figure 12.5**a** shows the mean annual net air-sea CO_2 flux due to impinging raindrops, F_R. The values of F_R were adjusted to correspond to salinity and local temperature using the NCEP/NCAR (National Centers for Environmental Prediction/National Center For Atmospheric Research) reanalyzed data [15]. To determine k_{LR}, we used the GPCP One-Degree Daily Precipitation Data Set [13, 14] for a period of one year from 1 Jan. to 31 Dec. 2001. The data are obtained by combining the precipitation data from SSM/I (Special Sensor Microwave/Imager), infrared (IR) sensor and TOVS (TIROS Operational Vertical Sounder) satellite data. For the distribution of the partial pressure difference ΔpCO_2, the data base provided by Takahashi et al. [16] was used. We assume that these ΔpCO_2 data are the same as in 2001. From Figure 12.5**a**, it is found that impinging raindrops promote CO_2 transfer from ocean to atmosphere in the tropics and from atmosphere to ocean in the mid-latitude region. In addition to CO_2 transfer by impinging raindrops, there is CO_2 absorption during the fall of raindrops. According to direct numerical simulation of air and water flows outside and inside a spherical droplet with mass transfer [9], the concentration field in a falling raindrop comes to equilibrium before impinging on the ocean surface. Therefore CO_2 flux due to raindrop absorption, F_D, was given by

$$F_D = R_n S_f (pCO_{2air}), \tag{12.6}$$

where S_f is the solubility of CO_2 in fresh water and pCO_{2air} is the partial pressure of CO_2 in the atmosphere. Figure 12.5**b** shows the mean annual net air-sea CO_2 flux due to the raindrop absorption, F_D. High absorption is seen in both equatorial and mid-latitude regions.

In order to compare these rainfall effects with the effect of wind shear, we estimate the CO_2 flux due to wind shear, F_W, by using the mass transfer velocity, k_{LW}(m/s), proposed by Wanninkhof [3]:

$$k_{LW} = 0.31 U_{10}^2, \tag{12.7}$$

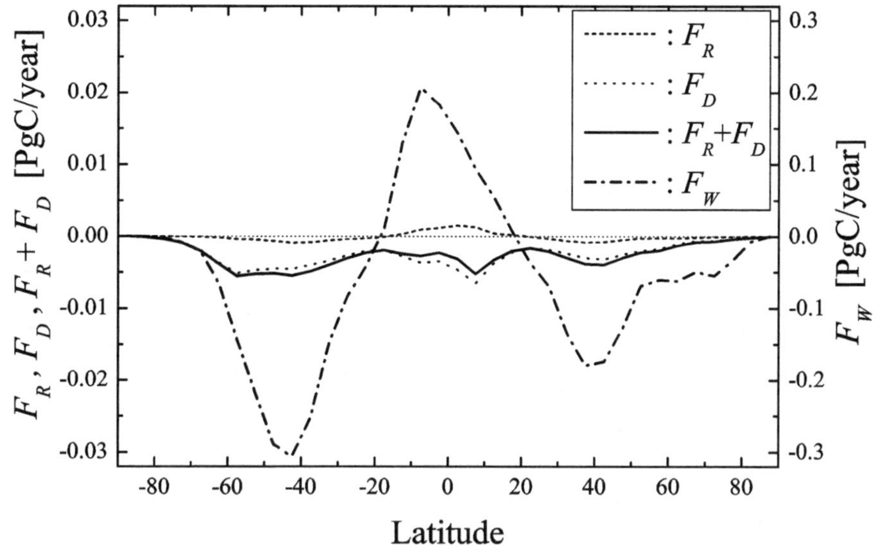

Figure 12.6. Distribution of the zone integrated values of mean annual net air-sea CO_2 fluxes, F_R, F_D, F_W and F_R+F_D against latitude.

where U_{10}(m/s) is the wind speed at 10m elevation from the ocean surface; we used the NCEP/NCAR reanalyzed wind data. The mean annual net air-sea CO_2 flux due to the wind shear is shown in Figure 12.5**c**. Figure 12.6 shows the distributions of the zone integrated values of annual net air-sea CO_2 fluxes, F_R, F_D, F_W and F_R+F_D against latitude. Compared to the wind shear effect, the rainfall effect is not so large. The impinging effect of raindrops is 2% of the wind shear effect at most in the tropics. The net effects of F_R+F_D are 6% and 3% at most, compared to the wind shear effect in the equatorial and mid-latitude region, respectively. On the other hand, the contribution of rainfall is locally big in the tropical region near 140 degrees of longitude, as shown in Figure 12.7, and the values of F_R/F_W and $(F_R+F_D)/F_W$ are about 7% and 35%, respectively.

Global air-sea CO_2 fluxes for 2001 were computed; F_W, F_R, F_D and F_R+F_D corresponded to -1.81, -0.003, -0.09 and -0.093PgC/Year, respectively. This shows that the global effect of rainfall is less than 5%. The above results also suggest that the rainfall effect on the CO_2 flux should be considered in terms of the local CO_2 budget, but it can be neglected for discussing the global CO_2 budget.

12.4 Conclusions

The water below the free surface is mixed by impinging raindrops and therefore, CO_2 transfer across the air-water interface is promoted. The

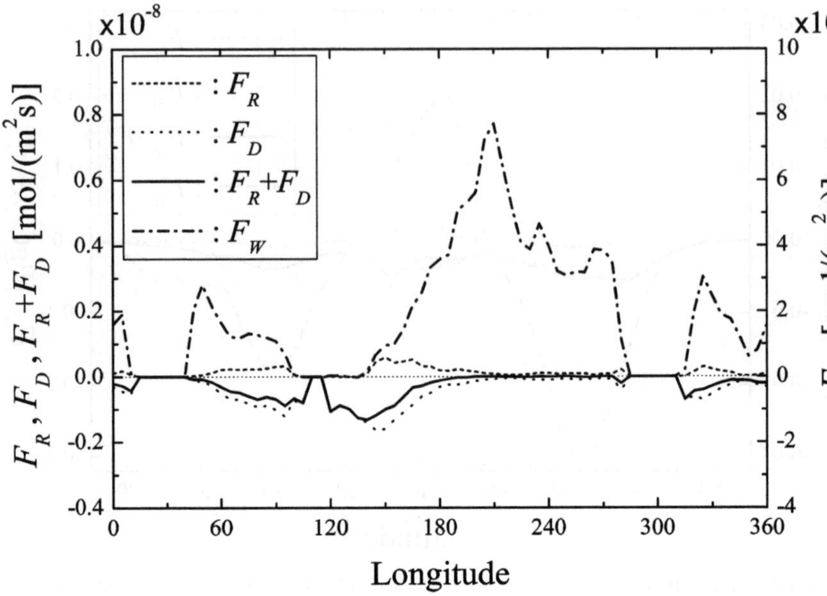

Figure 12.7. Longitudinal distribution of mean annual net air-sea CO_2 fluxes, F_R, F_D, F_W and F_R+F_D on the equator.

mass transfer velocity is determined by the mean vertical momentum flux of raindrops. The rainfall effect is significant for the local CO_2 budget between atmosphere and ocean in equatorial and mid-latitude regions, but not so important for discussing the global carbon budget.

Acknowledgement. This work was supported by the Ministry of Education, Science, Sports and Culture, Grant-in Aid (No.14102016). The part of the estimation of the global carbon budget was supported by the Core Research for Evolutional Science and Technology Program "Advanced Model Development and Simulations for Disaster Countermeasures" of Japan Science and Technology Agency. The authors acknowledge Profs. W. McGillis and T. Takahashi for their provision of the data sets of ΔpCO_2.

References

[1] Takagaki, N., Komori S. (2007) Effects of rainfall on mass transfer across the air-water interface. *J Geophys Res*, in press

[2] Liss, P.S., Merlivat L. (1986) Air-sea gas exchange rates : introduction and synthesis. In: The Role of Air–Sea Exchange in Geochemical Cycling (ed Buat–Menard P), 113–127

[3] Wanninkhof, R. (1992) Relationship between wind-speed and gas-exchange over the ocean. *J Geophys Res* 97:7373–7382

[4] Komori, S., Nagaosa R., Murakami Y. (1993) Turbulence Structure and Mass Transfer Across a Sheared Air-Water Interface in Wind-Driven Turbulence. *J Fluid Mech* 249:161–183

[5] Komori, S., Shimada T., Misumi R. (1999) Turbulence structure and mass transfer at a wind-driven air-water interface. In: *Wind-over-Wave Couplings: Perspective and Prospects*, Sajjadi SG, Thomas NH, Hunt JCR(eds), Oxford University Press, 273–285

[6] McGillis, W.R., Edson J.B., Hare J.E., Fairall C.W. (2001) Direct covariance air-sea CO2 fluxes. *J Geophys Res* 106:16729–16745

[7] Banks, R.B., Wickramanayake G.B., Lohani B.N. (1984) Effect of rain on surface reaeration. *J Environ Eng* 110(1):1–14

[8] Ho, D.T., Asher W.E., Bliven L.F., Schlosser P., Gordon E.L. (2000) On mechanisms of rain-induced air-water gas exchange. *J Geophys Res* 105:24045–24057

[9] Sugioka, K., Komori S. (2007) Drag and lift forces acting on a spherical water droplet in homogeneous linear shear air flow. *J Fluid Mech* 570:155-175

[10] Marshall, J.S., Palmer W.M. (1948) The distribution of raindrops with size. *J Meteor* 5:165–166

[11] Gunn, R., Kinzer G.D. (1949) The terminal velocity of fall for water droplets in stagnant air. *J Meteor* 6:243–248

[12] Ho, D.T., Bliven L.F., Wanninkhof R., Schlosser P. (1997) The effect of rain on air-water gas exchange. *Tellus* 49(B):149–158

[13] WCRP, Report of the workshop on global large scale precipitation data sets for the World Climate Research Program, WCP-111, WMO/TD - No.94, WMO, Geneva, 45 pp, (1986)

[14] Huffman, GJ, Adler RF, Morrissey MM (2001) Global precipitation at one-degree daily resolution from multisatellite observations. *J Hydrometeorology.* 2(1):36–50

[15] Jähne, B., Münnich K.O., Bösinger R., Dutzi A., Huber W., Libner P. (1987) On the Parameters influencing Air-Water Gas Exchange. *J Geophys Res* 92(C2):1937-1949

[16] Takahashi, T., Surherland S.C., Sweeney C., Poisson A., Metzl N., Tilbrook B., Bates N., Wanninkhof R., Feely R.A., Sabine C., Olafsson F., Nojiri Y. (2002) Global sea-air CO2 flux based on climatological surface ocean *p*CO2 and seasonal biological and temperature effects. *Deep-Sea Research II* 49:1601–1622

13

Air-Water Flux Reconciliation Between the Atmospheric CO_2 Profile and Mass Balance Techniques

Wade R. McGillis[1,2], John W. H. Dacey[3], Jonathan D. Ware[3], David T. Ho[1], Jonathan T. Bent[1], William E. Asher[4], Christopher J. Zappa[1], Peter A. Raymond[5], Rik Wanninkhof[6], and Satoru Komori[7]

[1] Lamont-Doherty Earth Observatory
Columbia University, Palisades, NY 10964 USA
[2] Earth and Environmental Engineering
Columbia University, New York, NY 10027 USA
[3] Woods Hole Oceanographic Institution, Woods Hole, MA 02543 USA
[4] Applied Physics Laboratory, University of Washington, Seattle, WA 98105 USA
[5] School of Forestry and Environmental Studies
Yale University, New Haven, CT 06511 USA
[6] NOAA Atlantic Oceanographic and Meteorological Laboratory, Miami, FL 33149 USA
[7] Department of Mechanical Engineering and Science
Kyoto University, Yoshida-Honmachi, Sakyo-ku, 606-8501 Japan

Abstract Studies deploying atmospheric flux-profile techniques in laboratory wind-wave tanks have been performed to demonstrate and verify the use of air-side turbulent transport models and micrometeorological approaches to accurately determine air-water gas transfer velocities. Air-water gas transfer velocities have been estimated using the CO_2 atmospheric flux-profile technique in laboratory wind-wave tanks both at the NASA Wallops Flight Facility, USA and Kyoto University, Japan. Gas fluxes using the flux-profile and the waterside mass balance techniques have been reconciled. Air-water fluxes of H_2O and momentum were also measured simultaneously in a linear wind-wave tank. The waterside mass balances used the evasion of SF_6. The CO_2, H_2O, and momentum fluxes were calculated using the atmospheric flux-profile technique over a wind speed range of 1 to 14 m s^{-1}. The CO_2 and H_2O atmospheric profile model uses airside turbulent diffusivities derived from momentum fluxes. These studies demonstrate that the quantification of air-water CO_2 fluxes using the atmospheric flux-profile technique can be implemented in the laboratory. The profile technique can be used to measure an air-water flux in much less time than a mass balance. Effects of surfactants, wind speed, and wind stress on air-water transfer are also explored using the flux-profile technique. Validation of the air-water CO_2 gas exchange in laboratory wind-wave tanks provides evidence and support that this technique may be used in field studies.

C.S. Garbe, R.A. Handler, B. Jähne (eds.): Transport at the Air Sea Interface
pp. 181-192, 2007,

13.1 Introduction

The air-water exchange, or gas transfer, of CO_2 depends on physical, chemical, and biological processes in aquatic and marine systems. Understanding the factors that control gas fluxes across the air-water interface is necessary to predict the fate and transport of climate relevant compounds in the environment. Gas transfer influences atmospheric and water concentrations on local, regional, and global scales. Therefore, the ability to measure air-water gas fluxes contributes to the understanding of ecosystem dynamics, contaminant transport, and climate change.

The fluxes of gases are controlled by the air-water concentration difference [11] and by the physics of both the air and water mass boundary layers [9]. The boundary layers are strongly forced by the wind and by the surface wave fields. Knowledge of wave-age and breaking, bubble intensity, surface films, spray, rain, chemical enhancement, mixed layer dynamics, biology and boundary layer instability is necessary for understanding the parameters that control air-water CO_2 fluxes.

To explore effects of physical processes on gas and water vapor transport both in water and air, fluxes were measured in the wind-wave tanks at the NASA Wallops Flight Facility and the Kyoto University wind-wave tanks. Fluxes of CO_2 and H_2O using an airside profile method were performed. This method employs a turbulent transport model in the air boundary layer determined from profiles of wind speed. Scalar fluxes are determined using the turbulent diffusivity with gradients of CO_2 and H_2O in the air boundary layer [10]. The accuracy of the profiling method is greatly enhanced by using a fixed-elevation reference for measurement of gas concentrations.

13.2 Methods

The flux of carbon dioxide, F, is often parameterized in terms of a gas transfer velocity, k, which is assumed to be a function of the environmental physical and chemical enhancement processes controlling gas exchange:

$$F_{CO_2} = k\Delta CO_2 \tag{13.1}$$

where ΔCO_2 is the concentration difference between the air and water.

The wind-wave tank at the NASA Wallops Flight Facility is 18.3-m long, 1.2-m high, and 0.9-m wide. The headspace is 0.5 m high. The maximum wind speed achievable in the facility is 25 m s^{-1}. The maximum current speed is 30 cm s^{-1}. The facility provided wave height measurements using capacitance probes and wave slope measurements using video imaging optical techniques. Airflow was measured using Pitot tubes. Waterflow was measured using acoustic Doppler velocimeters.

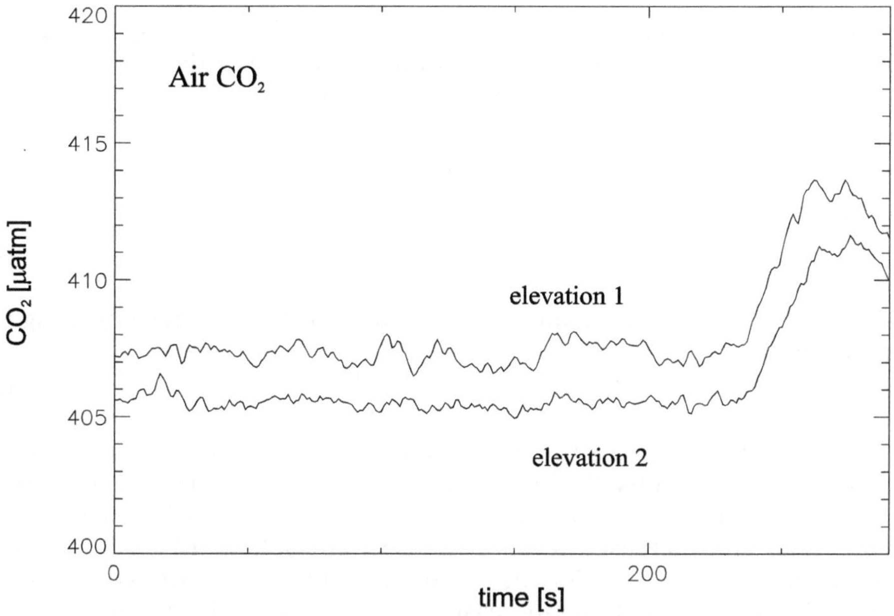

Figure 13.1. Comparison between two sample streams using two NDIR detectors. The timeseries shows how a dramatic jump in the ambient concentration is detected by both sensors. The differential concentration technique subtracts a profiling sample from a fixed sample to remove the mean bias. This technique results in a ±15 ppb concentration uncertainty over 10 minutes.

During these studies, the airside, or atmospheric measurements were also made of wind speed (U), H_2O (q), temperature (T), and CO_2 (C). To insure the best possible measurements of boundary layer turbulent transport, measurements always began approximately 2 cm from the top of the tallest wave height. Because wave height varies with wind speed, the height of the lowest measurement point in the profile increased as wave height increased. The mean water surface has a height of zero. 10-measurement heights where performed at 1-cm intervals. Because tanks have a turbulent boundary layer wake region above the wall-layer behavior, close attention was performed such that the boundary layer measurements were taken below the wake region, an artifact imposed in tanks.

Figure 13.1 shows two air sample streams using two NDIR detectors. The timeseries shows how a dramatic jump in the ambient concentration is detected by both sensors. The differential concentration technique subtracts a profiling sample from a fixed sample to remove the mean bias. This technique results in a ±15 ppb concentration uncertainty over 10 minutes.

In the laboratory, the friction velocity, $u_\star$, was determined from the atmospheric profiles of wind speed, $U(z)$. Carbon dioxide, sensible, and latent heat fluxes were determined using profiles of C, T, and q. Estimates of $u_\star$ under neutral atmospheric conditions can also be derived from the local wind speed through the neutral drag coefficient C_{dn}, given by:

$$u_\star = \sqrt{C_{dn}}U \tag{13.2}$$

where

$$C_{dn} = \left[\frac{\kappa}{ln(z/z_o)}\right]^2 \tag{13.3}$$

where the von Kármán constant, $\kappa = 0.4$, and z_o is the aerodynamic roughness height,

$$z_o = 0.11\frac{\nu}{u_\star} + a\frac{u_\star}{g} \tag{13.4}$$

where ν is viscosity, $u_\star$ is the friction velocity, $a = 0.011$, and g is gravity.

The NASA Wallops Flight Facility and Kyoto University wind-wave tanks use local ground water. The ground water carbonate levels are generally high and increased levels of pCO_2 were obtained by adding HCl to the tank water. The HCl addition decreased the pH in the system and shifts the equilibrium to an increase in pCO_2 through the reaction HCO_3^- + $H^+ \rightleftharpoons CO_2 + H_2O$ that governs the carbonate system [14]. Experiments performed at Kyoto University also increased the aqueous pCO_2 levels by sparging pure carbon dioxide into the water phase.

The aqueous pCO_2 was measured using an in situ equilibrator at the NASA facility. The aqueous pCO_2 was estimated using DIC and pH at the Kyoto University facilities. The in situ equilibrator was a dome placed at the water surface so that it pierced the air-water interface. This produced an air headspace and a water section. There were 3 ports on the 0.008 m^3 headspace. Water was pumped from the wind-wave tank centerline at 10 l min^{-1} and into the headspace through one port and through a nozzle that produced a spray. Effluent from the equilibrator was drained directly into the wind-wave tank. Headspace air was cycled through a Non Dispersive InfraRed (NDIR) gas analyzer at 3 l min^{-1} to measure aqueous pCO_2 concentrations. These flows provided a 30-second response time for pCO_2 measurements.

In the NASA facility, the scalar fluxes were determined using the differential concentration method to provide high precision. To increase the signal to noise in the air-water flux measurement, the flux, F, over a 10-minute interval is given as:

$$F_{\text{interval}} = \int F_{\text{instantaneous}}dt \tag{13.5}$$

The instantaneous flux, $F_{\text{instantaneous}}$, is derived from wall layer turbulence with corrections for atmospheric stability using Monin-Obukov theory. Here the instantaneous flux is given as:

$$F_{\text{instantaneous}} = u_{\star} c_{\star} \tag{13.6}$$

where $u_{\star}$ is the friction velocity and the relationship for $c_{\star}$ is given as:

$$\frac{\kappa z}{c_{\star}} \frac{dc'}{dz} = \phi_c \tag{13.7}$$

where c' the difference between the concentration at z and the concentration at a fixed height, and

$$\phi_c = (1 - \alpha\zeta)^{-\beta} \tag{13.8}$$

and ζ is z/L where L is the Monin-Obukhov length.

The stability function for flux-profile relationships have been tested and proved in laboratories and the field. While land flux-profile relationship studies have provided stability functions, recently, stability functions over water have been explored [4]. The parameters for α and β are 13.4 and 1/2 respectively.

The waterside mass balance technique is robust for quantifing the gas exchange rate from laboratory tanks. The method determines the air-water flux through the time rate of change of bulk concentration, which is conserved in a closed system. The transfer velocity using the mass balance technique is given as:

$$k = h \frac{dc_w/dt}{(c_w - c_a)} \tag{13.9}$$

where h is the depth of the tank, and c_w and c_a are the bulk and surface water concentrations, respectively.

Gas transfer velocities for gases with different mass diffusivities or temperature during exchange are normalized to a Schmidt number (Sc) corresponding to 600. The normalization for mass diffusion is given as:

$$k_{600} = k_{gas} \left(\frac{600}{Sc_{gas}} \right)^{-n} \tag{13.10}$$

where the exponent n is known to vary between 1/2 and 2/3 depending on the influence of surfactants. The mass balance equations were used to calculate the gas transfer velocity of SF_6 in the wind wave tank at NASA Wallops Flight Facility and the results were reported in Zappa et al. [19].

13.3 Results and Discussion

Figure 13.2 shows mass-balance derived gas transfer velocity data taken over a wide range of tanks spanning several decades [1, 2, 5, 6, 7, 12, 13, 15, 16]. Data using the flux-profile technique in the two Kyoto University

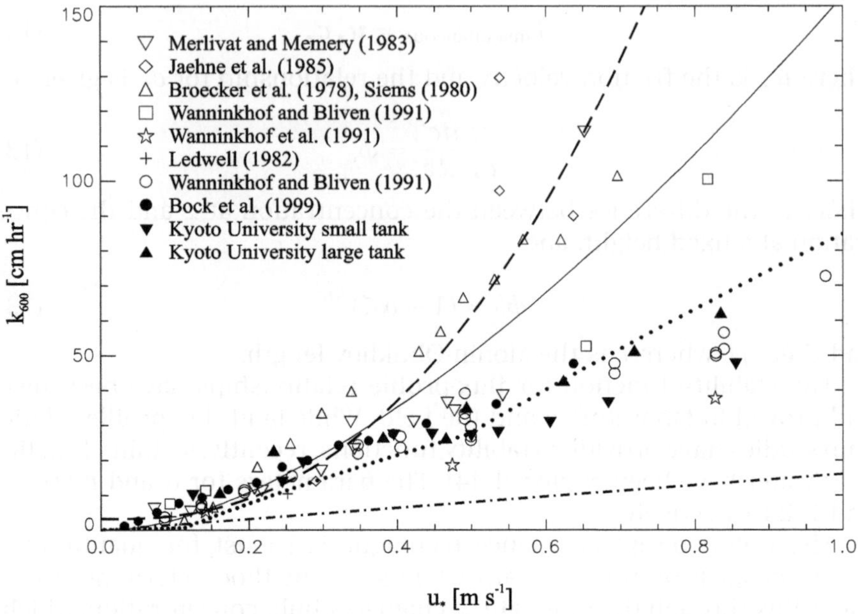

Figure 13.2. Comparison between tank gas transfer studies and models. The dashed line McGillis et al. [9], solid line Wanninkhof [18], dotted line Liss and Merlivat [8], and dashed-dotted line Deacon [3]. The Kyoto University studies are also shown for comparison.

laboratory wind-wave facilities are also shown. The results using the flux-profile technique are with-in the scatter of the historical gas exchange results of laboratory wind-wave tanks. The gas transfer measurements are also compared with in situ models of gas transfer velocity versus friction velocity.

The results from Kyoto University may be of very high quality because the water concentration of dissolved CO_2 was 30 times the concentration used in the NASA wind-wave tank. This corresponds to the same increase in flux resulting in large magnitudes of CO_2 profiles. The air-water pCO_2 difference was made to be greater than 90,000 μatm. This air-water difference causes a very large atmospheric gradient and very accurate results are obtained.

The scatter in the data from different tanks is not alarming. Tanks have different geometries, level of surfactants, and data processing details that result in a range of gas transfer velocities under different friction velocities. In fact, taking into account the maturation of the gas exchange field and the difficulties involved with the tank simulations and measurements, the results are useful. The comparison with the models developed from tank measurements, field measurements, fundamental theories, and theories adapted from the field are also promising. In general, tank mea-

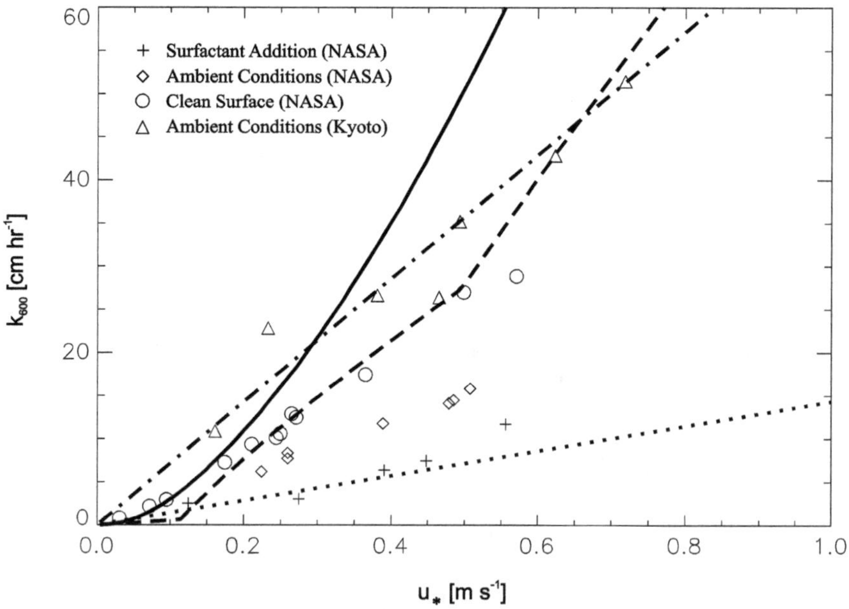

Figure 13.3. CO_2 transfer velocities versus friction velocity. The transfer velocity is calculated using the air-water CO_2 flux derived from atmospheric profiles. Models are also show for comparison. The solid line is the Wanninkhof [18] parameterization. The dashed line is the Liss and Merlivat [8] parameterization. The dashed-dot and dotted lines are the upper and lower limits, respectively, on clean and surfactant water gas exchange data. All data either use friction velocities measured in the tank or derived from drag coefficients at wind speeds measured during the tank experiments.

surements might be lower than field results because of the limited fetch, mitigation of breaking waves and bubbles, and increased surfactant levels. This might explain some of the lower laboratory data at higher friction velocities.

Figure 13.3 shows CO_2 transfer velocities versus friction velocity for the profile technique in the NASA and Kyoto wind-wave tanks. The results in Figure 13.3 show that under greater surface contamination, the gas transfer velocity is decreased.

Figure 13.4 shows the CO_2 transfer velocities are in good agreement with transfer velocities determined using SF_6 mass balance. Results are for ambient laboratory conditions and surfactant additions. The surfactant concentration was 1×10^{-6} mol l^{-1}. The R^2 value for the comparisons between the profile and mass balance method was 0.98 under both environmental surface conditions.

Figure 13.5 shows H_2O transfer velocities versus U_{10}. For a given wind speed, there was a decrease in the water vapor transfer velocity with sur-

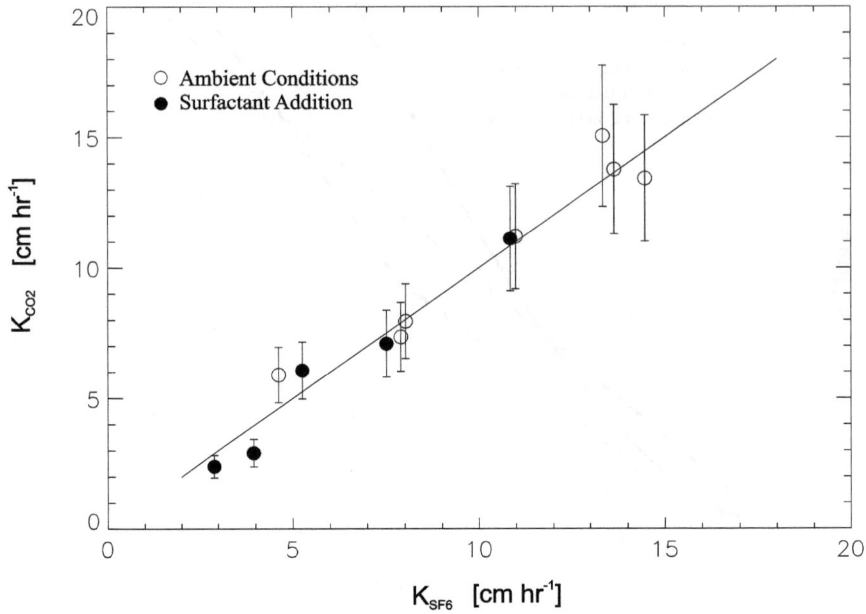

Figure 13.4. Comparison between the gas transfer velocities calculated using the CO_2 atmospheric profile technique and the SF_6 mass balance technique. The distinction between the Surfactant Addition (solid circles) and the Ambient Conditions (open circles) are shown.

factants. Figure 13.6 shows H_2O transfer velocities versus friction velocity. There was an indiscernible effect of surfactants on the H_2O transfer velocities with friction velocity. The fiction velocity takes into account a reduction of drag due to surfactants which may be why better agreement is shown.

Figure 13.7 is a comparison between the transfer velocities of water vapor and carbon dioxide. Because the fluxes of water vapor and carbon dioxide were measured simultaneously in the NASA wind wave tank, it provided a unique opportunity to compare the transfer rates under identical conditions. The fit between the exchange rates of water vapor, an airside controlled, and carbon dioxide, a waterside controlled gas is plotted in Figure 13.7. The line provides a relationship between the transfer velocity for water and carbon dioxide under the measurement conditions. The relationship plotted for comparison is

$$k_{water} = 800 + 12k_{CO_2}^2 \tag{13.11}$$

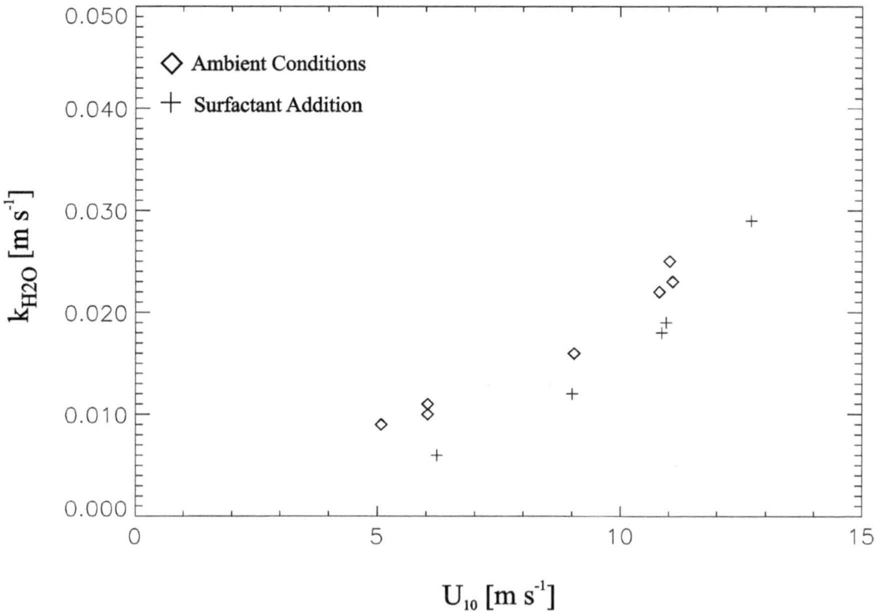

Figure 13.5. H_2O transfer velocities versus U_{10} wind speed. The transfer velocity is calculated using the air-water H_2O flux derived from atmospheric profiles.

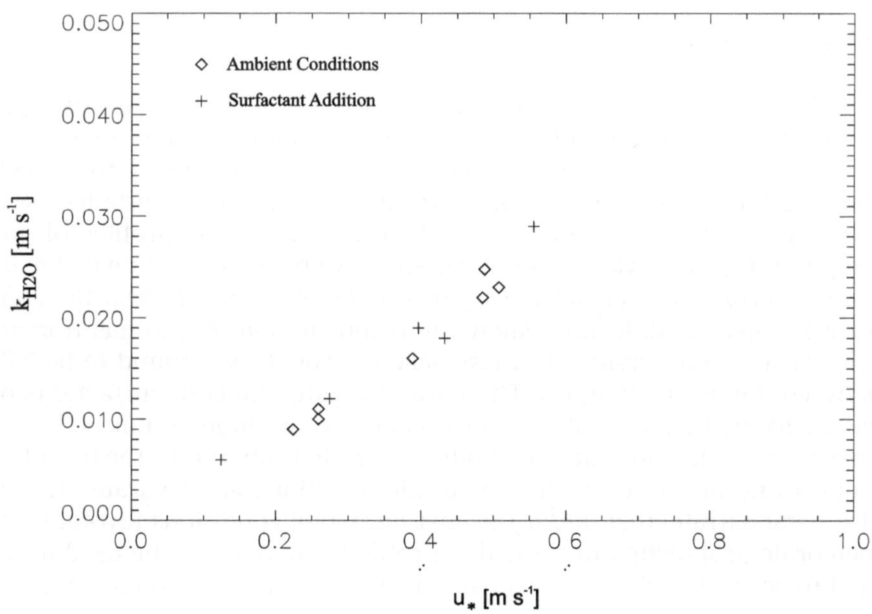

Figure 13.6. H_2O transfer velocities versus friction velocity. The transfer velocity is calculated using the air-water H_2O flux derived from atmospheric profiles.

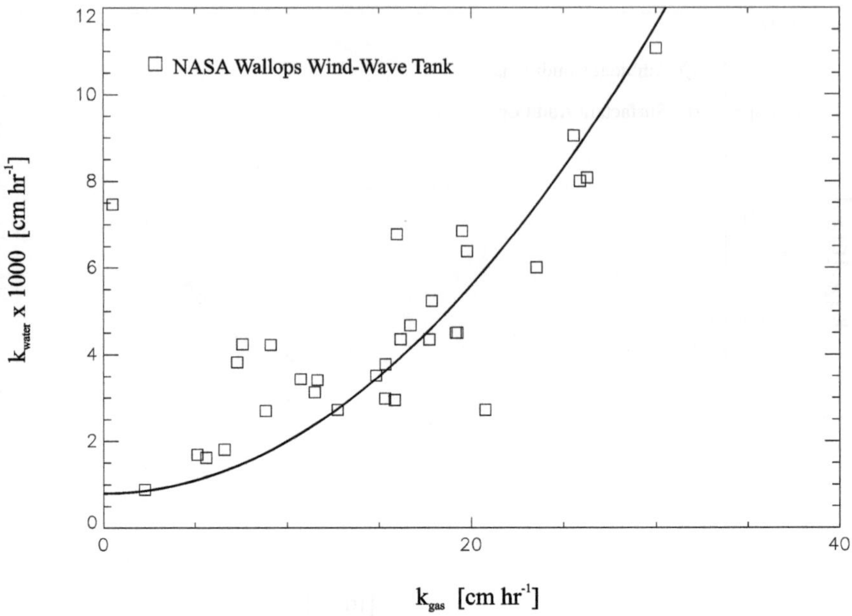

Figure 13.7. Comparison between the water and carbon dioxide transfer velocities.

13.4 Summary

The flux-profile technique was performed for CO_2 in the NASA Wallops Flight Facility (WFF) and the Kyoto University laboratory wind-wave tanks. These facilities are state-of-the-art linear wind-wave tanks and provide classical local atmospheric boundary layers to perform turbulent wall layer models to estimate air-water fluxes. Atmospheric profiles of gas, temperature, water vapor, and wind speed were performed near the air-water interface. The air-water CO_2 flux, H_2O flux, sensible heat flux, and $u_\star$ were measured. Results show the reconciliation of CO_2 gas transfer velocity with a waterside SF_6 mass balance. The R^2 was found to be 0.98 between the two techniques. There was a greater effect of surfactants on gas exchange than on water vapor exchange or on momentum exchange. These results demonstrate that under controlled conditions, the turbulent models that are used in the flux-profile technique are accurate. Results also demonstrate, that under the environmental conditions favorable for meteorological techniques, the flux-profile technique may be used in the field to accurately determine air-water CO_2 and other air-water fluxes.

Acknowledgement. The authors gratefully acknowledge Dr. Stephan Long of the NASA Wallops Flight Facility Air-Sea Interaction Research Facility for his generous assistance and use of the wind-wave tank. We thank Dr. Andrew Jessup from the

Applied Physics Laboratory at the Univeristy of Washington for leading the multi-institutional, multi-disciplinary FEDS-98 experiment. The authors also want to thank the three anonymous reviewers for their comments that greatly improving the manuscript. This research was performed and the manuscript prepared with support from: the Hudson River Foundation grant number 010/02A, the Japan Society for the Promotion of Science (JSPS); the National Science Foundation grant numbers OCE 03-27256, OCE 05-26677, and DEB 05-32075, the National Ocean and Atmospheric Administration grant number NA03OAR4320179, the Lamont-Doherty Earth Observatory at Columbia University, and the Woods Hole Oceanographic Institution. The Lamont-Doherty Earth Observatory contribution number is 7014.

References

[1] Broecker, H.C., Peterman J., Siems W. (1978) The influence of wind on CO_2 exchange in a wind-wave tunnel, including the effects of monolayers. *J Mar Res* 36: 595–610

[2] Broecker, H.C., Siems W. (1984) The role of bubbles for gas transfer from water to air at higher wind speeds; Experiments in the wind-wave facility in Hamburg. In: Brutsaert W, Jirka GH (eds) *Gas Transfer at Water Surfaces.* Reidel, Hingham, Massachusetts: 229–238

[3] Deacon, E.L (1977) Gas transfer to and across the air-water interace. *Tellus*, 29, 363–374

[4] Edson, J.B., C.J. Zappa, J.E. Hare, and W.R. McGillis (2004) Scalar flux profile relationships over the open-ocean, *J Geophys Res* 109: C08S09, doi:10.1029/2003JC001960

[5] Jähne, B., T. Wais, L. Memery, G. Caulliez, L. Merlivat, K.O. Munnich, and M. Coantic, He and Rn gas exchange experiments in the large wind-wave facility of IMST, *J. Geophys. Res.*, 90, 11989-11997, 1985.

[6] Ledwell, J. (1984) The variation of the gas transfer coefficient with molecular diffusivity, in Gas transfer at water surfaces. In: Brutsaert W, Jirka GH (eds) *Gas Transfer at Water Surfaces.* Reidel, Hingham, Massachusetts: 293–303

[7] Ledwell, J.R. (1982) Gas exchange across the air-water interface. Harvard, Cambridge, Massachusetts

[8] Liss, P.S., L. Merlivat.(1986) Air-sea gas exchange rates: Introduction and synthesis. In: P. Buat-Menard (ed) The Role of Air-Sea Exchange in Geochemical Cycling, NATO ASI Ser: C, Math. Phys. Sci., Vol. 185: 113–128

[9] McGillis, W.R., J.B. Edson, J.E. Hare, and C.W. Fairall (2001a) Direct covariance air-sea CO_2 fluxes. *J Geophys Res* 106: 16,729–16,745

[10] McGillis, W.R., J.B. Edson, J.D. Ware, J.W.H. Dacey, J.E. Hare, C.W. Fairall, and R. Wanninkhof (2001b) Carbon dioxide flux techniques performed during GasEx-98, *J Mar Chem*, 75: 267–280

[11] McGillis, W.R., Wanninkhof R. (2006) Aqueous CO_2 gradients for air-sea flux estimates. *J Mar Chem* 98:100–108

[12] Merlivat, L., Memery L. (1983) Gas exchange across an air-water interface: experimental results and modeling of bubble contribution to transfer. *J Geophys Res* 88: 707-724

[13] Siems, W. (1980) Modeluntersuchungen zur Verdunstung und zum Gasaustauch zwis en Wasser und Luft. Der Einfluss von Wellen und Oberflachenverundreinigungen, University of Hamburg

[14] Stumm, W., Morgan J.J. (1996) Aquatic Chemistry: chemical equilibria and rates in natural waters. John Wiley & Sons, New York

[15] Wanninkhof, R., Bliven L. (1991) Relationship between gas exchange, wind speed and radar backscatter in a large wind-wave tank. *J Geophys Res* 96: 2785–2796

[16] Wanninkhof, R., Bliven L.F., Glover D.M. (1991) Gas transfer velocities and radar backscatter from the water surface. In: Wilhelms, SC, Gulliver JS, (eds) *Proceedings of the second international symposium on gas transfer at water surfaces.* ASCE, New York, New York: 294–308

[17] Wanninkhof, R., McGillis, W.R. (1999) A cubic relationship between wind speed and gas exchange over the ocean, *Geophys Res Lett* 26:1889–1892

[18] Wanninkhof, R., (1992) Relationship between wind speed and gas exchange over the ocean, *J Geophys Res* 97: 7373–7382

[19] Zappa, C.J., W.E. Asher, A.T. Jessup, J. Klinke, and S.R. Long (2004) Microbreaking and the enhancement of air-water transfer velocity, *J Geophys Res* 109: (C08S16), doi:10.1029/2003JC001897.

14

Air-Water Interfacial Temperature Measurements

Brian Ward

Center for Coastal Physical Oceanography
Department of Ocean, Earth & Atmospheric Sciences
Old Dominion University, Norfolk, USA
bward@ccpo.odu.edu

Abstract Measurements at the air-water interface were made with a high spatial and temporal resolution microthermometer in the Air-Sea Interaction Saltwater Tank (ASIST). ASIST is a linear wind-wave tank of dimensions 15×1×1 m, and is constructed from transparent acrylic sheets. During the experiments, the air-water heat flux was varied by changing the air-water temperature difference (ΔT^{AW}), and the wind speed was varied with the ASIST fan. The microthermometers were mounted on a J-shaped support, which was attached to a linear servo motor which was driven in the vertical direction. This allowed profiles to be made from a depth of about 13 cm to the surface. Data was acquired for three ΔT^{AW} regimes (-5, +10, and +15°C) with wind speeds from 1 to 9 ms^{-1}. Data for a total of 8 runs were acquired and provided estimates of the boundary layer thickness (δ_c) and the skin-bulk temperature difference (ΔT^{SD}). We found that the relationship between ΔT^{SD} and wind depended on the ΔT^{AW} regime. There was a clear relationship between δ_c and wind (u), and we derived the empirical expression $\delta_c = 0.35 + 4.9e^{-u}$ based on the data. Comparison with previous estimates of δ_c based on measurements and rigid wall boundary layer theory showed that our values were lower by a factor of 2 at low wind speeds, but were in good agreement at 10 ms^{-1}.

14.1 Introduction

The ocean skin temperature is most commonly detected by infrared devices, which, due to absorption properties of water in the infrared region of the electromagnetic spectrum, measure only the top few micrometres of the temperature of the sea surface. There is a strong motivation to know the ocean skin temperature: it influences heat transfer between the ocean and atmosphere, and is therefore one of the most important parameters governing the climate [4, 16]. Polar-orbiting space-borne instruments provide global sea surface temperature (SST) measurements with infrared sensors, and hence a measurement of the skin temperature [2]. The skin temperature affects the exchange of gas across the air-sea interface be-

C.S. Garbe, R.A. Handler, B. Jähne (eds.): Transport at the Air Sea Interface
pp. 193-203, 2007,

cause of the temperature dependence of the solubility of gas in seawater [19].

The difference between the skin temperature T_{skin} and the temperature at depth T_{depth} is defined as $\Delta T^{SD} = T_{\text{skin}} - T_{\text{depth}}$. The first evidence of a surface temperature difference at a water surface was given by Woodcock and Stommel [23] who used a thermometer with a bulb of 1.5 mm outer diameter and 9 cm in length. Wave action was damped by a bottomless box and measurements were made under calm conditions. The ΔT^{SD} values ranged from 0 to 1 °C for wind speeds up to 4.7 ms^{-1}, and $T_{\text{depth}} - T_{\text{air}}$ temperature differences of -0.7 to 13.4 °C. Ewing and McAlister [3] were the first researchers to detect ΔT^{SD} using a radiometric device. They deployed from a pier an infrared radiometer with a spectral bandwidth from 6-20 µm allowing a measurement of the radiant flux originating within the top 0.1 mm. They showed a cool surface departure of up to 0.6 K from conventional temperature sensors placed 15 cm below the surface. They also investigated the effect of breaking waves on the skin layer by directing a jet of water with a pump placed below the surface. While the pump was running, the radiation temperature rose to approximately the value measured by the thermistors, and dropped to its normal value after cessation of the rupture of the surface. Less intense disturbances of the water below the surface had no effect on the radiometer's signal. Measurements coincident with actual wave breaking showed that the disturbance affected the signal for about 12 seconds before returning to its normal value. Since these initial measurements, there has been several studies related to ΔT^{SD} (see Donlon et al. [2] for a comprehensive list of references).

In contrast to the ΔT^{SD} measurements, there has been scant measurements of boundary layer thickness δ_c. McKeown and Asher [11] used an interferometric technique to measure the water temperature with two different wavelengths. The underlying principal was that radiation at different wavelengths emit from different depths. This was similar to the technique developed by McAlister and McLeish [10], who measured the temperature gradient with a two-channel radiometer having spectral bandwidths of 4.5-5.1 µm, and 3.5-4.1 µm, corresponding to optical depths of 0.025 mm and 0.075 mm. However, neither of these reports resolved the full depth of the thermal molecular boundary layer.

The boundary layer thickness has been estimated from theoretical models. Grassl [6] described the different sublayers at the air–water interface. Since turbulence is responsible for the transport of heat as well as momentum, the implication is that a viscous sublayer must also exist at the surface. Within the viscous layer δ_v is the conductive sub-layer δ_c. Grassl [6] calculated the ratio $\delta_v/\delta_c = 1.82 = Pr^{1/3}$, assuming that the one-third power of the Prandtl number for rigid boundaries is valid in this case. Wu [24] presented a theoretical estimate of the thicknesses of δ_v and δ_c. Wu defined the thickness of δ_v as $\delta_v = 11.6\nu/u_\star$, and argued

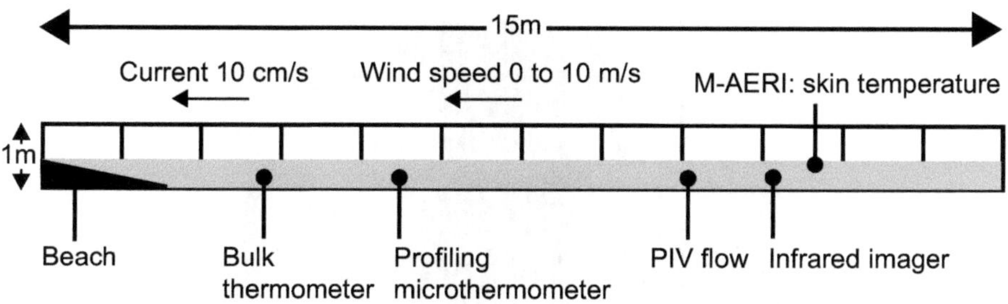

Figure 14.1. Location of instruments during the experiment in the ASIST wind-wave tank.

that the $\delta_\nu : \delta_c$ ratio was 2:1 (ν is the kinematic viscosity of water, and $u_\star$ is the friction velocity at the water surface).

In this study, we present measurements made in the Air–Sea Interaction Saltwater Tank (ASIST) with a high-resolution microthermometer. The objectives were to investigate the behavior of the molecular sublayer under varying heat flux and wind speed regimes. In section 14.2, we describe the experimental setup. Section 14.3 discusses the results with respect to δ_c and ΔT^{SD}, followed by conclusions from this study in section 14.4.

14.2 Measurements

The objective during the experiment was to resolve the molecular boundary layer at the sea surface by making profiles at intervals of $\sim$ 10 seconds during an experimental run i.e. under conditions of uniform wind speed, heat flux and wave frequency/amplitude.

The experiment was conducted in the ASIST wind-wave tank (Rosenstiel School of Marine and Atmospheric Sciences, University of Miami). This tank has a working section of 15 m and cross section of 1×1 m (see Figure 14.1). It is constructed with transparent acrylic panels to allow visualization of installed instrumentation and flow. During the experiment, fresh water was used and the wind tunnel was in the open mode to allow steady state fluxes to be obtained. This pumped air at the ambient external temperature into the tank, which remained fairly constant for the two week experiment. Air-water fluxes were controlled by changing the water temperature with a heat exchanger. Throughout the course of the experiment, air-water temperature differences ΔT^{AW} were varied from -15 K to

Figure 14.2. Photograph of the sensors mounted on the J-shaped support and its attachment to the linear servo motor.

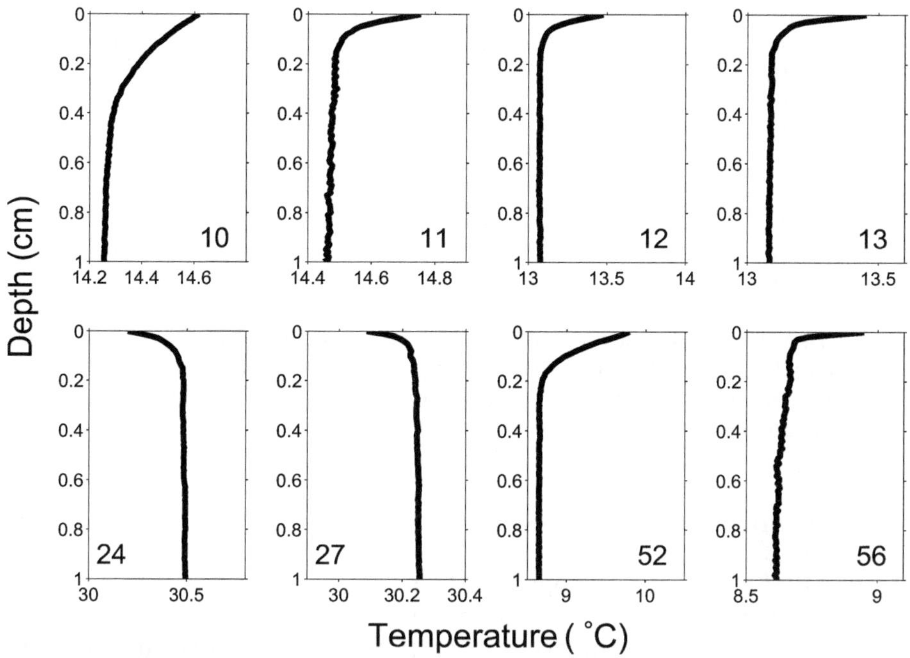

Figure 14.3. Average temperature profiles for each run.

+15 K in increments of 5 K, and for heat heat flux regime the wind speed u was varied from 0 to 10 ms^{-1} in increments of 1 ms^{-1}. Figure 14.1 shows the location of the instrument during the experiment.

Thermometric measurements of the molecular sublayer were conducted with a fine wire microthermometer. Absolute temperature was provided by a FP07 thermistor, and the surface was detected with a microconductivity sensor (see Ward et al. [20] for a full description of the sensors). The sensors were taken from the SkinDeEP (Skin Depth Experimental Profiler), an autonomous, self-contained, hydrodynamic instrument capable of making repeated, high resolution profiles of temperature and conductivity within the ocean's upper decameter [18].

The sensors were mounted on a vertical mast which was attached to a linear servo motor. Measurements were made from a depth of 13 cm to the surface at an ascent velocity of precisely 0.5 ms^{-1}, and at a rate of about 10 seconds. The sensors were mounted into J-shaped supports (see Figure 14.2). Profiling action was accomplished by mounting the sensors on the linear motor (originally deployed in the field as a "wave follower"). The data acquisition system was also mounted on the motor, but never immersed (Fig. 14.2).

Run No.	$T_a - T_w$ (°C)	u (ms^{-1})	$u_\star$ (cms^{-1})	$\overline{\delta_c}$ (mm)	$\overline{\Delta T^{SD}}$ (°C)
10	+10	1	0.16	2.16	0.36
11	+10	3	0.32	0.55	0.28
12	+10	5	0.58	0.39	0.40
13	+10	7	0.93	0.36	0.36
24	-5	4	0.44	0.44	-0.29
27	-5	5	0.58	0.40	-0.16
52	+15	3	0.33	0.85	1.14
56	+15	9	1.37	0.26	0.31

Table 14.1. Nominal conditions during the runs and the derived profile parameters. Here T_a and T_w the air and water temperatures, u is the wind speed, $u_\star$ the friction velocity, $\overline{\delta_c}$ is the mean sublayer thickness, and $\overline{\Delta T}$ is the mean temperature difference across δ_c.

14.3 Results and Discussion

Figure 14.3 shows the averaged profiles for each run. For each profile, it was necessary to extract the temperature drop across the boundary layer ΔT^{SD}, as well as the boundary layer thickness δ_c. This was carried out by fitting a model after Howard [8], and adjusting the parameters until the best fit was achieved [see 17].

Table 14.1 summarizes the conditions and parameters of interest. There was a total of 8 runs for three different ΔT^{AW} regimes, with winds ranging from 1 to 9 ms^{-1}. The mean values of ΔT^{SD} and δ_c are also presented in Table 14.1.

There has been some discussion in the literature about the relationship between ΔT^{SD} and wind speed. Fairall et al. [4] compare data to model estimates from Saunders [12] (the λ[1] value used was from Coppin et al. [1]), and Schlüssel et al. [14] (an empirical model based on field data). The Saunders' model showed a decreasing ΔT^{SD} with wind speed, while the empirical model showed an increasing ΔT^{SD} with wind.

Wick et al. [21] presented estimates of ΔT^{SD} from several models based on field data, where the ΔT^{SD} was calculated from measurements of wind speed and parameterized net heat flux. The results of the 6 different ΔT^{SD} models were striking, with "tremendous differences in the predicted behavior of ΔT^{SD} between the different models" [21]. Two of the six models predicted an increasing ΔT^{SD} with wind, while the remaining four predicted the opposite. There were also large discrepancies between the models at low wind speeds, ranging from 0.2 to 1.3°C.

Donlon et al. [2] used nighttime data from 6 independent field datasets to derive an expression for ΔT^{SD} based on wind only: $\Delta T^{SD} = -0.14 -$

[1] The λ coefficient was introduced by Saunders [12]. See Federov and Ginsburg [5] for a detailed discussion on λ.

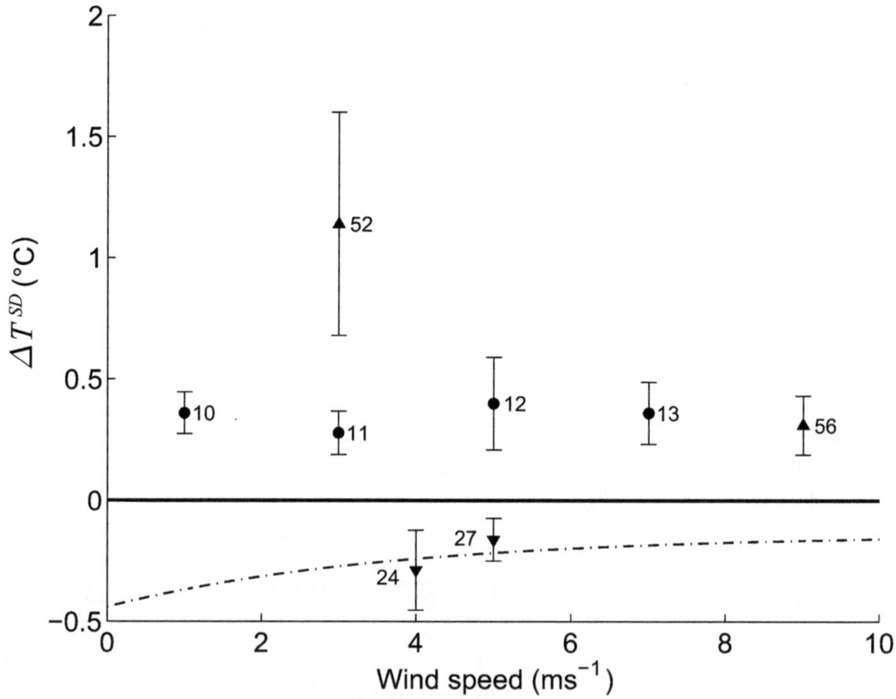

Figure 14.4. Mean values of ΔT^{SD} plotted against wind speed from the microthermometer data for the three different ΔT^{AW} regimes: -5°C (▼), +10°C (•), and +15°C (▲). The dashed line is the expression from Donlon et al. [2]: $\Delta T^{SD} = -0.14 - 0.3\exp(-u/3.7)$.

$0.3e^{(-u/3.7)}$. These results showed $|\Delta T^{SD}|$ decreasing with wind speed, where it approached a constant value of ~0.17°C above wind speeds of 6 ms^{-1}.

The measurements of ΔT^{SD} from this experiment (Fig. 14.4) did not show a strong relationship like the results from Donlon et al. [2]. Rather, there were 3 distinct relationships related to the air-water temperature difference. For ΔT^{AW} = -5°C, ΔT^{SD} decreased from 0.29°C to 0.16°C, which was in good agreement with the expression from Donlon et al. [2] (shown in Fig. 14.4 as the dashed line). However, for the other two heat flux regimes, the relationship between ΔT^{SD} and wind was very different. For ΔT^{AW} = +10°C, there was a constant ΔT^{SD} over 1 to 7 ms^{-1}. For ΔT^{AW} = +15°C, there was a large ΔT^{SD} of >1°C at 3 ms^{-1}, decreasing to 0.3°C at 9 ms^{-1}.

As mentioned in the introduction, there are few estimates of δ_c in the literature. The estimates from Wu [24] are based on 30 independent

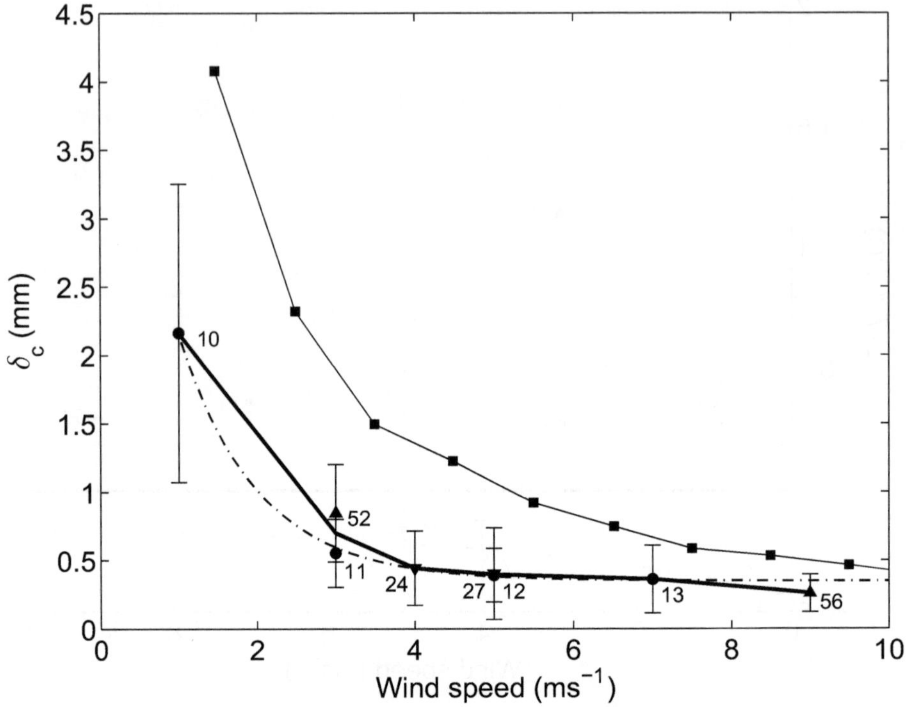

Figure 14.5. Microthermometer data showing the relationship between wind speed and δ_c for the three different ΔT^{AW} regimes: -5°C (▼), +10°C (•), and +15° (▲). A fit to the data yields $\delta_c = 0.35 + 4.9e^{-u}$ (dashed line). Also shown are the estimates of δ_c from Wu [24] (■).

datasets of measurements of the drag coefficient, which were then used to derive δ_c from boundary layer theory.

From our experimental data, we found a consistent relationship between δ_c and wind speed which was independent of ΔT^{AW} (Fig. 14.5). An empirical fit to the data yielded an expression for δ_c:

$$\delta_c = 0.35 + 4.9e^{-u} \tag{14.1}$$

The δ_c data from Wu [24] are also presented in Fig. 14.5. There is a factor of 2 difference between the estimates of Wu [24] and the measurements here for the lower wind speeds, but as the wind speed approaches 10 ms^{-1}, the estimates from Wu and our results are in good agreement.

14.4 Conclusions

We presented measurements with a high-resolution microthermometer in a wind-wave tank under various wind speed and air-water temperature differences. By fitting a model to the temperature profiles we were able to extract estimates of the bulk–skin temperature difference ΔT^{SD} and the thermal molecular boundary layer thickness δ_c based on these direct measurements. With these data we were able to study the relationship between wind speed and these two key variables.

There was no consistent relationship between ΔT^{SD} and wind speed alone, but rather it depended also on the ΔT^{AW} regime. This would indicate that ΔT^{SD} cannot be parameterized in terms of wind speed alone, but the heat flux must also be considered. However, the empirical parameterization from Donlon et al. [2] was in good agreement with the two cool skin runs in this experiment, indicating the validity of this model under certain conditions.

There exist many more physically–based parameterizations to predict the bulk-skin difference at the sea surface. The basic physical assumption underlying the models is that $\Delta T \propto Q_n/u_{10}$ or $\Delta T \propto Q_n/u_\star$. All of the models also assume that the heat in the microlayer is transferred by molecular conduction. However, the stagnant film models [e.g. 4, 7, 12] differ from the surface renewal models [e.g. 9, 15, 21], in that a parcel of water at the air-sea interface is intermittently replaced by bulk fluid, but in the former models is assumed to remain there. It is not possible to draw any conclusions over which model is more appropriate as only average values were considered for each run for this experiment.

The positive ΔT^{SD} values (i.e. a warm skin) measured in this experiment almost never occur in nature as the heat loss immediately at the water surface from latent, sensible, and longwave fluxes are greater than any shortwave radiation absorbed in the boundary layer. This point is further enhanced with the recent results from Wick et al. [22], who showed that previous estimates of shortwave absorption in the molecular boundary layer have been over–estimated.

Wu [24] concluded in his paper that estimates of δ_c and its functionality with wind speed was desirable, and that "laboratory experiments under controllable conditions would be ideal for studying the mechanisms involved in air–sea transfer processes". There was a clear relationship between δ_c and wind with an empirical fit yielding an exponential dependence.

The δ_c estimates from Wu [24] were based on boundary layer theory from Schlichting [13]. It would appear that the rigid boundary layer theory over–estimated the water surface molecular sublayers by a factor of 2 at lower wind speeds. However, these estimates of the drag coefficient were made several decades ago. There has been significant progress in measurements of the ocean surface drag coefficient since.

References

[1] P. A. Coppin, E. F. Bradley, I. J. Barton, and J. S. Godfrey. Simultaneous observations of sea surface temperature in the western equatorial pacific ocean by bulk, radiative and satellite methods. *J. Geophys. Res.*, 96:3410–3409, 1991.

[2] C. J. Donlon, P. J. Minnett, C. Gentemann, T. J. Nightingale, I. J. Barton, B. Ward, and J. Murray. Towards improved validation of satellite sea surface skin temperature measurements for climate research. *J. Climate*, 15:353–369, 2002.

[3] G. Ewing and E. D. McAlister. On the thermal boundary layer of the ocean. *Science*, 131:1374–1376, 1960.

[4] C. W. Fairall, E. F. Bradley, J. S. Godfrey, G. A. Wick, J. B. Edson, and G. S. Young. Cool-skin and warm-layer effects on sea surface temperature. *J. Geophys. Res.*, 101:1295–1308, 1996.

[5] K. N. Federov and A. I. Ginsburg. *The Near-Surface Layer of the Ocean.* VSP, Utrecht, 1992.

[6] H. Grassl. The dependence of the measured cool skin of the ocean on wind stress and total heat flux. *Bound.-Lay. Meteorol.*, 10:465–474, 1976.

[7] L. Hasse. The sea surface temperature deviation and the heat flow at the air-sea interface. *Bound.-Lay. Meteorol.*, 1:368–379, 1971.

[8] L. N. Howard. Convection at high rayleigh numbers. In H. Görtler, editor, *Proceedings of the 11th International Congress of Applied Mechanics*, pages 1109–1115, Munich, Germany, 1966. Springer.

[9] W. T. Liu and J. A. Businger. Temperature profile in the molecular sublayer near the interface of a fluid in turbulent motion. *Geophys. Res. Lett.*, 2:403–404, 1975.

[10] E. D. McAlister and W. L. McLeish. A radiometric system for airborne measurement of the total heat flow through the sea. *Appl. Opt.*, 9: 2697–2705, 1970.

[11] Walt McKeown and William Asher. A radiometric method to measure the concentration boundary layer thickness at an air-water interface. *J. Atmos. Oceanic Technol.*, 14:1494–1501, 1997.

[12] P.M. Saunders. The temperature at the ocean-air interface. *J. Atmos. Sci.*, 24:269–273, 1967.

[13] H. Schlichting. *Boundary-Layer Theory.* McGraw-Hill, New York, 1968.

[14] P. Schlüssel, W. J. Emery, H. Grassl, and T. Mammen. On the bulk skin temperature difference and its impact on satellite remote sensing of sea surface temperature. *J. Geophys. Res.*, 95:13341–13356, 1990.

[15] A. V. Soloviev and P. Schlüssel. Parameterization of the cool skin of the ocean and of the air-ocean gas transfer on the basis of modelling surface renewal. *J. Phys. Oceanogr.*, 24:1339–1346, 1994.

[16] B. Ward. Near-surface ocean temperature. *J. Geophys. Res.*, 111: C02005, 2006. doi: 10.1029/2004JC002689.

[17] B. Ward and M. A. Donelan. Thermometric measurements of the molecular sublayer at the air-water interface. *Geophys. Res. Lett.*, 33:L07605, 2006. doi: 10.1029/2005GL024769.

[18] B. Ward and P. J. Minnett. An autonomous profiler for near surface temperature measurements. In M. A. Donelan, W. M. Drennan, E. S. Saltzman, and R. Wanninkhof, editors, *Gas Transfer at Water Surfaces*, volume 127 of *Geophys. Monogr. Ser.*, pages 167–172. AGU, Washington, D.C., 2002.

[19] B. Ward, R. Wanninkhof, W. R. McGillis, A. T. Jessup, M. D. DeGrandpre, J. E. Hare, and J. B. Edson. Biases in the air-sea flux of CO_2 resulting from ocean surface temperature gradients. *J. Geophys. Res.*, 109:C08S08, 2004. doi: 10.1029/2003JC001800.

[20] B. Ward, R. Wanninkhof, P. J. Minnett, and M. Head. SkinDeEP: A profiling instrument for upper decameter sea surface measurements. *J. Atmos. Ocean. Technol.*, 21:207–222, 2004.

[21] G. A. Wick, W. J. Emery, L. H. Kantha, and P. Schlüssel. The behavior of the bulk-skin sea surface temperature difference under varying wind speed and heat flux. *J. Phys. Oceanogr.*, 26:1969—-1988, 1996.

[22] G. A. Wick, J. C. Ohlmann, C. W. Fairall, and A. T. Jessup. Improved oceanic cool-skin corrections using a refined solar penetration model. *J. Phys. Oceanogr.*, 35:1986–1996, 2005.

[23] A. H. Woodcock and H. Stommel. Temperatures observed near the surface of a fresh water pond at night. *J. Meteor.*, 4:102–103, 1947.

[24] J. Wu. An estimation of oceanic thermal-sublayer thickness. *J. Phys. Oceanogr.*, 1:284–286, 1971.

15

Observations of the Structure of the Surface Temperature Field at an Air-Water Interface for Stable and Unstable Cases

Geoffrey B. Smith, Robert A. Handler, and Nicholas Scott

Naval Research Laboratory
Remote Sensing Division, Code 7233
4555 Overlook Ave., SW
Washington, DC 20375, USA
geoffrey.smith@nrl.navy.mil

Abstract The thermal structure of an air-water interface is investigated by examining thermal imagery obtained from a high resolution infrared (IR) sensor. The experiments were performed at the ASIST facility at the University of Miami for wind speeds ranging from approximately 2 ms^{-1} to 10 ms^{-1} and for flux based Richardson numbers ranging from about 10^{-2} to 10^{-5}. Two cases were examined: (1) the so-called *cool-skin* case where the water surface was significantly cooler than the bulk water temperature and (2) the *warm-skin* case where the water surface was warmer than the bulk. In the cool-skin case, the low wind speed results reveal a cellular structure reminiscent of earlier results in which the lateral length scale of the cells (or *fish-scales*) varies as the inverse of the friction velocity. The imagery clearly reveals the progression from non-breaking gravity waves, to a system of omnidirectional breaking which seems to create a nearly isotropic surface temperature field. Though no wind waves were present at low wind speeds, the thermal imagery reveals the existence of persistent, highly coherent, Langmuir-like cell structures which were marked by surface convergent zones in which ambient surfactant may have accumulated. Imagery obtained for the case in which the water-side thermal boundary layer is stable constitutes a novel aspect of this work. In this *warm-skin* case, the cellular (fish-scale) structure appears as it does in the unstable case, strongly suggesting that these small scale features are due to shear instabilities in the surface layer. In addition, they are more clearly revealed as the natural convective instability of the thermal boundary layer is suppressed. This appears to reduce the appearance of the smallest scales of surface turbulence.

15.1 Introduction

Recently, high resolution infrared (IR) sensors have been used to study the complex hydrodynamical and thermal processes operating at the air-water

C.S. Garbe, R.A. Handler, B. Jähne (eds.): Transport at the Air Sea Interface
pp. 205-222, 2007, © Springer-Verlag Berlin, Heidelberg 2007

interface at small scales [5, 9, 24] and at large scales [14, 15, 16, 26]. One reason for the success of these sensors in examining the fluid dynamical motions at the interface is that heat diffuses slowly in water – the Prandtl number for heat in water is significantly greater than one – so that the temperature field can often be thought of as an accurate tracer of fluid motions. Thus, for example, one can use sequential images of IR data to determine the surface drift speed in a wind driven flow [3, 4, 5]. Such imagery allows for the determination of regions of high surface straining, as well as convergent and divergent zones [3, 22]. It is also possible to study the dynamics of surface waves, whose periodic motions are superimposed onto the surface current [5, 9]. In the case of breaking waves, for example, infrared methods reveal the details of the mixing of generally warmer bulk fluid with cooler surface fluid during a breaking event [8, 26]. Such information may be used as inputs to models that estimate the interfacial fluxes of momentum (wind shear), heat, gas, and related quantities of interest such as wave dissipation rates [7, 20, 27].

It is the purpose of this work to present results from an experiment performed at the ASIST (Air Sea Interaction Salt Water Tank) facility at the University of Miami. The experiment was designed primarily as a gas transfer study, with as many aspects of the sub-surface flow and cool skin being documented as possible. Instrumentation included down-looking radiometers (M-AERI [18]), thermal boundary layer temperature profilers [25], imaging slope gages, DPIV, hot films, and anemometers. Our main objective is to discuss in some detail, the kinds of processes revealed through high resolution infrared imagery, and to speculate on the microphysical processes which may give rise to the observed imagery. We have studied the thermal structure of the interface by systematically examining the IR imagery as the wind speed at the surface is increased from about 2 ms^{-1} to 10 ms^{-1}. We first investigate the classical *cool-skin* case where the interfacial heat flux is directed from the water to the atmosphere, and the thermal water-side boundary layer is clearly unstable. A novel aspect of this work is the investigation of IR imagery obtained from the stable case in which the heat flux was into the water column – the so-called *warm-skin* case. The stable situation allows for the study of purely shear induced instabilities since buoyancy induced motions are entirely suppressed.

15.2 Experimental Setup at the ASIST facility

The experiments were conducted in the Air-Sea Interaction Salt-water Tank (ASIST) located at the University of Miami (Rosenstiel School of Marine and Atmospheric Science) as part of a larger air-sea interaction study. The ASIST facility is a wind-wave water tunnel with a test section cross-section of 1m × 1m, and a length of 15 m. For this study the water depth was fixed at 40 cm, leaving 60 cm for the wind. For all the cases presented here a water pump was used to maintain a small, constant current (approximately 3 cms^{-1}) that enabled control of the water temperature. Water temperatures were varied to provide air-water temperature differences ranging from $-15°C$ to $+15°C$. Wind speeds were varied from 0 ms^{-1} up to 10 ms^{-1}. The cases in which the heat flux was directed *out* of the interface will be denoted as the *cool-skin* cases, and those in which the flux is directed *into* the boundary are referred to as *warm-skin* cases. In all cases, fresh water was used.

All the IR imagery described here was taken with an Indigo Systems Merlin Mid-Wave IR sensor, a snapshot imager with an array size of 320×256 pixels. For the current experiments this translates into image sizes of 23.9×29.9 cm for the cool-skin cases, and 23.6×29.5 cm for the warm skin. The array is cooled to increase sensitivity to better than $0.02°C$. The imager is sensitive to radiation in the 3–5 μm band. In this band the mean optical depth of the detected radiation is 35 μm [17]. The maximum frame rate of the IR imager is 60 Hz, although the actual acquisition rate was varied based on the flow being imaged. All the imagery was taken looking down at the water surface at an angle of approximately 60 degrees from the horizontal through a hole in the top of the tunnel. This was done for two reasons: first, to reduce the occurrence of reflections of the camera itself appearing in the data, and second, to accommodate a second IR imager viewing through the same port in the roof of the wind-wave tunnel.

The tunnel was instrumented with a number of sensors which allowed for the determination of the wind shear, and the sensible and latent heat fluxes at the interfaces. The experimental parameters for both the cool and warm skin cases are given in the Tables 15.1 and 15.2 in the Appendix. In the tables we list only the total heat flux which was computed as the sum of the sensible and latent fluxes. We also list in each case the so-called flux based Richardson number (Ri) which is defined as:

$$Ri = \frac{\beta g Q \nu}{{u_\star}^4 \rho c}, \tag{15.1}$$

where β, c, ρ, and ν are, respectively, the coefficient of expansion, specific heat, density, and kinematic viscosity for water. The gravitational acceleration is denoted by g, Q is the total flux (positive values designate flux from the water to the air), and $u_\star = \sqrt{\tau/\rho}$ is the friction velocity where τ is the wind-induced shear stress at the air-water interface. This flux based Richardson number can be interpreted as the ratio of turbulence production by buoyancy to that by shear [5, 22, 23]. We also list in each case the viscous length scale $l^+ = \nu/u_\star$. We note that the Richardson numbers for the cool skin cases range from about 10^{-2} at the lowest wind speed, to values of order 10^{-5} at high wind speeds – a range of about three orders of magnitude. Thus, our experiments represent a range of conditions in which shear production is always larger than buoyancy production. In the warm-skin cases, the surface flow is clearly stable with respect to buoyancy forces since the heat flux is always directed into the water column, as indicated in the Table 15.2.

In interpreting the IR imagery presented in this work, we follow the surface strain model described in detail in Smith et al. [22] and Leighton et al. [13]. In brief, the surface strain model (SSM) states that, when the heat flux at the surface is fixed, the local thermal boundary layer thickness controls the temperature at the surface. In the cool-skin case, for example, when the subsurface motions cause the thermal boundary layer to become thinner (a divergent surface straining) the surface temperature appears warmer than the average surface temperature. In a similar manner the surface will appear cooler when subsurface motions cause the thermal boundary layer to thicken (a convergent straining). It is interesting to note that according to this model, the surface temperature can change in the absence of any water mass transport. An excellent example of this is given by Marmorino et al. [15] where surface temperature fluctuations were observed during the passage of a subsurface internal wave.

15.3 Infrared Imagery

15.3.1 Unstable Cases

In Figures 15.1-15.5 we show IR images for the cool thermal boundary layer case over the range of wind speeds 1.9 ms^{-1} to 10.1 ms^{-1}. The temperature differences represented in these images are on the order of 1.5°C. Note that the descriptions of the phenomena in these figures are based upon detailed viewing of movies of the IR images and not solely upon these still images. Thus, some stated conclusions may not be readily evident by viewing just these snapshots. In particular, only the viewing of such movies can reveal phenomena such as wave motion or surface convergent-divergent motions. With this caveat, we proceed to describe the processes observed as the wind speed is systematically increased.

For the lowest wind speed case two snapshots are shown: one very early in the run (Fig. 15.1**a**) and one taken about sixteen minutes later (Fig. 15.1**b**). At this low wind speed there is no evidence of gravity-capillary waves in the IR images. It is therefore assumed that in this case the surface was sufficiently smooth that curvature and roughness effects due to any waves could be ignored. Early in time (Fig. 15.1**a**), a number of significant features are evident. For example, we note the existence of *warm spots* (labeled A) which were dubbed *fish-scales* in earlier work [5]. These structures seem to be characterized by a *head-tail* structure, such that a vector drawn from head to tail is approximately parallel to and in the same direction as the mean wind speed direction. Also indicated in the figure is the scale $100l^+$, which has been found in classic experiments on wall bounded turbulence ([12] [21]) to characterize the observed streaky

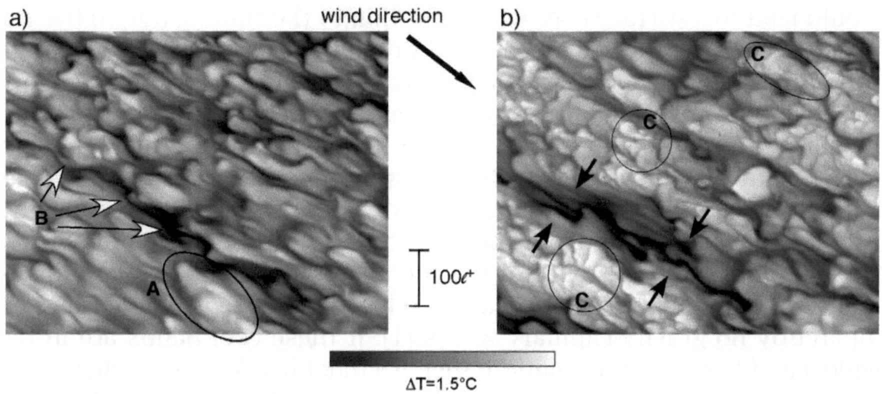

Figure 15.1. Infrared images of the water surface with a wind speed of 1.9 ms^{-1} (Run 31). **a** This image is taken very early in the run. Region designated by *A* indicates the typical thermal cellular structures termed *fishscales* and *B* indicates a cool band. **b** A thermal image of the water surface taken (16 minutes later). The region indicated by *C* shows the cellular structures with embedded small scale features absent in image (A). These small scale features are assumed to be associated with the absence of surfactants. Arrows indicate the convergent surface flow pattern associated with the cool band. This flow pattern was discerned from viewing movies of the IR imagery. Indicated in this figure, and in all subsequent figures, is (1) The length scale $100l^+$ (2) The wind direction, and (3) The greyscale color scheme in which brighter regions are always warmer than darker regions.

structures. This scale has also been found in previous work to be nearly equal to the lateral scale size of the fish-scale cellular structure [5]. Here, the lateral scale size is defined as the distance along a line perpendicular to the wind direction between any two successive peaks in the temperature field. A more general discussion of the scale-size distribution of the surface temperature field may be found in Schimpf et al. [20].

Also indicated in Fig. 15.1**a** is a cold band (indicated by the arrows labeled 'B') which at this instant stretches diagonally across almost the entire image. In a movie sequence at this wind speed such cool bands come and go apparently in a random manner, but are probably visible in half the images. In addition, viewing movie sequences clearly indicates that the cooler fluid associated with these bands moves faster than the speed of the surrounding warmer cells. At this time the authors have no definitive explanation as to the nature of this phenomenon, but one possible idea is that the signature of a Langmuir-like circulation is being observed. If it is supposed for the moment that a Lamgmuir-like circulation is present, it should generate a surface convergence region. In such a region, it is expected [10, 22] that a local down-welling and therefore a local thickening of the thermal boundary layer should result. If it is further assumed that the surface heat flux is constant, then such a thickening should lead to a surface cooling. In addition to the cooling due to the surface convergence, there may be an additional effect due to the accumulation of naturally occurring surfactant in the convergence zone itself. The presence of any surfactant in the convergence region will tend to damp out surface renewal eddies, thus further increasing the thermal boundary layer thickness which leads to further surface cooling [6]. Here it is assumed that the existence of surfactant does not significantly change the rate of evaporative flux from the surface [19]. Although this may be an explanation for the existence of the cool band, it does not explain the existence of Langmuir circulation at such low wind speeds where there are apparently no gravity-capillary waves [1]. If these cool bands are in fact evidence of Langmuir circulation, then we must invoke some other mechanism by which the surface shear-induced spanwise (perpendicular to the mean wind direction) vorticity is reoriented into the streamwise direction. It seems therefore, that this observation will require further explanation.

In Fig. 15.1**b**, a snapshot taken later in time, it is evident that much brighter (warmer) cellular-like structures seem to erupt onto the surface. Close examination of these structures shows that they consist of temperature variations possessing length scales significantly smaller than those observed earlier in time in Fig. 15.1**a**. A plausible explanation for this is that certain local regions of the surface are swept clean of surfactant as the wind continues to blow. Thus the brightest patches on the surface are interpreted as the cleanest regions. This is very reminiscent of the observations of Saylor et al. [19] who showed, using thermal convection experiments, that surfaces which are free of surfactants are warmer than contaminated surfaces and that such regions of the surface exhibit very small scale spatial variations in temperature. Again, this is almost certainly due to the removal of the damping effect associated with the surface elasticity provided by the surfactant. In addition, we observe again the cool dark band extending diagonally over the lower half of the image. Examination of the inner structure within the band reveals that the small scale temperature structures appear to be almost entirely eliminated and the band appears wider and more robust than the one shown earlier in Fig. 15.1**a**. This is certainly consistent with the supposition that naturally occurring ambient surfactant continues to accumulate in convergent regions as the wind continues to stress the surface. Also, as indicated by the vectors drawn in the figure, viewing the movie sequence associated with these images reveals that the flow pattern discerned from the movement of the thermal structures is clearly that of a convergent motion associated with the cool band.

In Fig. 15.2, a typical IR image from the 4.0 ms^{-1} case is shown. The so-called fish-scale structure is again apparent but appears at a smaller lateral scale compared to similar structures observed at 1.9 ms^{-1}. This reduction in scale is consistent with the increase in the friction velocity. In addition, a cold dark band appears as it did in the 1.9 ms^{-1} case. The

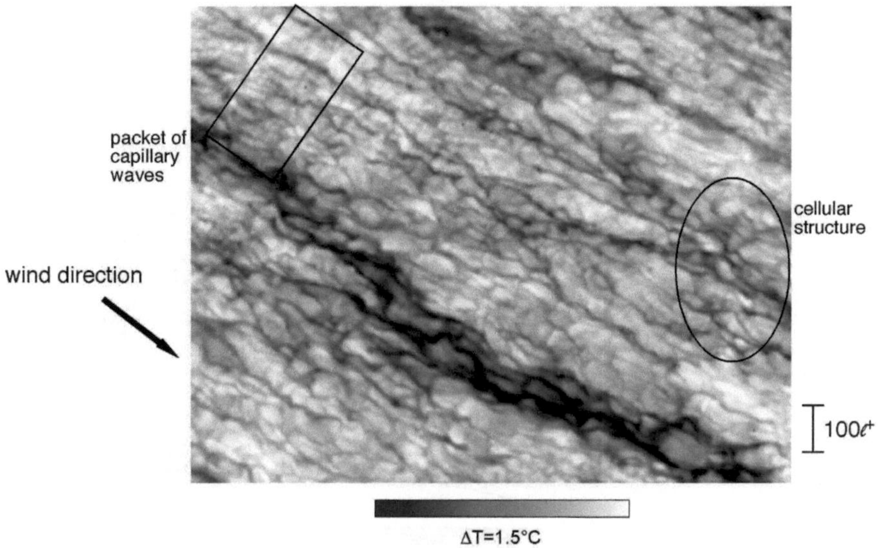

Figure 15.2. Infrared images of the water surface with a wind speed of 4.0 ms^{-1} (Run 32). Indicated in the figure are regions where capillary wave activity is apparent. A cool band is evident, and the scale of the cellular structure is clearly smaller compared to the scales at the lower wind speed of Figure 15.1.

newest aspect emerging in the IR images at this wind speed is the clear appearance of gravity waves which are relatively hard to see in the still image of Fig. 15.2, but are easily recognized in the movies. Their phase speed is markedly larger than the surface drift speed, and we expect their phase speed and wavelength to be consistent with deep water gravity wave systems [5]. As was noted in previous work [5], the most likely reasons for the visibility of gravity waves in IR imagery are: (1) The stretching and compression of the surface cellular structure at troughs and crests respectively; (2) The possible changes in surface emisivity as the orientation of the free surface changes, and (3) The tilting of surface, which may tend to reduce the *apparent* length of the surface structures as their true length scale is projected onto the plane of the IR imager. Although not easily discerned, we have also pointed out small warm bands oriented in a direction perpendicular to the wind direction which we believe are capillaries. No evidence for any wave breaking was found at this wind speed.

In Fig. 15.3 an IR snapshot from the 5.0 ms^{-1} wind speed case is shown. Here an indication of where we believe the wave crests are located in the image is given. The wavelength is longer than that for 3.997 ms^{-1}, and it also appears that the wave crests have substantially steepened, which is revealed very clearly in movies for this case. Also, there is no clear evidence of wave-breaking at this wind speed. At 7.1 ms^{-1}, (Fig. 15.4) the

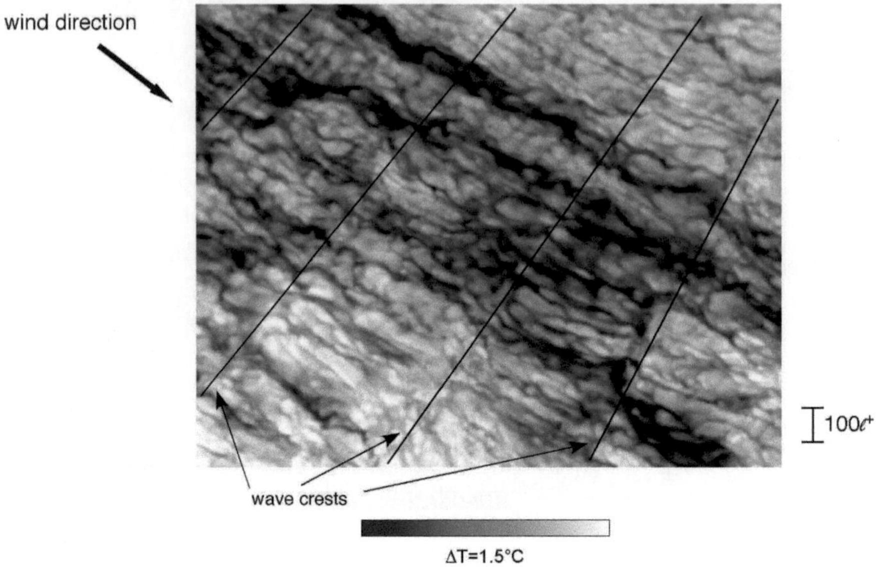

Figure 15.3. Infrared images of the water surface with a wind speed of 5.0 ms^{-1} (Run35). The location of wave crests is indicated.

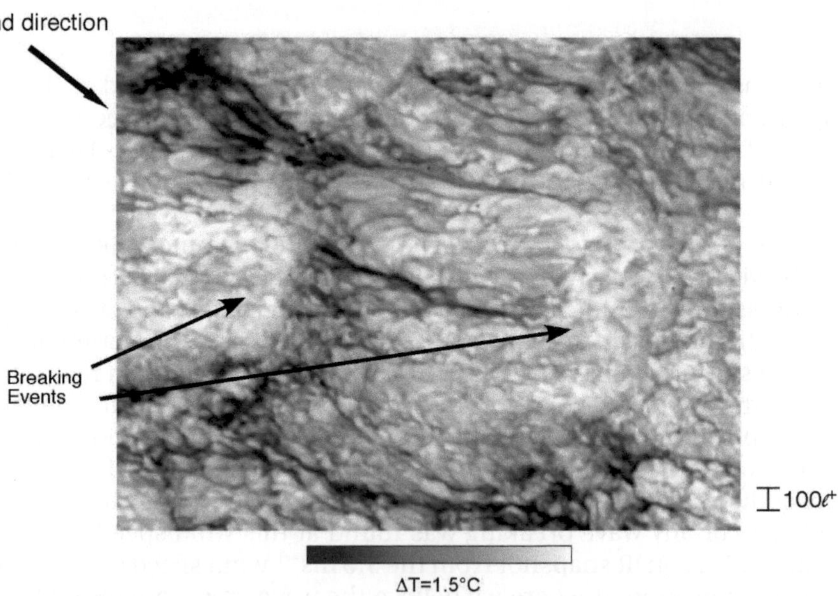

Figure 15.4. Infrared images of the water surface with a wind speed of 7.0 ms^{-1} (Run 33). The location of breaking events is indicated.

Figure 15.5. Infrared images of the water surface with a wind speed of 10.1 ms^{-1} (Run 34).

first evidence of wave-breaking appears, with two such events visible in the image. These are assumed to be breaking waves since warmer patches of fluid are seen behind the crests, indicating that the micro-breaker has left behind turbulence which exposes fluid whose temperature is near that of the bulk [27]. At the highest wind speed (10.1 ms^{-1}) shown in Fig. 15.5, the surface temperature pattern seems nearly isotropic in its appearance. There seems to be little or no indication in the still images, or in the movies, of the wind direction. It seems reasonable to suppose that this is due to the fact that wave breaking events are so numerous at this wind speed that the surface temperature pattern is simply a reflection of the isotropic nature of the turbulence generated behind such events.

15.3.2 Stable Cases

In these experiments, we also explored the case where the air temperature was substantially warmer than the water so that the heat flux was into the water column. This is the so-called *warm-skin* case. We note that the temperature differences in the stable cases are on the order of 0.25°C, substantially smaller than for the the unstable cases. In Figures 15.6-15.8 snapshots from three such cases alongside cool-skin cases at exactly the same wind speed and friction velocity are displayed. In Fig. 15.6, the case of 2.95 ms^{-1}is shown. Several features of the warm-skin image are immediately apparent as follows: (1) The cellular structures, or fish-scales, are cool relative to the surrounding fluid and are very clearly defined; (2) Capillaries appear clearly as warm lines perpendicular to the flow direction, and (3) Long warm bands appear as indicated in Fig. 15.6**a**. The existence of the cellular fish-scale structure in the stable case is strong evidence that such structures are due to a shear driven instability. This observation also directly confirms numerical results [5] in which such structures were shown to exist in the absence of buoyancy. We note in passing that one may get the impression when viewing movies of the thermal structure of the interface at low wind speeds that buoyant plumes appear to rise, and are then stretched in the direction of the wind as they encounter surface shear [5]. While this may perhaps be an explanation for the appearance of the cellular structures at very low wind speeds where Ri may be near zero or somewhat larger than one, the present observations definitively show that shear alone can create *fish scale* structure. This should not be too surprising since in all cool and warm skin cases presented here, the Richardson number was always significantly below one, indicating that buoyancy effects were not appreciable.

In the corresponding cool-skin case (Fig. 15.6**b**) similar features appear but there are obvious differences: (1) The cellular structures are not as clearly defined and close examination reveals that the warmest cells have an inner structure (see Fig. 15.1**b**) consisting of very fine-scale turbulence and (2) The elongated bands are cool as we have noted above. The most important difference between these two cases is their stability. The cool skin case is unstable in the sense that if there were no shear, the cool thermal boundary layer would form an unstable turbulent layer [2, 11, 13] in which cold fluid would sink into the warmer bulk, and warmer plumes would rise to the surface. On the other hand, the warm-skin is stable, since a fluid particle at the interface, if displaced downward from its initial position, would return back to this position as it encounters cooler, denser sub-surface fluid. In the absence of shear, the warm skin layer could not spontaneously break into convectively driven turbulence. This explains the absence of the fine scale thermal structure in the warm thermal boundary layer, which is clearly evident in the cool-skin case as described above. In addition, the elongated warm bands in the warm-skin case are the analogues of the cool bands in the cool-skin case: they are due to a region of convergence in which the thermal boundary boundary layer is thickened which, assuming the heat flux *into* the surface from the air remains fixed, leads to a warmer surface temperature.

In Fig. 15.7 the two cases are shown side by side for a wind speed of 5.0 ms^{-1}. In the warm skin case (Fig. 15.7**a**), the wave crest locations are very evident as they are in the companion cool skin example. The cellular structure has become noticeably smaller in lateral scale as the friction velocity has increased, and the same *warm* bands described in Fig. 15.6**a** are evident. At 7.1 ms^{-1} (Fig. 15.8**a**) there are two wave crests which are

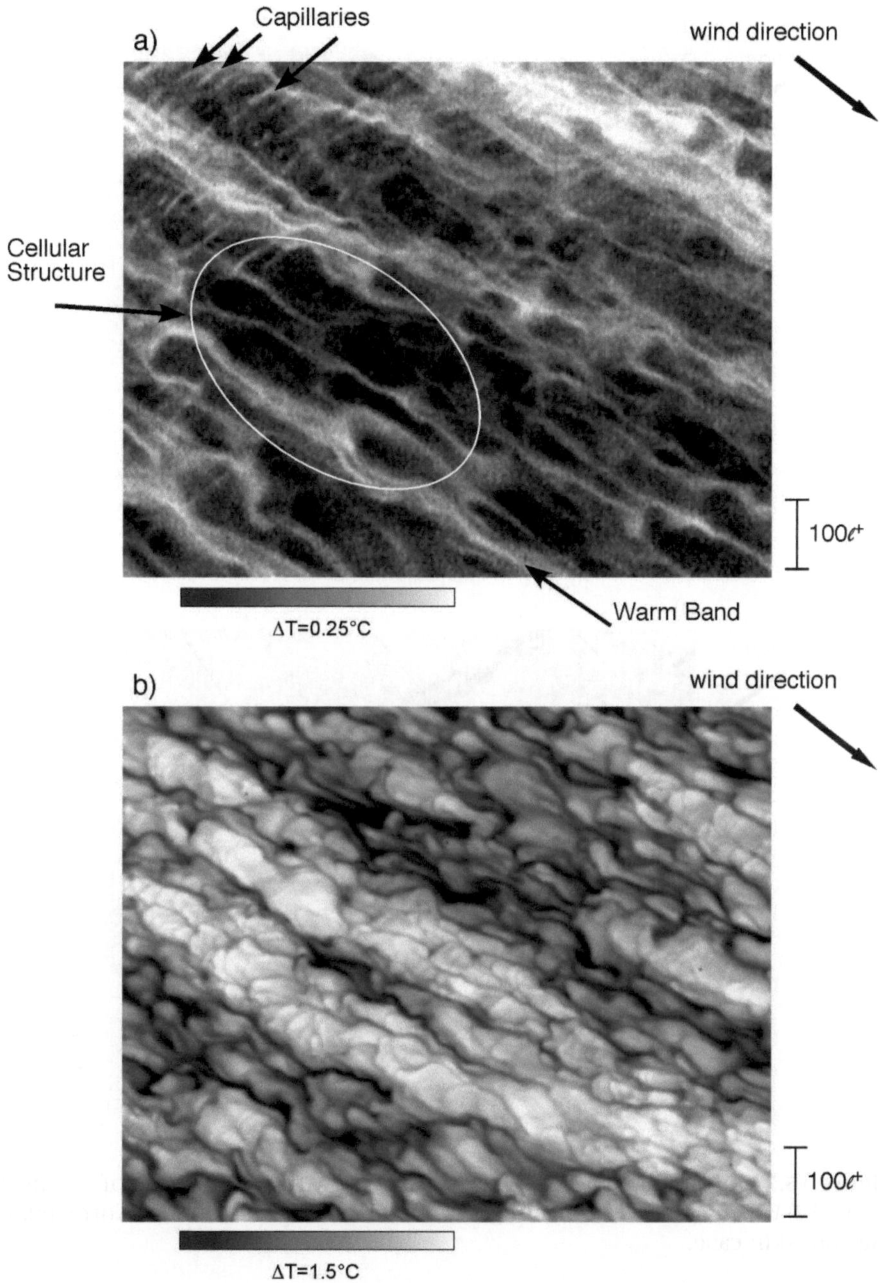

Figure 15.6. Infrared images of the water surface with a wind speed of 3.0 ms^{-1}. **a** (Run 11) Warm-skin case indicating capillary wave structure and cellular structure. **b** (Run 36) Corresponding cool-skin case.

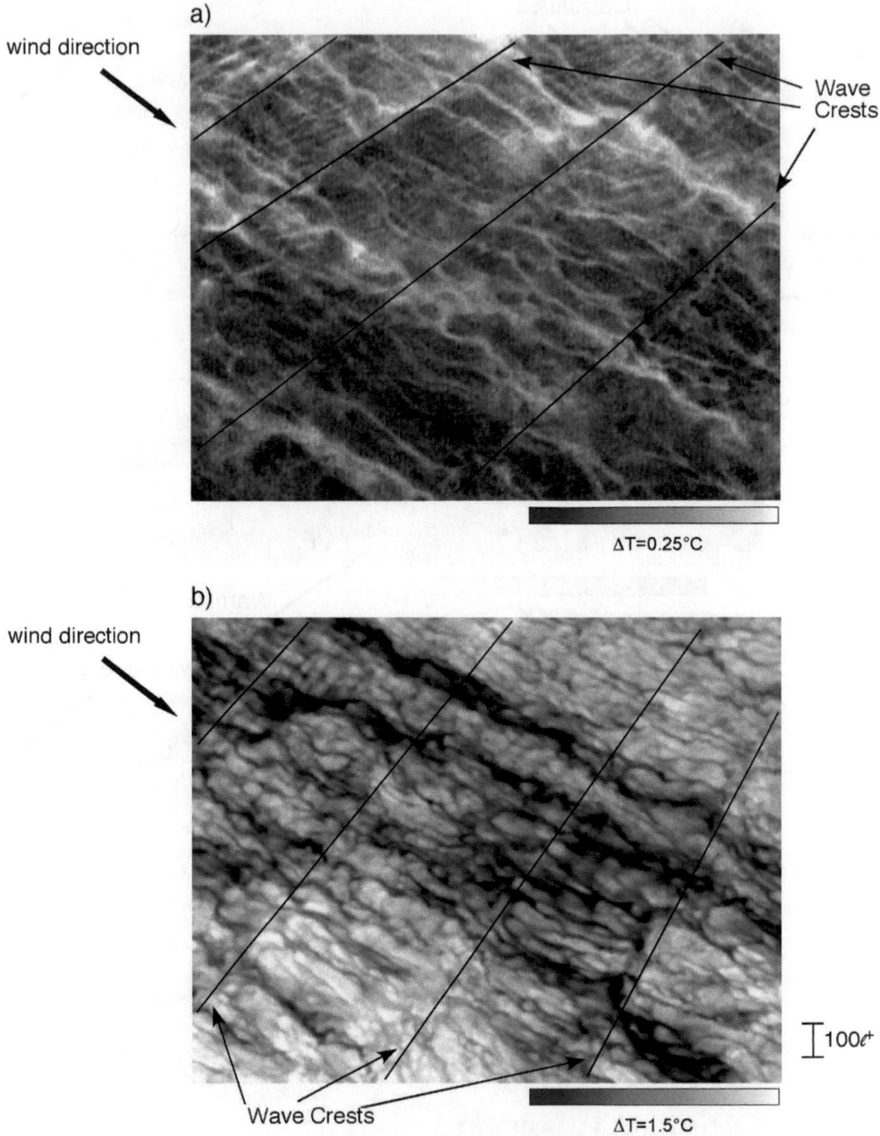

Figure 15.7. Infrared images of the water surface with a wind speed of 5.0 ms^{-1}. **a** (Run12) Warm-skin case indicating wave crest locations. **b** (Run 35) Corresponding cool-skin case.

readily apparent in the warm skin case. Note that these crests seem to be marked by a series of dark *spots* which we have delineated in the figure. These dark spots are almost always present, and in reviewing movies of this case the spots were observed to *move with the wave* which indicates that they are not associated with fluid particle motion: instead we speculate that they may be due to specular reflections in the vicinity of wave crests. In addition, there are clearly regions behind wave crests which are noticeably cooler than the surrounding fluid. This is evidence of micro-breaking in which cooler bulk fluid is being mixed with surface fluid. An interesting aspect of these images, which is very apparent in the warm-skin case, is the appearance of cellular structure *between* the wave crests. The size of these structures is not noticeably different from those in the 5.0 ms^{-1} case in which the friction velocity is much smaller. This leads us to believe that this may be due to a wave sheltering effect. If this is the case, the shear stress in the troughs is expected to be substantially lower than at the crests. Therefore, if the lateral scale of the thermal cells is inversely proportional to the friction velocity, we would expect larger cells compared to what we might expect from the average value of the friction velocity. Future experiments will be required to confirm these speculations.

15.4 Summary and Conclusions

Observations of the thermal structure of an air-water interface were made using a high resolution IR imager at the University of Miami ASIST facility for wind speeds from about 2 to 10 ms^{-1}. We have made observations for both the *unstable* case in which the heat flux was directed from the water to the atmosphere, and the *stable* case in which the heat flux was directed into the water column. For the unstable case, for which the flux based Richardson number varied between approximately 10^{-2} to 10^{-5}, these observations seemed to be consistent with those of previous investigations which revealed thermal cells with lateral length scales which varied inversely with the friction velocity. At low wind speeds a Langmuir-like cell structure was clearly observed even though no gravity waves were present. As the wind speed was increased, the surface thermal pattern revealed an increasingly complex structure: (1) Gravity-capillary waves at about 3 ms^{-1} (2) Steep gravity waves at 5 ms^{-1} (3) Breaking events at 7 ms^{-1} and (4) Isotropic thermal structure at 10 ms^{-1}.

For the stable cases, for the wind speeds between 3 and 7 ms^{-1}, the thermal structure was in many ways the mirror image of the temperature field for the unstable cases. We find, for example, that the small cellular structures which are warm in the unstable case, are cool in the stable case, that the cool bands which we identify with a Langmuir-like circulation in the unstable case, appear as warm bands in the stable case, and that

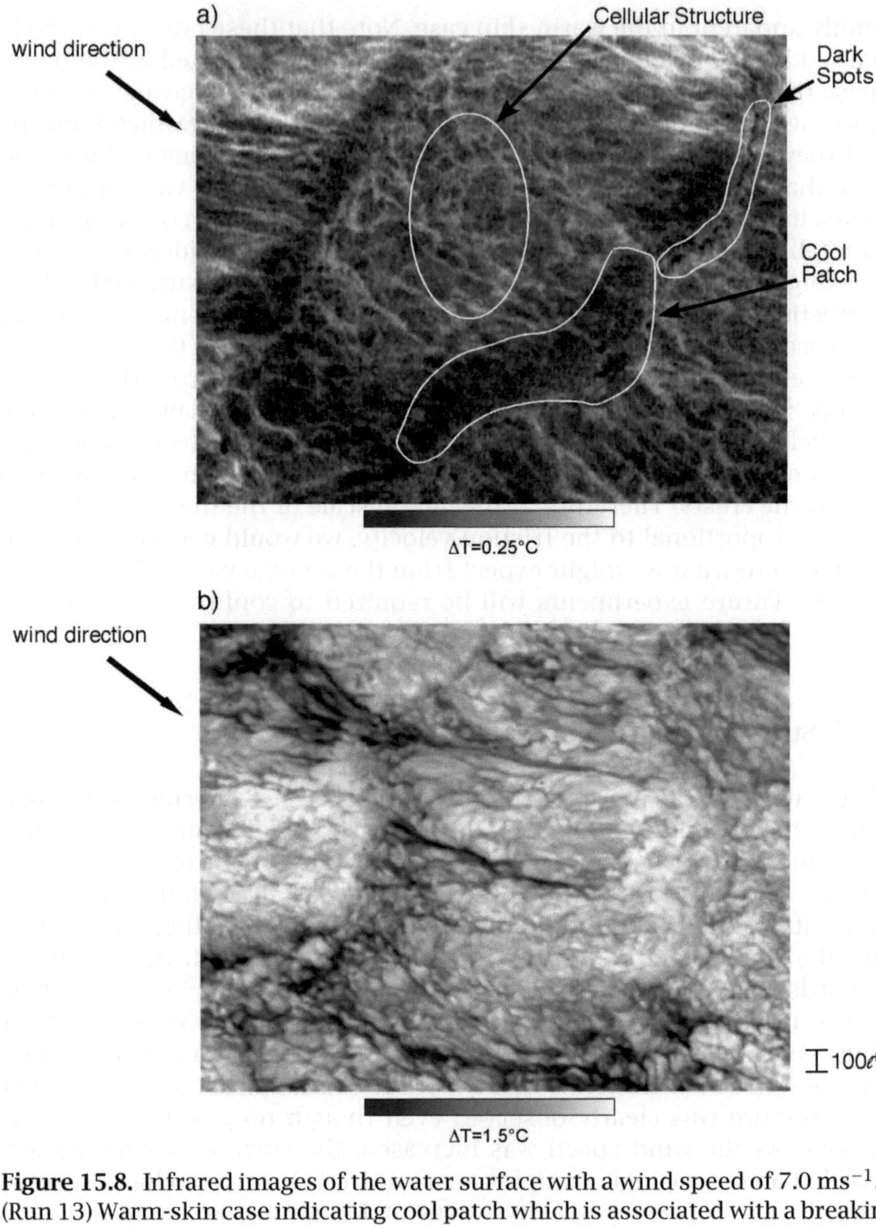

Figure 15.8. Infrared images of the water surface with a wind speed of 7.0 ms^{-1}. **a** (Run 13) Warm-skin case indicating cool patch which is associated with a breaking event. Also noted are the cellular structures between wave crests and dark spots whose appearance is discussed in the text. **b** (Run 33) Corresponding cool-skin case.

breaking events which appear as warm patches in the unstable case, are cool regions in the stable case. However, in the stable case the thermal structures associated with the so-called fish-scale pattern are more clearly defined compared to the same structures in the unstable case, despite the fact that the magnitude of the temperature fluctuations in the imagery for the former case were smaller than those in later. This is likely due to the absence of buoyancy induced small scale turbulence in the stable case which we expect, and apparently observe, in the unstable case. Also, we observed *dark spots* in the stable case which seemed to very closely follow the wave crests. The cause of these spots is presently unknown.

The existence of fish-scale cellular structure in the stable case is strong evidence that the origin of such structures is the shear stress at the surface itself, with buoyancy effects playing a negligible role in their formation and maintenance. These observations corroborate and strengthen earlier conclusions [5] that the lateral scale of these structures can be used with some confidence to estimate wind stress. Since one objective of this research is to develop possible means to remotely infer such quantities as surface shear, heat flux, and small scale wave statistics, these results also imply that a better chance of achieving this goal may be to examine more closely the thermal structure in the stable case where buoyancy driven small scale turbulence is suppressed, thereby leading to less *noisy* thermal signals.

Acknowledgement. This work was funded by the Office of Naval Research through the Fluid Dynamics Task Area at the Naval Research Laboratory. Nicholas Scott also acknowledges the support of the National Research Council through the Research Associateship Program at the Naval Research Laboratory. We also acknowledge assistance of Prof. Peter Minnett and other staff members at the Rosenstiel School (RSMAS) of the University of Miami. Funding to conduct the experiments was provided to RSMAS by the Office of Naval Research under award N00014-03-1-0384.

Appendix: Table of Parameters for Both the Warm and Cold Skin

Table 15.1. Cool-skin cases: In both Table 15.1 and Table 15.2 below, U,T_A, T_W, $u_\star$, l^+, Q, and Ri are the wind speed in meters per second, the air and water temperatures respectively in degrees centigrade, the friction velocity on the water side in meters per second, the viscous length scale on the water side (as defined in section 15.2) in meters, the net heat flux out of the water surface in Watts per square meter, and the Richardson number (as defined in section 15.2).

Run	U	T_A	T_W	$u_\star$	l^+	Q	Ri
31	1.925	25.98	40.22	2.18×10^{-3}	4.59×10^{-4}	393	7.36×10^{-3}
36	2.951	25.13	40.45	3.12×10^{-3}	3.20×10^{-4}	601	2.66×10^{-3}
32	3.977	25.54	40.32	4.31×10^{-3}	2.32×10^{-4}	850	1.04×10^{-3}
35	5.0031	25.64	40.27	5.74×10^{-3}	1.74×10^{-4}	1144	4.45×10^{-4}
33	7.0552	25.59	40.28	9.32×10^{-3}	1.07×10^{-4}	1764	9.88×10^{-5}
34	10.133	25.52	40.24	16.5×10^{-3}	6.06×10^{-5}	2592	1.48×10^{-5}

Table 15.2. Warm-skin cases.

Run	U	T_A	T_W	$u_\star$	l^+	Q	Ri
11	2.951	23.962	14.231	3.12×10^{-3}	3.20×10^{-4}	-140	-6.22×10^{-4}
12	5.003	23.085	10.988	5.74×10^{-3}	1.74×10^{-4}	-343	-1.33×10^{-4}
13	7.055	—	—	9.32×10^{-3}	1.07×10^{-4}	—	—

References

[1] A. D. D. Craik and S. Leibovich. Rational model for langmuir circulations. *Journal of Fluid Mechanics*, 73:401–426, 1976.

[2] T. D. Foster. Intermittent convection. *Geophysical Fluid Dynamics*, 2: 201–217, 1971.

[3] C. S. Garbe, H. Spies, and B. Jähne. Estimation of surface flow and net heat flux from infrared images sequences. *Journal of Mathematical Imaging and Vision*, 19:159–174, 2003.

[4] C. S. Garbe, U. Schimpf, and B. Jähne. A surface renewal model to analyze infrared image sequences of the ocean surface for the study of air-sea heat and gas exchange. *Journal of Geophysical Research*, 109(C08S15), 2004. doi:10.1029/2003JC001802.

[5] R. A. Handler, G. B. Smith, and R. I. Leighton. The thermal structure of an air-water interface at low wind speeds. *Tellus Series a-Dynamic Meteorology and Oceanography*, 53(2):233–244, 2001.

[6] R. A. Handler, R. I. Leighton, G. B. Smith, and R. Nagaosa. Surfactant effects on passive scalar transport in a fully developed turbulent flow. *International Journal of Heat and Mass Transfer*, 46(12):2219–2238, 2003.

[7] B. Jähne, P. Libner, R. Fischer, T. Billen, and E. J. Plate. Investigating the transfer processes across the free aqueous viscous boundary layer by the controlled flux method. *Tellus*, 41B:177–195, 1989.

[8] A. T. Jessup and C. J. Zappa. Defining and quantifying microscale wave breaking with infrared imagery. *Journal of Geophysical Research*, 102(C10):23,145–23,153, 1997.

[9] A. T. Jessup, C. J. Zappa, M. R. Loewen, and V. Hesany. Infrared remote sensing of breaking waves. *Nature*, 385(6611):52–55, 1997.

[10] K. P. Judd, S. Phongikaroon, G. B. Smith, and R. A. Handler. Thermal structure of clean and contaminated free-surfaces subject to an impinging gas jet. *Experiments in Fluids*, 38(1):99–111, 2005.

[11] K. B. Katsaros, W. T. Liu, J. A. Businger, and J. E. Tillman. Heat-transport and thermal structure in interfacial boundary-layer measured in an open tank of water in turbulent free convection. *Journal of Fluid Mechanics*, 83:311–335, 1977.

[12] S. J. Kline, W. C. Reynolds, J. A. Schraub, and P. W. Runstadler. The structure of turbulent boundary layers. *Journal of Fluid Mechanics*, 30:741–773, 1967.

[13] R. I. Leighton, G. B. Smith, and R. A. Handler. Direct numerical simulations of free convection beneath an air-water interface at low rayleigh numbers. *Physics of Fluids*, 15(10):3181–3193, 2003.

[14] G. O. Marmorino and G. B. Smith. Bright and dark ocean whitecaps observed in the infrared. *Geophysical Research Letters*, 32(11):L11604, 2005.

[15] G. O. Marmorino, G. B. Smith, and G. J. Lindemann. Infrared imagery of ocean internal waves. *Geophysical Research Letters*, 31(11):L11309, 2004.

[16] G. O. Marmorino, G. B. Smith, and G. J. Lindemann. Infrared imagery of large-aspect-ratio langmuir circulation. *Continental Shelf Research*, 25(1):1–6, 2005.

[17] E. D. McAlister and W. McLeish. Heat transfer in the top millimeter of the ocean. *Journal of Geophysical Research*, 74:3408–3414, 1969.

[18] P. J. Minnett, R. O. Knuteson, F. A. Best, B. J. Osborne, J. A. Hanafin, and O. B. Brown. The marine-atmospheric emitted radiance interferometer: A high-accuracy, seagoing infrared spectroradiometer. *Journal of Atmospheric and Ocean Technology*, 18(6):994–1013, 2001.

[19] J. R. Saylor, G. B. Smith, and K. A. Flack. The effect of a surfactant monolayer on the temperature field of a water surface undergoing evaporation. *International Journal of Heat and Mass Transfer*, 43 (17):3073–3086, 2000.

[20] U. Schimpf, C. S. Garbe, and B. Jähne. Investigation of transport processes across the sea surface microlayer by infrared imagery. *Journal of Geophysical Research*, 109(C08S13), 2004. doi:10.1029/2003JC001803.

[21] C. R. Smith and S. P. Metzler. The characteristics of low-speed streaks in the near-wall region of the turbulent boundary layer. *Journal of Fluid Mechanics*, 129:27–54, 1983.

[22] G. B. Smith, R. J. Volino, R. A. Handler, and R. I. Leighton. The thermal signature of a vortex pair impacting a free surface. *Journal of Fluid Mechanics*, 444:49–78, 2001.

[23] A. V. Soloviev and P. Schlüssel. Parameterization of the skin of the ocean and of the air-ocean gas transfer on the basis of modeling surface renewal. *Journal of Physical Oceanography*, 24:1339–1346, 1994.

[24] F. Veron and W. K. Melville. Experiments on the stability and transition of wind-driven water surfaces. *Journal of Fluid Mechanics*, 446: 25–65, 2001.

[25] B. Ward and M. Donelan. Thermometric measuremnets of the molecular sublayer at the air-water interface. *Geophysical Research Letters*, 33(L07605), 2006. doi:10.1029/2005GL024769.

[26] C. J. Zappa, A. T. Jessup, and H. Yeh. Skin layer recovery of free-surface wakes: Relationship to surface renewal and dependence on heat flux and background turbulence. *Journal of Geophysical Research*, 103(C10):21,711–21,722, 1998.

[27] C. J. Zappa, W. E. Asher, A. T. Jessup, J. Klinke, and S. R. Long. Microbreaking and the enhancement of air-water transfer velocity. *Journal of Geophysical Research*, 109(C08S16), 2004. doi:10.1029/2003JC001897.

16

Estimating the Viscous Shear Stress at the Water Surface from Active Thermography

Christoph S. Garbe[1], Kai Degreif[2], and Bernd Jähne[1,2]

[1] Interdisciplinary Center of Scientific Computing (IWR)
University of Heidelberg, Germany
Christoph.Garbe@iwr.uni-heidelberg.de
[2] Institute for Environmental Physics (IUP)
University of Heidelberg, Germany

Abstract A novel technique is presented that makes it possible to measure the viscous shear stress τ_μ from active thermography. With a CO_2 laser, patterns are written to the sea surface. This temperature structure is distorted by the linear velocity profile in the viscous boundary layer. Due to the non-zero penetration depth of both the laser and the infrared camera, this vertical velocity profile can be resolved. By resolving the velocity profile, the viscous shear stress can be extracted from the recorded image sequences. At the same time, the flow field at the water surface can be measured accurately. Estimating both quantities is only possible by modeling the imaging process as well as the velocity profile in the boundary layer. The model parameters can then be computed in a standard parameter estimation framework. This novel technique was tested both on simulated data and on measurements conducted in a small annular wind-wave flume. The friction velocity computed in this fashion compared favorably to independent measurements. Although not tested yet, this technique should be equally applicable to field measurements.

16.1 Introduction

The transport of energy, momentum and mass across the air-sea interface are central questions in the study of air-sea interaction and ocean atmosphere modeling. Previously, techniques relying on thermographic image sequence analysis for measurung the transport of energy, or heat, have been presented [11, 12]. A long-standing effort has been put into measuring the transfer of mass across the air water interface [5, 6, 24, 29, 42] as well as parameterizing this transfer [28, 38, 39, 40, 41]. Due to the advantages of using heat as a proxy for mass transfer, research has been undertaken to scale from the transfer velocity of heat to that of gas [1, 16, 18, 25, 34]. A review on the subject of air-water mass transfer can be found in Jähne and Haußecker [23].

C.S. Garbe, R.A. Handler, B. Jähne (eds.): Transport at the Air Sea Interface
pp. 223-239, 2007,

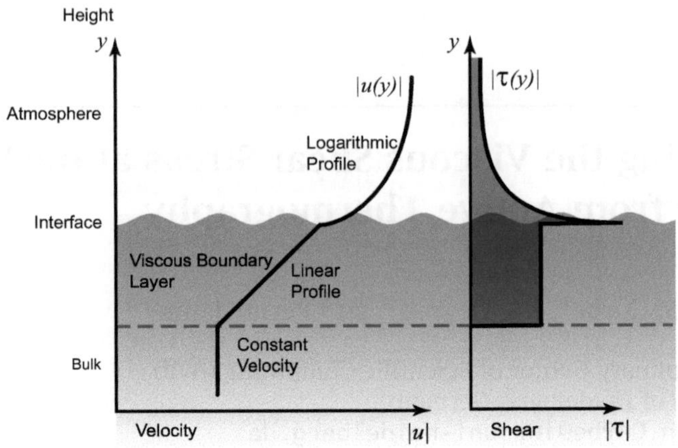

Figure 16.1. Sketch of the velocity profile $u(z)$ and the shear τ at the sea surface. In the bulk, the velocity is considered constant with vanishing shear τ. In the viscous boundary layer, the velocity changes linearly as seen in plane Couette flow, leading to a constant shear. Above the sea surface, the wind profile is logarithmic and the resulting shear $\tau \sim 1/z$.

Apart from the transport of energy and mass, the transport of momentum is of great importance for ocean-atmosphere modeling as well as for understanding the processes at the air-sea interface. It is the driving force in a number of interfacial processes.

The transport of momentum is expressed by the momentum flux j_m which is equivalent to the shear τ at the interface. The momentum flux j_m is defined by

$$j_m = \tau = \rho u_\star^2 = \mu \cdot \left.\frac{\partial u}{\partial z}\right|_0 , \tag{16.1}$$

where $u_\star$ is the friction velocity, ρ is the density of water, μ is the viscosity and $\partial u/\partial z|_0$ indicates the horizontal velocity gradient with respect to depth, directly at the air-water interface.

Generally, the momentum flux τ is partitioned into viscous stress τ_μ and wave-induced stress τ_w. The wave induced stress, also known as form stress, is due to pressure force acting on the slope of waves. This stress τ_w is strongly connected to the wave age β, which is defined as the momentum transfer rate from wind to waves per unit wave momentum. A recent analysis of this quantity in the air-sea momentum flux budget analysis can be found in Kukulka and Hara [27]. Banner and Peirson [2] conducted experiments to measure the partition of the total momentum transfer into the viscous and the wave-induced stress. Contrary to Okuda et al. [33] they found a large partition of more than 50% of the total momentum transfer to be made up by the wave-induced stress. More recently Uz et al. [36] found the partition of total momentum flux to wave induced stress to be 64% for typical conditions.

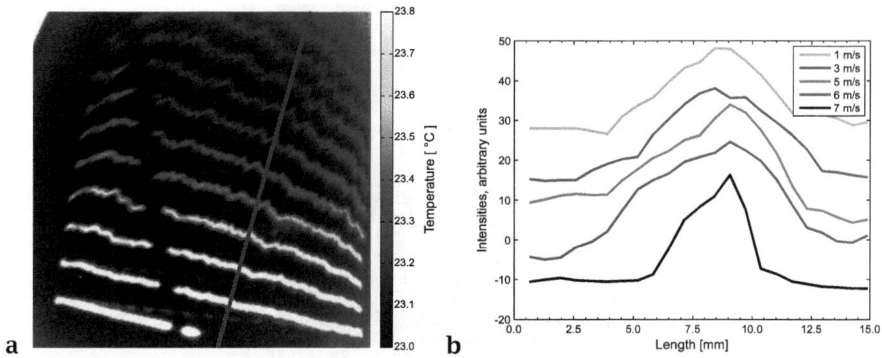

Figure 16.2. In **a** a thermal image at 5.0 m/s wind speed directed from bottom to top is shown. Straight lines are written with a CO_2 laser at the bottom of the frame. Indicated by the blue circle is a section of the profile of the blue line for different wind speeds, which are presented in **b**. The profiles are shifted for better visualization and thus presented in arbitrary units of image intensities.

While Okuda et al. [33] measured the viscous stress τ_μ from hydrogen bubbles, Banner and Peirson [2] performed their tangential stress measurements with a PIV technique using particles of diameters ranging from 20-60 μm. The discrepancy in the results between Okuda et al. [33] and Banner and Peirson [2] can be attributed to a large part in uncertainties in the use of hydrogen bubbles for the flow visualization. The resulting bubbles in measurements by Okuda et al. [33] and in previous measurements by McLeish and Putland [31] are quite large which leads to a number of difficulties, as discussed in detail by Banner and Peirson [2].

While Banner and Peirson [2] produced some excellent results to the extent of measuring viscous stress in relation to wave phase, their technique relying on PIV measurements is not applicable to the field. This represents a significant drawback and opens room for further investigations. In this chapter, a novel technique will be presented that relies on active thermography for measuring the viscous stress τ_μ [13]. Currently, only laboratory measurements have been conducted. They will be presented in Section 16.5. Due to the simple experimental set-up, which is presented in Section 16.4, this technique should be well adapted to field use.

This technique is based on the assumption of plane Couette flow in the viscous boundary layer. The resulting linear velocity profile is sketched in Figure 16.1. Information concerning the velocity structure with depth cannot be recovered from passive thermography. Therefore, active thermography is necessary, as shall be described in the next section.

Figure 16.3. A simulation of the intensities recorded by the IR camera a fixed time t after the pattern has been written to the surface. Wind induces a shear along the x-axis (from top to bottom). From left to right, the velocity gradient dx/dz increases. It can clearly be seen that the image intensity is smeared over a wide area for strong velocity gradients. Also, it should be noted that the intensity maximum remains at a fixed location at the trailing edge of the written structure.

16.2 Motion Estimation for Active Thermography

Estimating motion from image sequences represents an inverse problem [14]. As such, it is important to derive a motion model that takes into account the underlying fundamental processes for measuring motion accurately from active thermography. In Figure 16.2**a** an image from active thermography is shown. Here, a line pattern is written onto the interface which is advected away from the heated location downwind. Principally any conceivable pattern can be written to the interface. Here a line pattern was chosen for illustration purposes. For applications in the field a simple line pattern is disadvantageous, since velocities tangential to the lines cannot be estimated due to the aperture problem [19, 20].

As can be seen from Figure 16.2**a**, the appearance of the structures written with the laser are distorted as they are advected downstream. The leading edge appears to remain sharp, while the trailing edge blurs increasingly with time. This change of shape of the written temperature structure becomes more pronounced at stronger wind-speeds, associated with a higher shear at the interface. This can be seen in Figure 16.2**b** where profiles of the temperature structure are shown for different wind speeds. In these experiments, a surface film was present, effectively suppressing waves. The whole wind induced shear stress is thus transferred to viscous stress. This same effect can be observed in Figure 16.3. Here the results of a simulation of this effect is visualized for different velocity depth gradients related to different shear rates.

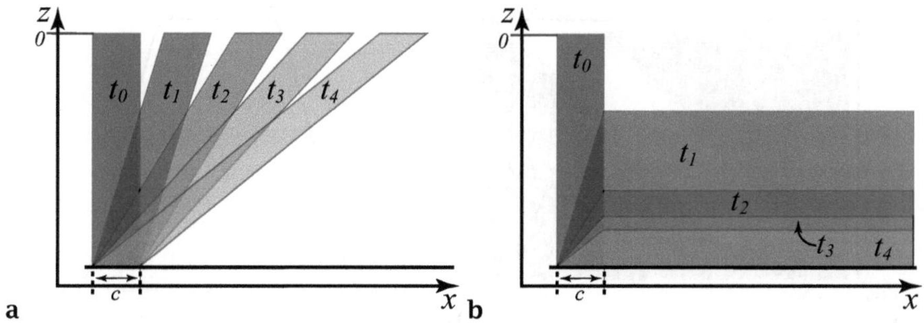

Figure 16.4. Sketch of a marker in plane Couette Flow written with at time t_0 in subsequent time steps $t_1 - t_4$ in **a**. The marker is sheared due to the flow. Shown in **b** is the depth integration of the marker, as visualized with the camera.

These blurring effects have to be incorporated into the motion model. Otherwise inaccurate motion fields will be estimated. First, the equations of motion will be derived not taking into account the decay of infrared radiation with depth due to absorption in water. This is of course a very crude approximation, as the penetration depth in water is only a few µm. Hence in a second step, this absorption will be modeled explicitly. Only from this motion model, the viscous shear stress τ_μ can be estimated from the image sequences.

16.2.1 Motion Model without Infrared Absorption in Water

The velocity profile in the viscous boundary layer is considered to decrease linearly with depth. This type of flow can thus be approximated by plane Couette flow. In this configuration, the flow in between two plates is driven by the relative velocity of the boundary plates.

In active thermography a laser is used for heating up patches of water. Due to its penetration depth, the laser heats up a three dimensional structure inside the viscous boundary layer. This structure becomes sheared due to the velocity profile. An infrared camera is used for imaging these heated patches of water. The projection of the three dimensional structure leads to an integration with depth in the boundary layer. The shearing of the thermal structure and the subsequent integration leads to an apparent smearing of the structure with an accompanying decrease in intensity. This process is sketched in Figure 16.4. Results of a computer simulation of this process are shown in Figure 16.5 **a** and **b**.

In the following, it is assumed that the laser heats up the thermal boundary layer instantaneous at the time $t = t_0 = 0$. Once the structure is written at time t_0, it is sheared due to the velocity profile as indicated in Figure 16.4 in successive time steps $t_1 - t_4$. In the imaging process the dimension of depth z is lost through integration. The projection of

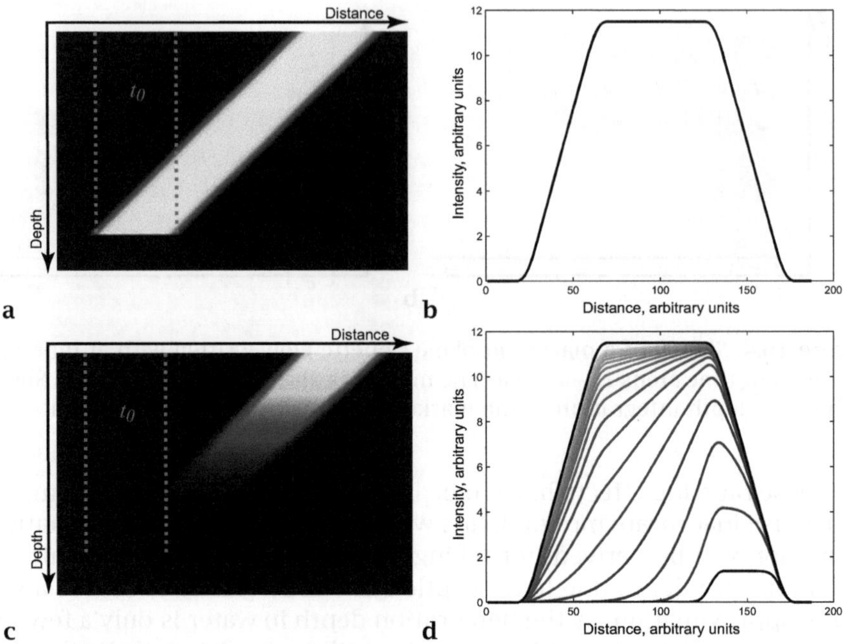

Figure 16.5. Simulation of the viscous boundary layer heated up locally by a CO_2 laser at time t_0 (indicated by blue lines in **a** and **c**. Shown is the shearing of the temperature structure due to the velocity gradient. In **b** and **d** the depth integrated intensities are shown. In **a** and **b** the intensities remain constant with depth, in **c** and **d** the intensities are attenuated due to the Lambert-Beer law. In **d** plots are presented for different penetration depths. In all plots the wind direction is from left to right.

intensities T onto the surface at $z = \delta$ is given by

$$T(x,t) = \int_{\frac{\delta\cdot(x-c)}{U\cdot t}}^{\frac{\delta\cdot x}{U\cdot t}} 1\,dz = \frac{x\cdot\delta}{t\cdot U} - \frac{(x-c)\cdot\delta}{t\cdot U} = \frac{c\cdot\delta}{t\cdot U} = \frac{\mu\cdot c}{t\cdot\tau} \qquad (16.2)$$

Here μ is the viscosity, τ is the shear stress, δ is the boundary layer thickness, U is the flow at the top of the boundary layer and c denotes the width of the area marked, as can be seen in Figure 16.4.

Differentiating Equation (16.2) with respect to time leads to

$$\frac{dT}{dt} = \frac{d}{dt}\left(\frac{c\cdot\delta}{t\cdot U}\right) = -\frac{1}{t}T. \qquad (16.3)$$

Estimating the velocity of the intensity structures subject to a plane Couette type shear flow with a linear velocity gradient can thus be computed by solving the differential equation $dT/dt = -(t)^{-1}T$ which can be written as the motion constraint equation

$$\frac{dT}{dt} = u_1 \frac{\partial T}{\partial x} + u_2 \frac{\partial T}{\partial y} + \frac{\partial T}{\partial t} = -\frac{1}{t} T. \tag{16.4}$$

Here $\mathbf{u} = [u_1, u_2]^\top$ indicates the flow field at the water surface. Rewriting this equation in vector notation leads to

$$\frac{dT}{dt} = d^\top \cdot p = \left[\tfrac{1}{t} T \ \tfrac{\partial T}{\partial x} \ \tfrac{\partial T}{\partial y} \ \tfrac{\partial T}{\partial t} \right] \cdot \left[1 \ u_1 \ u_2 \ 1 \right]^\top = 0\,. \tag{16.5}$$

This motion constraint equation can be extended to incorporate isotropic conduction of the heat pattern. The combination of these two processes can be expresses as the following differential equation

$$\frac{dT}{dt} = u_1 \frac{\partial T}{\partial x} + u_2 \frac{\partial T}{\partial y} + \frac{\partial T}{\partial t} = \kappa \Delta T - \frac{1}{t} T, \tag{16.6}$$

where κ is the thermal diffusivity and t is the time since the thermal structures were written. Rewriting this equation in vector notation leads to

$$\frac{dT}{dt} = d^\top \cdot p = \left[\tfrac{\partial^2 T}{\partial x^2} + \tfrac{\partial^2 T}{\partial y^2} \ \tfrac{1}{t} T \ \tfrac{\partial T}{\partial x} \ \tfrac{\partial T}{\partial y} \ \tfrac{\partial T}{\partial t} \right] \cdot \left[-\kappa \ 1 \ u_1 \ u_2 \ 1 \right]^\top = 0\,. \tag{16.7}$$

This type of equation can be solved straight forward with the local gradient based approach that will be presented in Section 16.2.3. If the exact time t since the thermal structures were written is not available, one might also estimate this parameter. This will reduce the accuracy but an estimation of the velocities u_1 and u_2 should still be more accurate than not taking this term into account. The motion constraint equation for this case is given by

$$\frac{dT}{dt} = d^\top \cdot p = \left[\tfrac{\partial^2 T}{\partial x^2} + \tfrac{\partial^2 T}{\partial y^2} \ T \ \tfrac{\partial T}{\partial x} \ \tfrac{\partial T}{\partial y} \ \tfrac{\partial T}{\partial t} \right] \cdot \left[-\kappa \ \tfrac{1}{t} \ u_1 \ u_2 \ 1 \right]^\top = 0. \tag{16.8}$$

16.2.2 Absorption with Depth

In the previous section, the motion constraint equation for heat structures was derived without taking absorption in the water body into account. For active thermography, a CO_2 laser is used for heating up water parcels. These parcels are visualized with an IR camera sensible in the spectral range from 3-5 μm. In this range, the penetration depth for radiation in water is only approximately 10 μm. Equally, the penetration depth of the CO_2 laser is 11.5μm. This strong absorption needs to be considered in order to gain accurate estimates of velocity.

The absorption of radiation with depth is modeled by Lambert-Beer's law. Integration with depth thus results in

$$T(x,t) = \int_{\frac{\mu(x-c-U_0 t)}{t\tau}}^{\frac{\mu(x-U_0 t)}{t\tau}} e^{-\kappa z}\, \mathrm{d}z = \frac{1}{\kappa} e^{\frac{\kappa\mu(tU_0 - x)}{t\tau}} \cdot \left(e^{\frac{c\kappa\mu}{t\tau}} - 1 \right). \tag{16.9}$$

Differentiating this equation with respect to time yields the following differential equation:

$$\frac{dT}{dt} = \left(\frac{\kappa \cdot \mu \cdot x}{t^2 \cdot \tau} + \frac{\kappa \cdot \mu \cdot c}{t^2 \cdot \tau} \left(\frac{1}{e^{-\frac{\kappa\mu c}{t\tau}} - 1} \right) \right) T. \tag{16.10}$$

When not considering absorption, the motion constraint equation was derived to $dT/dt = t^{-1}T$ in the previous section. By noting that

$$\lim_{x \to 0} \frac{x}{e^{-\alpha x} - 1} = -\frac{1}{\alpha}. \tag{16.11}$$

Equation (16.10) reduces to $dT/dt = -t^{-1}T$, which is exactly the constraint equation that was found in the setting without absorption. This formulation is thus consistent to previous findings.

While it is possible to use Equation (16.10) to solve the motion problem, it is generally preferred to approximate that solution. The measurement is conducted by heating up water parcels with a CO_2 laser. The width c of the structure is small compared to the distance x the water parcels traverses during the measurement. Therefore, Equation (16.10) can be approximated by requiring $\lim_{c \to 0}$. This results in

$$\lim_{c \to 0} \frac{dT}{dt} = \left(\frac{\kappa \mu x}{t^2 \tau} - \frac{1}{t} \right) T = \left(\frac{\alpha x}{t^2} - \frac{1}{t} \right) T \quad \text{with} \quad \alpha = \frac{\kappa \mu}{\tau}, \tag{16.12}$$

which can also be written as

$$\frac{dT}{dt} = u_1 \frac{\partial T}{\partial x} + u_2 \frac{\partial T}{\partial y} + \frac{\partial T}{\partial t} = \frac{T}{t} \left(\frac{\alpha x}{t} - 1 \right) = \frac{T}{t} \left(\alpha \sqrt{u_1^2 + u_2^2} - 1 \right). \tag{16.13}$$

The parameters of this constraint equation can again be estimated by writing it in vector notation, leading to

$$\left[\tfrac{T}{t} \; \tfrac{\partial T}{\partial x} \; \tfrac{\partial T}{\partial y} \; \tfrac{\partial T}{\partial t} \right] \cdot \left[\xi_{c \to 0} \; u_1 \; u_2 \; 1 \right]^\top = \mathbf{d}^\top \cdot \mathbf{p} = 0, \tag{16.14}$$

$$\text{with} \quad \xi_{c \to 0} = 1 - \alpha \sqrt{u_1^2 + u_2^2}.$$

The parameter vector $\mathbf{p}$ can now be estimated from the data with the algorithm presented in Section 16.2.3. From the parameters $\xi_{c \to 0}$ and $\mathbf{u} = [u_1, u_2]^\top$ the surface stress $\tau_{c \to 0}$ and thus the friction velocity $u_\star = \sqrt{\tau/\rho}$ can be computed from

$$\tau_{c \to 0}(\xi_{c \to 0}, ||\mathbf{u}||_2) = \frac{\kappa \cdot \mu}{1 - \xi_{c \to 0}} \sqrt{u_1^2 + u_2^2} = \frac{\kappa \cdot \mu}{1 - \xi_{c \to 0}} ||\mathbf{u}||_2 \tag{16.15}$$

$$u_{\star; c \to 0}(\xi_{c \to 0}, ||\mathbf{u}||_2) = \sqrt{\frac{\tau}{\rho}} = \sqrt{\frac{\kappa \cdot \mu}{\rho(1 - \xi_{c \to 0})} ||\mathbf{u}||_2}, \tag{16.16}$$

where $|| \cdot ||_2$ is the L-2 norm. The subscript $c \to 0$ indicates that this solution only holds for the limit case $\lim_{c \to 0}$.

In this section, motion constraint equations were developed that make it possible to estimate the viscous shear stress τ_μ from active thermography. At the same time, accurate estimates of surface flow are also achieved from the same technique. This is due to the fact that both τ and the parameters of motion u_1 and u_2 are estimated as parameters of the motion model.

16.2.3 Gradient Based Approach for Motion Estimation

The motion constraint equation derived for active thermography and isotropic diffusion is given by Equation (16.8). This equation represents an ill-posed problem in the sense of Hadamard [14]. In order to estimate the model parameters, additional constraints are necessary. These constraints can be constancy of parameters on a very local support in the sense of Lucas and Kanade [30] or global smoothness as proposed by Horn and Schunk [21]. The problem of estimating fluid motion at the air-water interface can thus be formulated in a local framework as presented by Garbe et al. [11] with a robust extension such as the approach of Garbe and Jähne [9]. This local approach of motion estimation can be embedded in a global variational approach as shown by Spies and Garbe [35]. A possible alternative is using a combined local global approach as proposed by Bruhn et al. [7] which can also be extended to include brightness change models.

For the performance analysis presented here, dense flow fields are not of utmost importance. Therefore, only a total least squares (TLS) approach, also known as structure tensor approach [3, 4], was chosen here. It should be noted that accuracy could be improved by employing a more elaborate approach at the cost of additional computing complexity. The technique of simultaneously estimating optical flow and change of image intensity is well known in literature [15, 17, 32, 43]. Details of the technique employed in the context of this paper have been explained previously by Garbe et al. [11]. Accuracy improvements were introduced in Garbe and Jähne [9] and Garbe et al. [10]. Therefore, only a brief overview of the technique shall be presented here.

Most gradient based techniques for the estimation of optical flow from image sequences $I(\mathbf{x}, t)$ assume brightness constancy along trajectories. Here, the gray-values are denoted by I; $\mathbf{x}$ and t denote spatial and temporal positions, respectively. This brightness constancy can then be formulated as

$$T(\mathbf{x}, t) = c, \tag{16.17}$$

for an arbitrary constant c. Taking the total temporal derivative on both sides of this equation leads to

$$\frac{dT}{dt} = (\nabla T)^\top \mathbf{u} + T_t = 0, \tag{16.18}$$

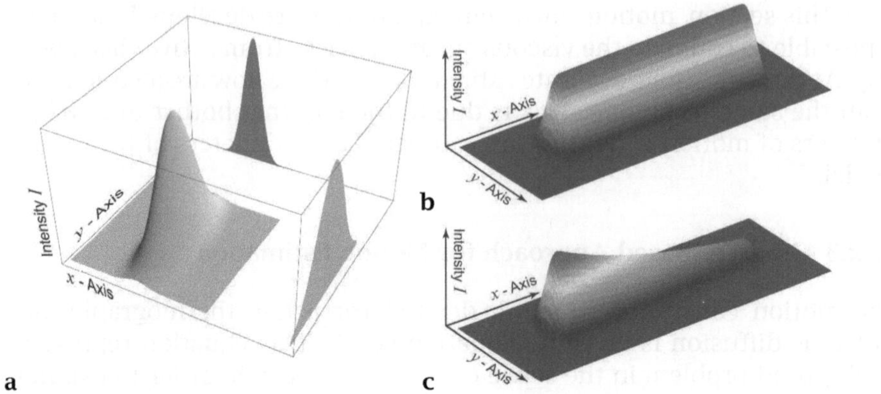

Figure 16.6. In **a** the Gaussian profile of the incident radiative heat flux is shown, in **b** frame 124 of the intensity of the simulated IR image sequence is shown without Lambert Beer's law and in **c** with decay as model by Lambert Beer's law.

where $\mathbf{u} = [u_1, u_2]^\top$ is the sought optical flow field. Subscripts indicate partial derivatives with respect to the coordinate and ∇T indicates the spatial gradient. Since this equation provides one constraint in two unknowns, one has to find additional constraints. A commonly made assumption is that of a locally smooth motion field. Therefore, Equation (16.18) can be pooled over a local neighborhood, leading to an overdetermined system of equations. This system can be solved for the parameter **u** using a weighted total least squares approach [37]. More refined motion models can also be used, incorporating brightness change [15, 17]. It is readily observed, that the Equation (16.8) derived for motion in the presence of active thermography and isotropic diffusion is very similar to Equation (16.18). It is also a partial differential equation linear in its parameters. Therefore, the same extended structure tensor approach can be used for estimating the model parameter.

16.3 Synthetic Data

In order to analyze the validity of the proposed technique it was tested on simulated data. Also, by varying the level of noise, the robustness of the algorithm to sensor noise was analyzed. The laser was simulated to be a 2D Gaussian with different widths in the x_1 and x_2 direction. An image of the simulated laser profile is shown in Figure 16.6**a**. The shearing of the heated structure due to the velocity profile was then simulated. Two simulations were run, one taking absorption of radiation in water into account, one without absorption. These two simulations are presented in Figure 16.6**b** and 16.6**c**. This transport equation was discretized on a $[x, y, z] = 256 \times 64 \times 480$ grid in 125 time steps. Different noise levels were added to this sequence. The parameter estimation was performed on this sequence. The result of the velocity vector field was very accurate. The relative error was well below 1% for noise levels equivalent to the IR camera. It is also possible to give a rough estimate for the relative accuracy of this novel technique. From Gaussian error propagation, it turns out that the relative accuracy of the shear stress should be estimated better than 5%. This is in good agreement to our findings based on the simulated data. Detailed information regarding this simulation and results are presented in Garbe et al. [13].

16.4 Experimental Set-Up

Infrared thermography represents an excellent tool for measuring boundary layer processes directly at the air water interface. This is due to the high temperature resolution of modern IR cameras as well as to the high absorption of water in the two spectral windows where midwave (3-5 μm) and longwave (8-10 μm) IR cameras are sensitive. This leads to a very shallow penetration depth of around 10 μm into the boundary layer. Added to this is an easy geometric and radiometric calibration of these cameras. More in-depth information about the IR properties of water and calibration of IR cameras can be found in Garbe [8].

Generally, two classes of thermographic techniques exist. The first class relies on imaging temperature fluctuations at the air-water interface due to a "natural" net heat flux at the interface. From this technique the temperature difference ΔT, the flow velocity directly at the interface, the net heat flux j_{heat}, as well as the transfer velocity k and parameters of the underlying transport model and their statistics can be estimated [12]. This class is termed *passive* thermography, since no external sources of heat are used for conducting these measurements. The downside of this technique is that it can only be used for conditions with sufficiently strong heat fluxes present. Otherwise the surface temperature will be homogeneous and no patterns are visible. Under these circumstances, an external heat source such as a CO_2 laser can be used for applying an external heat flux to the topmost surface layer. If this heat flux is spatially homogeneous, turbulences will become visible, as is shown in Figure 16.7. Alternatively, patterns can be written to the interface, marking individual patches of water. This class of techniques, relying on an external heat flux are called *active* thermography. A subclass of this technique is also termed *controlled flux technique* (CFT), since this external heat flux can be controlled precisely. This allows to analyze the response of the sea surface to this controlled heat flux in a linear system theoretical approach [25, 26].

For *active* thermography, a heat source is needed in addition to the set-up for passive thermography. For the active techniques presented here,

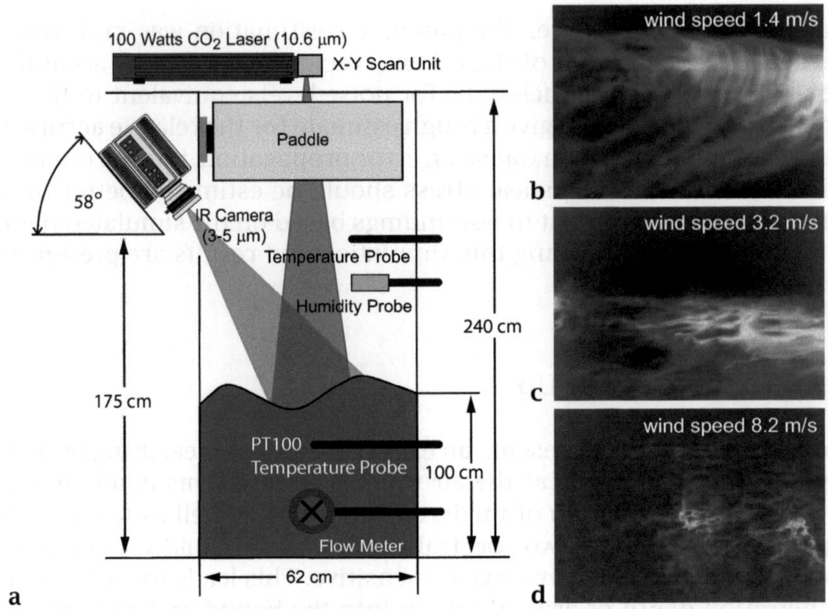

Figure 16.7. Images of typical thermal structures as seen with the active IR technique. The wind velocities are 1.4 m/s, 3.2 m/s and 8.2 m/s in **a**, **b** and **c** respectively. In all cases, the direction of the wind is from right to left.

this laser needs to be synchronized with the infrared imager. This synchronization is achieved by triggering the laser with an integral reduction of the frame rate of the camera. In the context of this work, a 25W Synrad 48-2 laser was used. Generally, the water is heated by less than one Kelvin. Stronger heating of the water is not necessary due to the high sensitivity of the IR cameras. This is important, since a strong heating would lead to a stable stratification and an associated change in turbulent structures close to the interface. The set-up for active thermography is shown in Figure 16.7.

16.5 Laboratory Measurements of Viscous Stress from Active Thermography

Apart from proving its applicability on simulated data, our novel technique for estimating the viscous shear stress τ_μ from active thermography was also tested in a laboratory setting. In the small Heidelberg annular wind wave facility measurements were conducted on surfactant covered water. As a surfactant, stearic acid (octadecanoic acid, or $CH_3(CH_2)_{16}COOH$) was used. This insoluble surfactant was applied to effectively suppress surface waves. The reason for this setting is, that the

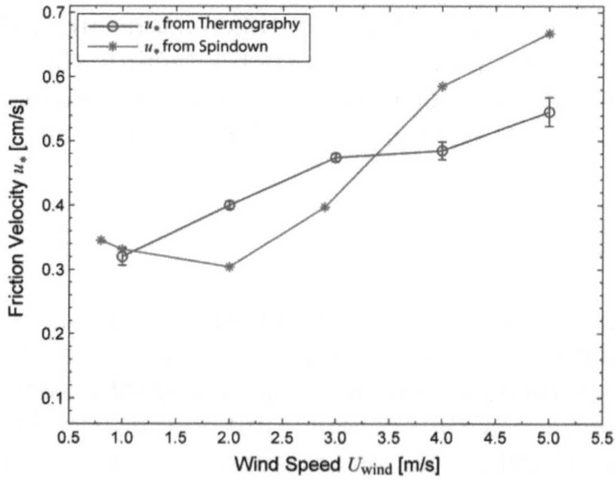

Figure 16.8. Results for measurements conducted at the small wind wave facility in Heidelberg.

form drag cannot be resolved with this technique. Comparison with independent measurements of τ are thus difficult to achieve in the presence of waves, because it is not known exactly how much of the shear stress is due to form drag. This uncertainty can thus be resolved by measuring on flat water surfaces. The wind speed range covered was from 1.0 m/s up to 5.0 m/s in the annular tank. At higher wind speeds, the surfactant would break up in certain areas and sporadic waves growth was visible. Measurements in this regime would thus have lead to bigger uncertainties in the independent measurement of τ.

Ground truth measurements of τ were achieved with the "spin down" technique. For this technique, flow velocity of the bulk is measured and at a known point in time the wind generating force is turned off. From the following exponential decay of the bulk velocity, the drag of the walls can be computed. In the presence of wind, this drag is in equilibrium with the drag of the wind on the water body and hence the shear stress τ. More details concerning this technique can be found in Ilmberger [22].

From the measurement of τ_μ the friction velocity $u_\star = \sqrt{\tau/\rho}$ can be computed. The results of these measurements are presented in Figure 16.8. Shown are the results of $u_\star$ from the novel technique based on active thermography and the results from the spin down technique. It can clearly be seen that both the results from the spin down technique and from thermography are very close to each other. At low wind speed of 1 m/s both measurements agree within the error bars. The maximum discrepancy can be seen at 5 m/s where the measurements differ by as much as 10-15%. This is a bigger discrepancy than would be expected based on the error analysis or the measurements on synthetic data in Section 16.3. More measurements in different facilities are needed in order to better assess the accuracy of this technique. For this reason, currently a linear wind wave facility is being constructed. In such a facility, the shear stress can also be computed from determining the wind profile close to the boundary. This would thus present an alternative measurement of τ.

16.6 Conclusion

A novel technique was presented for measuring the viscous shear stress τ_μ directly at the water surface from active thermography. By modeling the flow visualization process of active thermography, a motion constraint equation was derived. With this constraint equation, the flow velocity directly at the water surface can be measured very accurately. At the same time, the linear velocity profile in the viscous boundary layer is recovered. This makes it possible to estimate the viscous shear stress τ_μ.

Tests were conducted on simulated data, resulting in a relative error of 1% for noise levels encountered with modern infrared cameras. This finding is in agreement with Gaussian error propagation, from which also a relative error of less than 5% is expected. Results of test measurements in an annular wind wave facility also exhibit promising results. It should be noted that these reported measurements are the first test measurements. A thorough series of measurements will be reported in a subsequent publication. While not tested at this point, the technique should also be applicable to the sea surface in the field. These measurements will be conducted in the near future.

Acknowledgement. The aurhots acknowledge Kerstin Richter of the IUP, University of Heidelberg, for her help in proof reading the manuscript. This work has partly been funded by the German Science Foundation (DFG) within the priority program SPP 1114 "Mathematical methods for time series analysis and digital image processing" as well as by the German BMBF within the SOPRAN inititiative (Surface Ocean Processes in the Anthropocene).

References

[1] M. A. Atmane, W. Asher, and A. T. Jessup. On the use of the active infrared technique to infer heat and gas transfer velocities at the air-water interface. *Journal of Geophysical Research*, 109:C08S14, 2004. doi: 10.1029/2003JC001805.

[2] M. L. Banner and W. Peirson. Tangential stress beneath wind-driven air-water interfaces. *Journal of Fluid Mechanics*, 364:115–145, 1998.

[3] J. Bigün and G. H. Granlund. Optimal orientation detection of linear symmetry. In *ICCV*, pages 433–438, London, UK, 1987.

[4] J. Bigün, G. H. Granlund, and J. Wiklund. Multidimensional orientation estimation with application to texture analysis and optical flow. *IEEE Transactions on Pattern Analysis and Machine Intelligence*, 13 (8):775-790, 1991.

[5] W. S. Broecker and T. H. Peng. Gas exchange rates between air and sea. *Tellus*, 24:21-35, 1974.

[6] W. S. Broecker and T. H. Peng. *Tracers in the Sea.* Lamont-Doherty Geological Observatory, Columbia University, Palisades, New York, 1982.

[7] A. Bruhn, J. Weickert, and C. Schnörr. Lucas/kanade meets horn/schunck: combining local and global optic flow methods. *International Journal of Computer Vision*, 61(3):211-231, 2005.

[8] C. S. Garbe. *Measuring Heat Exchange Processes at the Air-Water Interface from Thermographic Image Sequence Analysis.* PhD thesis, University of Heidelberg, Heidelberg, Germany, December 2001.

[9] C. S. Garbe and B. Jähne. Reliable estimates of the sea surface heat flux from image sequences. In B. Radig, editor, *Mustererkennung 2001, 23. DAGM Symposium, München*, number 2191 in LNCS Lecture notes on computer science, pages 194-201. Springer-Verlag, 2001.

[10] C. S. Garbe, H. Spies, and B. Jähne. Mixed ols-tls for the estimation of dynamic processes with a linear source term. In L. Van Gool, editor, *Pattern Recognition*, volume LNCS 2449 of *Lecture Notes in Computer Science*, pages 463-471, Zurich, CH, 2002. Springer-Verlag.

[11] C. S. Garbe, H. Spies, and B. Jähne. Estimation of surface flow and net heat flux from infrared image sequences. *Journal of Mathematical Imaging and Vision*, 19(3):159-174, 2003. doi: 10.1023/A:1026233919766.

[12] C. S. Garbe, U. Schimpf, and B. Jähne. A surface renewal model to analyze infrared image sequences of the ocean surface for the study of air-sea heat and gas exchange. *Journal of Geophysical Research*, 109(C08S15):1-18, 2004. doi: 10.1029/2003JC001802.

[13] C. S. Garbe and B. Jähne. Viscous stress measurements from active thermography on a free air-water interface. *Experiments in Fluids*, 2007. In preparation.

[14] J. Hadamard. Sur les problèmes aux dérivées partielles et leur signification physique. *Princeton University Bulletin*, pages 49-52, 1902.

[15] H. Haußecker and D. J. Fleet. Computing optical flow with physical models of brightness variation. *PAMI*, 23(6):661-673, June 2001.

[16] H. Haußecker, S. Reinelt, and B. Jähne. Heat as a proxy tracer for gas exchange measurements in the field: Principles and technical realization. In B. Jähne and E. C. Monahan, editors, *Air-Water Gas Transfer - Selected Papers from the Third International Symposium on Air-Water Gas Transfer*, pages 405-413, Heidelberg, 1995. AEON Verlag & Studio Hanau.

[17] H. Haußecker, C. Garbe, H. Spies, and B. Jähne. A total least squares framework for low-level analysis of dynamic scenes and processes. In *DAGM*, pages 240–249, Bonn, Germany, 1999. Springer.

[18] H. Haußecker, U. Schimpf, C. S. Garbe, and B. Jähne. Physics from IR image sequences: Quantitative analysis of transport models and parameters of air-sea gas transfer. In E. Saltzman, M. Donelan, W. Drennan, and R. Wanninkhof, editors, *Gas Transfer at Water Surfaces*, Geophysical Monograph. American Geophysical Union, 2001.

[19] E. C. Hildreth. Computations underlying the measurement of visual motion. *Artificial Intelligence*, 23:309–354, 1984.

[20] E. C. Hildreth. The analysis of visual motion: from computational theory to neural mechanisms. *Annual Review of Neuroscience*, 10: 477–533, 1987.

[21] B. K. P. Horn and B. Schunk. Determining optical flow. *Artificial Intelligence*, 17:185–204, 1981.

[22] J. Ilmberger. Impulsübertrag und Strömungsverhältnisse in einem runden Wind-Wasser Kanal. Master's thesis, University of Heidelberg, 1980.

[23] B. Jähne and H. Haußecker. Air-water gas exchange. *Annual Reviews Fluid Mechanics*, 30:443–468, 1998.

[24] B. Jähne, K. O. Münnich, R. Bösinger, A. Dutzi, W. Huber, and P. Libner. On the parameters influencing air-water gas exchange. *Journal of Geophysical Research*, 92(C2):1937–1949, 1987.

[25] B. Jähne, P. Libner, R. Fischer, T. Billen, and E. J. Plate. Investigating the transfer process across the free aqueous boundary layer by the controlled flux method. *Tellus*, 41B(2):177–195, 1989.

[26] B. Jähne, C. Popp, U. Schimpf, and C. S. Garbe. Analysis of the heat transfer process across the aqueous heat boundary layer by active thermography: mean transfer velocities and intermittence. In *Transport at the Air Sea Interface-Measurements, Models and Parameterizations*. Chapter 18, pages 255–274, Springer, this volume, 2007.

[27] T. Kukulka and T. Hara. Momentum flux budget analysis of wind-driven air-water interfaces. *Journal of Geophysical Research*, 110: C12020, 2005. doi: 10.1029/2004JC002844.

[28] P. S. Liss and L. Merlivat. Air-sea gas exchange rates: Introduction and synthesis. In P. Buat-Menard, editor, *The role of air-sea exchange in geochemical cycling*, pages 113–129. Reidel, Boston,MA, 1986.

[29] P. S. Liss and P. G. Slater. Flux of gases across the air-sea interface. *Nature*, 247:181–184, 1974. doi: 10.1038/247181a0.

[30] B. Lucas and T. Kanade. An iterative image registration technique with an application to stereo vision. In *DARPA Image Understanding Workshop*, pages 121–130, 1981.

[31] W. McLeish and G. E. Putland. Measurements of wind-driven flow profiles in the top millimeter of water. *Journal of Physical Oceanog-*

raphy, 5(3):516–518, 1975. doi: 10.1175/1520-0485(1975)005<0516: MOWDFP>2.0.CO;2.

[32] S. Negahdaripour and C. H. Yu. A generalized brightness chane model for computing optical flow. In *International Conference in Computer Vision*, pages 2–7, Berlin, 1993.

[33] K. Okuda, S. Kawai, and Y. Toba. Measurement of skin friction distribution along the surface of wind waves. *Journal of Oceanography*, 33(4):190–198, 1977. doi: 10.1007/BF02109691.

[34] U. Schimpf, C. S. Garbe, and B. Jähne. Investigation of transport processes across the sea surface microlayer by infrared imagery. *Journal of Geophysical Research*, 109(C08S13), 2004. doi: 10.1029/2003JC001803.

[35] H. Spies and C. S. Garbe. Dense parameter fields from total least squares. In L. Van Gool, editor, *Pattern Recognition*, volume LNCS 2449 of *Lecture Notes in Computer Science*, pages 379–386, Zurich, CH, 2002. Springer-Verlag.

[36] B. M. Uz, M. A. Donelan, T. Hara, and E. J. Bock. Laboratory studies of wind stress over surfacewaves. *Boundary-Layer Meteorology*, 102 (2):301–331, 2002. doi: 10.1023/A:1013119313063.

[37] S. Van Huffel and J. Vandewalle. *The Total Least Squares Problem: Computational Aspects and Analysis.* Society for Industrial and Applied Mathematics, Philadelphia, 1991.

[38] R. Wanninkhof. Relationship between gas exchange and wind speed over the ocean. *Journal of Geophysical Research*, 97(C5):7373–7382, 1992.

[39] R. Wanninkhof. The impact of different gas exchange formulations and wind speed products on global air-sea co_2 fluxes. In C.S. Garbe, R.A. Handler, and Jähne B., editors, *Transport at the Air Sea Interface - Measurements, Models and Parametrizations.* Chapter 1, pages 1–23, Springer, this volume, 2007.

[40] R. Wanninkhof and W. R. McGillis. A cubic relationship between gas transfer and wind speed. *Geophysical Research Letters*, 26(13):1889–1892, 1999. doi: 10.1029/1999GL900363.

[41] R. Wanninkhof, W. Asher, R. Weppernig, C. Hua, P. Schlosser, C. Langdon, and R. Sambrotto. Gas transfer experiment on georges bank using two volatile deliberate tracers. *Journal of Geophysical Research*, 98(C11):20237–20248, 1993.

[42] A. J. Watson, R. C. Upstill-Goddard, and P. S. Liss. Air-sea exchange in rough and stormy seas measured by a dual tracer technique. *Nature*, 349(6305):145–147, 1991.

[43] D. Zhang and M. Herbert. Harmonic maps and their applications in surface matching. In *CVPR'99*, Fort Collins, Colorado, June 1999.

17

Estimation of Air-Sea Gas and Heat Fluxes from Infrared Imagery Based on Near Surface Turbulence Models

Tetsu Hara[1], Eric VanInwegen[1], John Wendelbo[1], Christoph S. Garbe[2], Uwe Schimpf[2], Bernd Jähne[2], and Nelson Frew[3]

[1] University of Rhode Island, USA thara@uri.edu
[2] University of Heidelberg, Germany
[3] Woods Hole Oceanographic Institution, USA

Abstract Water surface infrared images were obtained during the GASEX2001 experiment in the South Equatorial Pacific waters and during the laboratory experiment in the AEOLOTRON wind wave tank at University of Heidelberg in October 2004. The infrared imagery during these experiments reveals coexistence of roller type turbulence and intermittent breaking events. Previous interpretations of the infrared images relied on the surface renewal model, in which the water surface is assumed to be occasionally renewed by bursts of turbulent eddies reaching the water surface. A new complementary model (eddy renewal model) based on stationary and spatially periodic turbulent eddies is developed to reinterpret the infrared images of near surface turbulence. The model predicts warm elongated patches bounded by cold streaks aligned with mean wind, being consistent with field and laboratory infrared images. The model yields bulk temperature estimates and mean heat flux estimates that are very close to those based on the surface renewal model.

17.1 Introduction

In any air-sea climatological model, the importance of the transfer between ocean and atmosphere of gasses (especially greenhouse gasses such as CO2) cannot be understated. However, the physical processes responsible for such transfers are still poorly understood. Compounding the issue is that the transfers are affected by many variables, including, but not limited to, wave action, wind speed, temperatures of atmosphere and seas, concentrations of gasses, coefficients of molecular diffusivity, and condition of surface waters (e.g. presence of surfactants).

Recently, methods have been developed to study air-sea gas transfer using heat as a proxy tracer, since water surface temperature can be measured accurately using infrared imaging. Thus, much of the research of

C.S. Garbe, R.A. Handler, B. Jähne (eds.): Transport at the Air Sea Interface
pp. 241-254, 2007,

air-sea gas transfer has been focused on understanding the mechanisms of heat transfer. In order to understand the mechanisms, mathematical models have been developed to explain heat transfer due to turbulent eddies. The surface renewal model (e.g., Garbe et al. [2]) has been commonly used to describe the near surface turbulence statistically and to estimate the resulting heat flux. Atmane et al. [1] have proposed a random eddy model to explain the near surface process. The model is an extension of the surface renewal model to include the eddy approach distance as a new parameter. The model was applied to experimental results using the active controlled flux technique and yielded improved transfer velocity estimation.

These models assume that water surface is periodically renewed by turbulence; the renewal occurs instantaneously in time and spatial variability of the turbulence is negligible. Such assumptions may be appropriate to simulate passage of breaking wave fronts that occur intermittently. However, these models are not consistent with visual observations of near surface temperature fields away from breaking events. Infrared images of the surface temperature distribution are typically quite stationary in time but are variable (often streaky) in space, suggesting that near surface turbulence eddies are also quasi stationary and are elongated in the mean wind direction.

This study proposes a simple model of scalar transfer due to stationary two dimensional wind generated turbulent eddies. The objective here is to provide a more realistic theoretical framework to interpret infrared images of air sea interfaces under low to moderate winds. In Sections 17.2 and 17.3 the model is presented and is compared with the surface renewal model. The new model is then applied to infrared images obtained in laboratory and field conditions in Section 17.4.

17.2 Eddy Renewal Model and Surface Renewal Model

In observing the infrared images of the air-water interface under low to moderate winds, it appears that distinct roller type turbulence patterns exist. The images often show elongated patches of warm water alternating with cold water streaks, with their long axes aligned with the predominant wind direction (e.g., Figure 17.4**a**). These eddies exhibit the characteristics of Langumir turbulence that is distorted by the wave induced Stokes drift, although the origin of these eddies are not clear. If near surface turbulent eddy motions are detected from such images, it is in principle possible to solve for heat transfer caused by such motions explicitly and estimate the resulting air-water heat transfer velocity. Furthermore, if the turbulence field is taken to be time independent, it is possible to construct a relatively simple model where bulk water is constantly being advected up, spreading

Figure 17.1. Schematic of near surface turbulence - cross section perpendicular to mean wind.

out at the surface, and sinking down near the downwelling regions, during which the advected water is continuously cooled by molecular diffusion.

Let us assume that the surface turbulence is visualized as two-dimensional eddies whose cross section looks like Figure 17.1; this gives the general physical description of long "rollers" spinning next to each other but in alternating directions, creating alternating regions of upwelling and downwelling, as first described by Lamont and Scott [3]. When temperature is added to the model, this equates to regions of warm water upwelling, cooled water downwelling and regions of fairly constant temperature between. While Lamont and Scott [3] studied such a model with eddies bounded by both top and bottom boundaries, we examine eddies that are bounded by the top free surface boundary only, in order to simulate roller type turbulence interacting with the air-sea interface. We will call our model an "eddy renewal model" in contrast to the existing "surface renewal model".

Let us define a coordinate system such that x and y are in horizontal and z is in vertical (upward) from the mean water surface. The mean wind and wave direction is set to be in positive y. The equation governing the temperature field T is written as

$$\frac{dT}{dt} = \frac{\partial T}{\partial t} + u\frac{\partial T}{\partial x} + v\frac{\partial T}{\partial y} + w\frac{\partial T}{\partial z} = \kappa\left(\frac{\partial^2}{\partial x^2} + \frac{\partial^2}{\partial y^2} + \frac{\partial^2}{\partial z^2}\right)T, \quad (17.1)$$

where (u, v, w) are water velocities in (x, y, z) directions, and κ is the molecular diffusivity. The surface boundary condition is set to

$$j_H = -k\frac{\partial T}{\partial z} = -\kappa\rho C_P\frac{\partial T}{\partial z}, \qquad z = 0, \quad (17.2)$$

where j_H is the surface heat flux, k is the thermal conductivity, ρ is the water density, and C_P is the specific heat. It is assumed that j_H is mainly controlled by the atmospheric turbulence and is speficied a priori for the water side problem considered here. The temperature is assumed to be uniform (T_b, called "bulk temperature") below the diffusive sublayer;

$$T = T_b, \qquad z = -\infty. \tag{17.3}$$

If the turbulent eddies are assumed to be uniform in y and stationary in t, the equation is simplified to

$$u\frac{\partial T}{\partial x} + w\frac{\partial T}{\partial z} = \kappa\left(\frac{\partial^2}{\partial x^2} + \frac{\partial^2}{\partial z^2}\right)T. \tag{17.4}$$

It is further assumed that the water velocity at the surface is periodic with a wavenumber $\tilde{k}$ such that

$$u = u_0 \sin(\tilde{k}x), \qquad z = 0. \tag{17.5}$$

Then, the Taylor expansion in vertical around $z = 0$ yields

$$u = u_0 \sin(\tilde{k}x) + \left[\frac{\partial u}{\partial z}\right]_{z=0} z + \ldots, \tag{17.6}$$

and the continuity requires that

$$\frac{\partial w}{\partial z} = -\frac{\partial u}{\partial x} = -\tilde{k}u_0 \cos(\tilde{k}x) - \left[\frac{\partial^2 u}{\partial z \partial x}\right]_{z=0} z + \ldots, \tag{17.7}$$

hence,

$$w = -z\tilde{k}u_0 \cos(\tilde{k}x) - \left[\frac{\partial^2 u}{\partial z \partial x}\right]_{z=0} \frac{z^2}{2} + \ldots. \tag{17.8}$$

Let us introduce the following nondimensional variables:

$$\tilde{x} = \tilde{k}x, \qquad \tilde{z} = \frac{z}{\delta}, \qquad \tilde{T} = \frac{(T - T_b)k}{\delta j_H}, \tag{17.9}$$

with

$$\delta^2 = \frac{\kappa}{\tilde{k}u_0}, \tag{17.10}$$

where δ is the depth scale of the diffusive sublayer and is determined by the diffusivity κ and the surface divergence scale $\tilde{k}u_0$. Then, the normalized governing equation and boundary conditions become

$$\sin\tilde{x}\frac{\partial \tilde{T}}{\partial \tilde{x}} - \tilde{z}\cos\tilde{x}\frac{\partial \tilde{T}}{\partial \tilde{z}} = \frac{\partial^2 \tilde{T}}{\partial \tilde{z}^2} + O(\tilde{k}\delta), \tag{17.11}$$

$$\frac{\partial \tilde{T}}{\partial \tilde{z}} = -1, \qquad \tilde{z} = 0, \tag{17.12}$$

$$\tilde{T} = 0, \qquad \tilde{z} = -\infty. \tag{17.13}$$

Therefore, provided the diffusive sublayer depth is much smaller than the wavelength of the eddies ($\tilde{k}\delta \ll 1$), heat diffusion is governed by a single set of governing equation and boundary conditions, regardless of the size and intensity of the turbulent eddies. (Here, it has been assumed that the surface viscous shear stress does not affect the scaling of $\partial u/\partial z$, since the stress applies in y direction.)

The solution of Equations (17.11) to (17.13) can be obtained numerically. Before discussing the solution, however, let us contrast this problem to the surface renewal model. The assumptions of the surface renewal model are that the water velocity is zero (except during the instantaneous renewal event) and that the horizontal gradient of the temperature is negligible. These assumptions lead to the governing equation,

$$\frac{\partial T}{\partial t} = \kappa \frac{\partial^2 T}{\partial z^2}, \qquad 0 \leq t \leq \alpha\tau, \tag{17.14}$$

where τ is the renewal time scale and $\alpha\tau$ is the actual interval between the renewal events (the coefficient α is determined later), and the boundary conditions,

$$j_H = -k\frac{\partial T}{\partial z} = -\kappa\rho C_P \frac{\partial T}{\partial z}, \qquad z = 0, \tag{17.15}$$

and

$$T = T_b, \qquad z = -\infty. \tag{17.16}$$

The initial condition at $t = 0$ is given as

$$T = T_b, \qquad \text{for all } z. \tag{17.17}$$

If we introduce the following nondimensional variables

$$\tilde{t} = \frac{t}{\tau}, \qquad \tilde{z} = \frac{z}{\delta}, \qquad \tilde{T} = \frac{(T - T_b)k}{\delta j_H}, \tag{17.18}$$

with

$$\delta^2 = \kappa\tau, \tag{17.19}$$

the governing equation becomes

$$\frac{\partial \tilde{T}}{\partial \tilde{t}} = \frac{\partial^2 \tilde{T}}{\partial \tilde{z}^2}, \qquad 0 \leq \tilde{t} \leq \alpha, \tag{17.20}$$

and the boundary conditions become

$$\frac{\partial \tilde{T}}{\partial \tilde{z}} = -1, \qquad \tilde{z} = 0, \tag{17.21}$$

$$\tilde{T} = 0, \qquad \tilde{z} = -\infty, \tag{17.22}$$

with the initial condition,

$$\tilde{T} = 0, \qquad \text{for all } \tilde{z}. \tag{17.23}$$

Again, the temperature field is determined by a single set of governing equation and boundary/initial conditions. The solution of Equations (17.20) to (17.23) can be obtained analytically as

$$\tilde{T} = 2\sqrt{\tilde{t}}\left[\frac{-\tilde{z}}{2\sqrt{\tilde{t}}}\operatorname{erfc}\left(\frac{-\tilde{z}}{2\sqrt{\tilde{t}}}\right) - \frac{1}{\sqrt{\pi}}e^{-\left(\frac{-\tilde{z}}{2\sqrt{\tilde{t}}}\right)^2}\right]. \tag{17.24}$$

By comparing Equations (17.10) and (17.19), it is evident that the renewal time scale τ in the surface renewal model is equivalent to the inverse of the surface divergence $\tilde{k}u_0$ in the eddy renewal model. In fact, the inverse of the surface divergence can be interpreted as a time scale over which a water parcel travels from the upwelling region to the downwelling region.

In Figure 17.2 the solution of our eddy renewal model from Equation (17.11) to (17.13) is compared with the solution of the surface renewal model presented in Equations (17.20) to (17.23). In order to make the comparison easier, the coefficient α is set 5.54 so that both the mean surface temperature and the surface heat flux are identical between the two models. (That is, the actual (dimensional) interval between the renewal events in the surface renewal model is 5.54 times the inverse of the surface divergence $\tilde{k}u_0$ in the eddy renewal model if the two eddies yield the same heat transfer velocity.) In addition, the temperature scale is adjusted so that the normalized mean surface temperature is equal to -1.

The result of the eddy renewal model clearly shows that water is advected up toward the surface, along (and just below) the surface, and then down, while it is continuously cooled. Rather than the motion of the water being described by an instantaneous event, the motion is described as a continuous flow from upwelling to downwelling. Once the water has moved down sufficiently far from the surface, it is assumed that the turbulent motions of the mixed layer will blend the downwelled water with the existing mixed layer.

One obvious difference between the eddy renewal model and the surface renewal model is that the bulk temperature is never observed on the surface of the eddy in the former. This difference can be easily explained. In the surface renewal model, the process that transfers water to the surface is assumed to be instantaneous. That is, no time is given for the bulk water advected to the surface to cool. The eddy renewal model allows for the bulk water to be cooled as it is advected closer to the surface. The strength of the cooling is based on the strength of the vertical motions of the eddy.

Another way to interpret the difference between the models is to compare the histograms of surface temperature (Figure 17.3). In the surface renewal model, the water reaches colder temperatures more quickly than

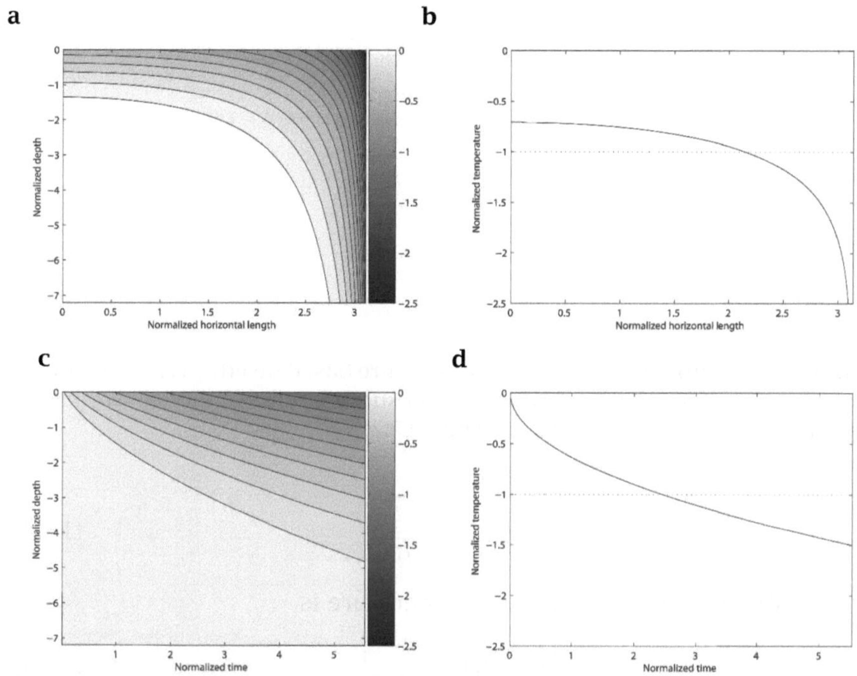

Figure 17.2. **a** Normalized temperature distribution (indicated by gray scales and contours every 0.125) based on eddy renewal model. Upwelling is near $\tilde{x} = 0$ and downwelling is near $\tilde{x} = \pi$. **b** Normalized surface temperature distribution (solid line) and its mean (dashed line) based on eddy renewal model. **c** Normalized temperature distribution (indicated by gray scales and contours every 0.125) based on surface renewal model. **d** Normalized surface temperature distribution (solid line) and its mean (dashed line) based on surface renewal model. All results have been rescaled such that the surface mean is -1 and the bulk mean is 0.

it does in the eddy renewal model. However, the maximum value of the bulk temperature is never observed at the surface in the eddy renewal model. This leads to two very different temperature distributions. With the eddy renewal model, the histogram increases with temperature up to the maximum temperature, which is still lower than the bulk temperature, and then suddenly decreases to zero. The histogram of the surface renewal model, on the other hand, decreases with temperature but remains above zero all the way to the bulk temperature.

17.3 Heat Flux Estimates from the Two Models

The solution of the surface renewal model in Equation (17.24) shows that the temperature at the water surface varies like

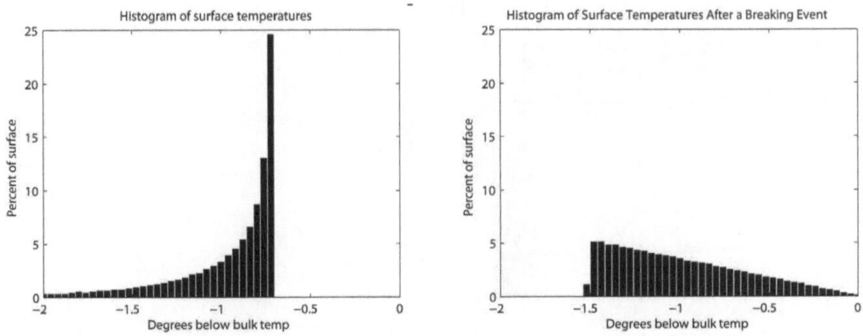

Figure 17.3. **a** Histogram of surface temperature based on eddy renewal model. **b** Histogram of surface temperature based on surface renewal model. Normalized temperature scale is the same as in Figure 17.2.

$$\tilde{T} = -\frac{2\sqrt{\tilde{t}}}{\sqrt{\pi}}, \tag{17.25}$$

and the rate of change of surface temperature is

$$\frac{d\tilde{T}}{d\tilde{t}} = \frac{\partial \tilde{T}}{\partial \tilde{t}} = -\frac{1}{\sqrt{\pi \tilde{t}}}. \tag{17.26}$$

The product of $\tilde{T}$ and $d\tilde{T}/d\tilde{t}$ is

$$\tilde{T}\frac{d\tilde{T}}{d\tilde{t}} = \frac{2}{\pi}, \tag{17.27}$$

or

$$\frac{k^2}{\kappa j_H^2}(T - T_b)\frac{dT}{dt} = \frac{2}{\pi} \tag{17.28}$$

in a dimensional form. Therefore, if simultaneous measurements are made of both $(T - T_b)$ and dT/dt for any pixel in an infrared image, it is possible to estimate the heat flux at this pixel as

$$\frac{j_H}{k} = \sqrt{\frac{\pi}{2}}\sqrt{\frac{1}{\kappa}(T - T_b)\frac{dT}{dt}} \tag{17.29}$$

as shown by Garbe et al. [2].

With the eddy renewal model, the rate of change of surface temperature following a fluid parcel is

$$\frac{d\tilde{T}}{d\tilde{t}} = \sin\tilde{x}\frac{\partial \tilde{T}}{\partial \tilde{x}}. \tag{17.30}$$

Therefore, the heat flux is expressed as

$$\frac{j_H}{k} = \sqrt{\frac{1}{\tilde{T}\sin\tilde{x}\frac{\partial\tilde{T}}{\partial\tilde{x}}}}\sqrt{\frac{1}{\kappa}(T - T_b)\frac{dT}{dt}}. \tag{17.31}$$

Unlike the surface renewal model, the product $(T-T_b)\,{}^{dT}/_{dt}$ is not uniquely related to the heat flux but the relationship depends on $\tilde{x}$, that is, depends on where the pixel is located relative to the eddy below. However, if the estimates of $\sqrt{(T-T_b)\,{}^{dT}/_{dt}}$ are averaged over many pixels, the results can be related to the averaged heat flux as

$$\frac{\overline{j_H}}{k} = \left(\overline{\sqrt{\tilde{T}\sin\tilde{x}\frac{\partial\tilde{T}}{\partial\tilde{x}}}}\right)^{-1}\overline{\sqrt{\frac{1}{\kappa}(T - T_b)\frac{dT}{dt}}}, \tag{17.32}$$

provided the heat flux j_H is uncorrelated with $\tilde{x}$, i.e., the location relative to the upwelling and downwelling points. Here, overbars denote averages over many pixels. The coefficient $\left(\overline{\sqrt{\tilde{T}\sin\tilde{x}\,{}^{\partial\tilde{T}}/_{\partial\tilde{x}}}}\right)^{-1}$ is calculated to be 1.33, while $\sqrt{{}^{\pi}/_{2}} = 1.25$ in the surface renewal model. Therefore, the mean heat flux estimates are only 6% higher with the eddy renewal model compared to those with the surface renewal model.

17.4 Application to Real Infrared Images

We now apply the eddy renewal model to actual infrared images of air-water interfaces. First the images obtained during the GASEX2001 experiment in the South Equatorial Pacific waters are used. An example of the surface image (128 by 128 pixels corresponding to roughly 0.5 m by 0.5 m) is shown in Figure 17.4**a**. The mean wind is from right to left and the image shows cold streaks that tend to be parallel to the wind. The first step of the analysis is to examine the temperature variation along a vertical column of pixels (perpendicular to the wind) as shown by the + symbols in Figure 17.4**e**. Next, local temperature maxima (red + symbols) and minima (blue + symbols) are identified, and each interval between a maximum point and a neighboring minimum point is defined as one eddy. For each eddy, the theoretical temperature variation curve of the eddy renewal model (green lines) is fitted (with the least square error method) to estimate the bulk temperature below the eddy (red lines). Eddies consisting of 4 pixels or less tend to yield less accurate estimates of the bulk temperature and are excluded from this analysis. Since our eddy renewal model solution is not strictly valid at the downwelling location, the minimum temperature point within each eddy is not used in the best fit analysis.

Figure 17.4**e** clearly shows that the theoretical temperature profile fit to each eddy is generally consistent with the observed temperature profile. The observed crosswind variations of the surface temperature field exhibit

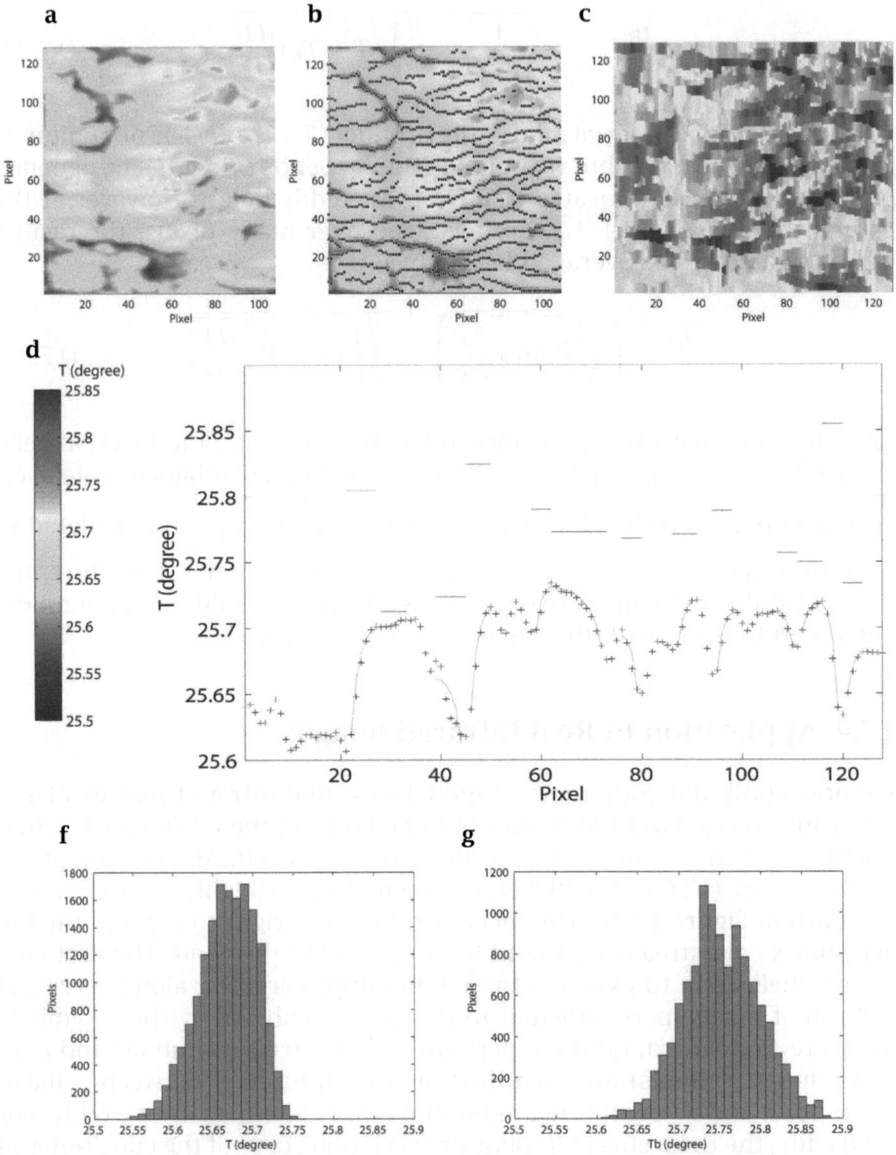

Figure 17.4. Analysis of infrared image from GASEX2001. The 10 meter wind speed is 5.1 m/s. **a** Observed surface temperature distribution. **b** Modeled surface temperature distribution. Black pixels indicate local temperature minima corresponding to downwelling zones. **c** Estimated bulk temperature distribution. **d** Temperature color scale for **a**, **b**, and **c**. **e** Observed surface temperature along a single column (+), local maxima (red +), local minima (blue +), fit to eddy renewal model (green line), and estimated bulk temperature (red line). **f** Histogram of observed temperature. **g** Histogram of estimated bulk temperature.

typical signatures with round high temperature "crests" and sharp low temperature "troughs", which are very similar to the evolution of the surface temperature predicted by the eddy renewal model. The local estimate of the bulk temperature varies from eddy to eddy but is consistently above the observed maximum temperature of each eddy.

After the same analysis is performed along all 128 columns, the resulting fitted temperature field is shown in Figure 17.4**b**, in which the local minima are identified by black pixels. Our method is able to identify cold streaks clearly because the locations of the temperature minima tend to vary slowly in the wind direction. This suggests that the near surface eddies are indeed elongated in the mean wind direction. The model also reproduces the observed temperature field very well (compare Figure 17.4**a** and Figure 17.4**b**). The corresponding (eddy by eddy) estimates of the bulk temperature are presented in Figure 17.4**c**. Again, the variation from one column to the next is relatively small.

The mean of the estimated bulk temperature is 25.75 degree and the mean observed surface temperature is 25.67 degree. If the surface renewal model is applied and the bulk temperature of this image is estimated from the distribution of the observed temperature using the method proposed in Garbe et al. [2], the bulk temperature is estimated to be 25.74 degree, which is almost identical to our estimate based on the eddy renewal model. The major difference, however, is that our approach yields not only the mean but the variability of the bulk temperature estimates. The histogram of the observed surface temperature distribution and the histogram of the estimated bulk temperature exhibit similar overall variability as seen in Figure 17.4**f** and 17.4**g**. This is somewhat surprising since the variation of the bulk temperature is expected to be much less than that of the surface temperature. It is likely that our analysis overestimates the variability of the bulk temperature because fitting the theoretical temperature profile to each eddy is never perfect and introduces some error in the bulk temperature estimate.

When our analysis is applied to a sequence of images, the variations of the mean surface temperature and the bulk temperature may be investigated as shown in Figure 17.5. Overall, the variation of the bulk temperature is very small (less than 0.05 degree over 500 images corresponding to 5 seconds). The bulk temperature estimates with our eddy renewal model (average of the estimates from eddy by eddy fit to the theoretical curve) are surprisingly close to the estimates with the surface renewal model (estimates from the surface temperature distribution of each image) using the method of Garbe et al. [2].

If smaller eddies (4 pixels or less) are included in our analysis, the eddy by eddy bulk temperature estimates show more scatter but the mean estimate per image remains very similar to the estimate without smaller eddies, suggesting that our method is not sensitive to filtering of small scale eddies.

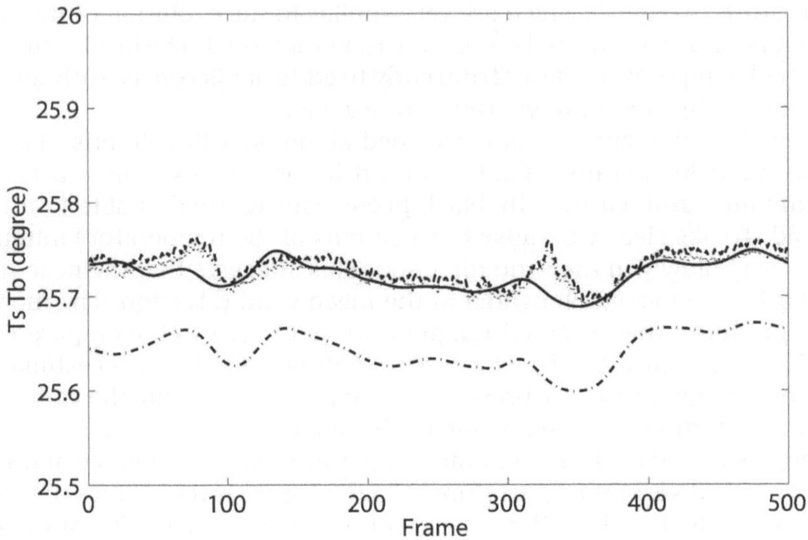

Figure 17.5. Analysis of sequence of infrared images from GASEX2001. The 10 meter wind speed is 5.1 m/s. Dashed line is bulk temperature estimates from eddy renewal model. Dotted line is bulk temperature estimates from eddy renewal model including small eddies. Solid line is bulk temperature estimates from surface renewal model. Dash-dot line is mean surface temperature.

Finally, the same analysis based on the eddy renewal model is applied to laboratory data obtained during the laboratory experiment at AEOLOTRON wind wave tank at University of Heidelberg in October 2004. An example of the infrared image and the results of the analysis are presented in Figure 17.6**a** to 17.6**g**. Overall, the results are qualitatively very similar to those with the field data. The estimated mean surface and bulk temperatures are 25.21 and 25.31 degrees, respectively. The bulk temperature estimate based on the surface renewal model is 25.32 degrees, which is again very close to our estimate.

17.5 Conclusion

A new model of the near surface turbulence and resulting air-water heat transfer has been developed to explain infrared images of air-water interfaces under low to moderate winds. The model (called an eddy renewal model) assumes that near surface eddies are stationary, uniform in the mean wind direction, and periodic in the cross wind direction. The model predicts surface temperature patterns with warm elongated

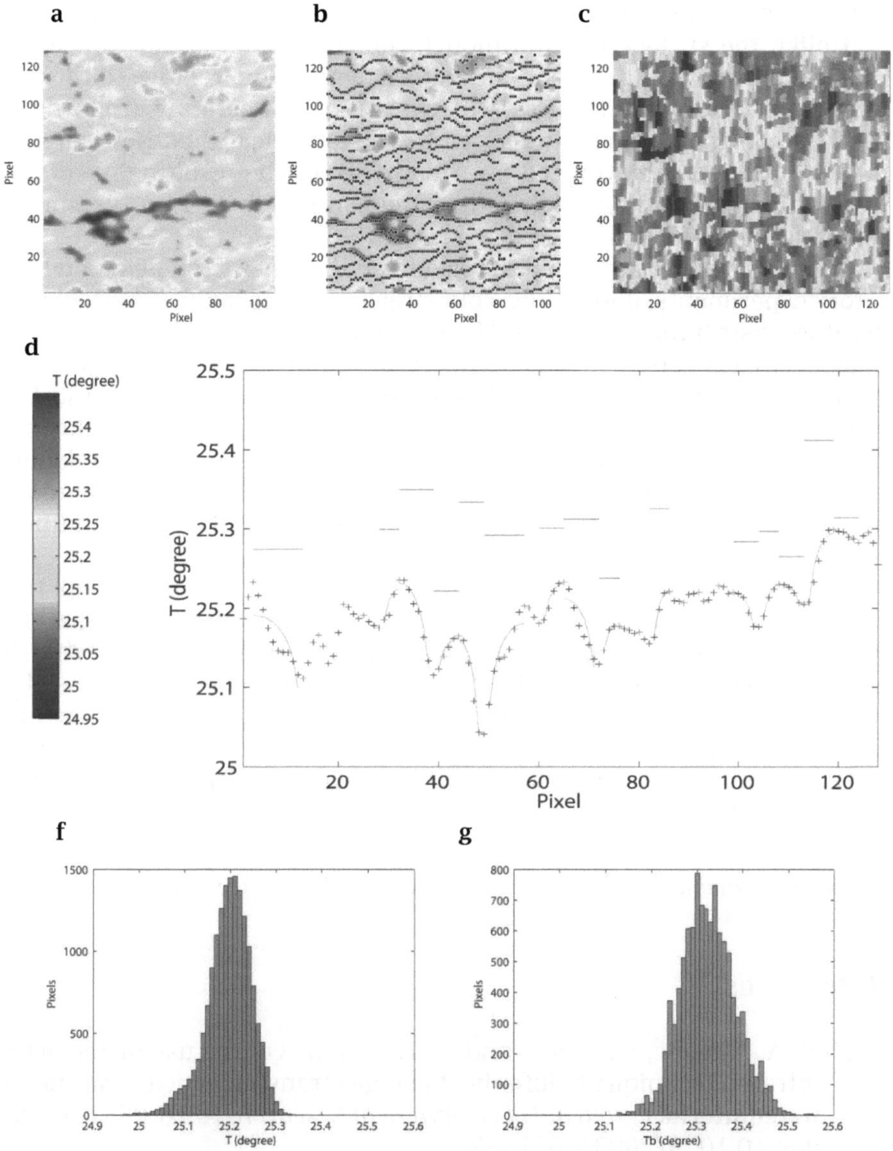

Figure 17.6. Analysis of infrared image from AEOLOTRON wind wave tank. The 10 meter wind speed is roughly 5 m/s. **a** Observed surface temperature distribution. **b** Modeled surface temperature distribution. Black pixels indicate local temperature minima corresponding to downwelling zones. **c** Estimated bulk temperature distribution. **d** Temperature color scale for **a**, **b**, and **c**. **e** Observed surface temperature along a single column (+), local maxima (red +), local minima (blue +), fit to eddy renewal model (green line), and estimated bulk temperature (red line). **f** Histogram of observed temperature. **g** Histogram of estimated bulk temperature.

patches bounded by cold streaks, both being parallel to the wind direction.

Unlike the surface renewal model, the eddy renewal model predicts that the product of the local temperature depression (relative to the bulk temperature) and the local rate of change of temperature (following a fluid parcel) is not uniquely related to the local heat flux. Nevertheless, if such products are averaged over many pixels, the resulting mean heat flux estimate is only 6% higher than the estimate based on the surface renewal model.

When the model is applied to real infrared images from field and laboratory experiments, it identifies cold streaks and reproduces surface temperature distribution very well. The model yields a bulk temperature estimate over each eddy. The bulk temperature estimates averaged over individual infrared images are very close to the estimates based on the surface renewal model using the method of Garbe et al. [2]. These conclusions, however, are preliminary since only two cases (one field and one laboratory with similar wind speeds) have been examined. The model should be applied to more data sets over a wide range of wind and wave conditions to determine its applicability.

Our method provides a new approach to estimate the mean bulk temperature and the mean heat flux from the surface infrared images. The method is more based on actual physical processes compared with the existing surface renewal model. In addition, the model may be used to investigate the near surface turbulence field in detail; it yields the distribution of eddy sizes and eddy intensities (surface divergence) under different wind and wave conditions. These will be subjects of our future investigation.

Acknowledgement. We thank NOAA CPO/GCC for supporting this research.

References

[1] M. A. Atmane, W. Asher, and A. T. Jessup. On the use of the active infrared technique to infer heat and gas transfer velocities at the air-water interface. *Journal of Geophysical Research*, 109,(C08S14),2004. doi: 10.1029/2003JC001805

[2] C. S. Garbe, U. Schimpf, and B. Jähne. A surface renewal model to analyze infrared image sequences of the ocean surface for the study of air-sea heat and gas exchange. *Journal of Geophysical Research*, 109(C08S15):1–18, 2004. doi: 10.1029/2003JC001802.

[3] J. C. Lamont and D. C. Scott. An eddy cell model of mass transfer into the surface of a turbulent liquid. *A.I.Ch.E.J*, 16(4):513–519, 1970. doi: 10.1002/aic.690160403.

18

The Influence of Intermittency on Air-Water Gas Transfer Measurements

Bernd Jähne, Christopher Popp, Uwe Schimpf, and Christoph S. Garbe

Institute for Environmental Physics and
Interdisciplinary Center for Scientific Computing
University of Heidelberg, 69120 Heidelberg, Germany
Bernd.Jaehne@iwr.uni-heidelberg.de

Abstract This paper theoretically investigates the influence of intermittency on determining average transfer velocities using different measuring techniques. It is shown that all measuring techniques can significantly be biased by intermittency. Mass balance and eddy correlation measurements are only biased when the concentration difference between the air and the water is spatially or temporally inhomogeneous over the measurement interval. Mean transfer velocities calculated either from mean boundary layer thicknesses or from thermographic techniques, which compute the mean transfer velocity either from concentration differences of from time constants, are biased toward lower values. The effects can be large and a simple stochastic bimodal model is used to estimate the effect.

18.1 Introduction

The exchange of inert and sparingly soluble gases including climate-relevant species such as carbon dioxide, methane, and fluorochlorocarbons between the atmosphere and oceans is controlled by a thin 20-200 μm thick mass boundary layer at the top of the ocean. For a long time, mass balance techniques, which infer the mean flux density across the water interface from temporal concentration changes, were the only techniques to investigate air-sea gas transfer experimentally, both in the field and in laboratory facilities. These techniques do not give any detailed insight into the mechanisms themselves. The last two decades have shown significant progress in developing new techniques for a more detailed investigation of the underlying transport mechanisms. Nowadays three further experimental approaches are available.

Eddy correlation and profile measurements are elegant micrometerological measuring technique. These are techniques that allow direct measurements of the flux density of important species such as CO_2 in the field. The measurements, however, are conducted at the "wrong" side in

C.S. Garbe, R.A. Handler, B. Jähne (eds.): Transport at the Air Sea Interface
pp. 255-274, 2007,

the air space. Because the gas transfer of sparingly soluble gases is controlled by the transport across the water sided mass boundary layer, only a small fraction of the total concentration difference between air and water remains in the air side. Therefore it is experimentally very difficult to resolve these small concentration differences and to measure fluxes reliably. Early eddy correlation measurements gave transfer velocities that were one or even two orders of magnitude higher than geochemical mass balance methods and thus stirred a lot of controversy [4]. Meanwhile, the reliability of measurements using these techniques has been demonstrated although the accuracy is still low [16].

With the pioneering work of Asher and Pankow [2], techniques became available that use fluorescent dyes to visualize concentration fields close to the water interface in small laboratory facilities. Concentration fields can either be made visible by using fluorescent pH-indicators and acid or alkaline gases [11] or dyes that are sensitive to oxygen quenching [14]. These techniques are suitable to measure instantaneous and averaged vertical concentration profiles. A mean transfer velocity can be inferred from the mean boundary layer thickness and the known diffusion coefficient of the transported species.

A forth technique, known as the controlled flux technique (CFT, [9]), applies a known flux density and determines the transfer velocity either by measuring the concentration difference across the mass boundary layer or the time constant of transfer across the boundary layer. This technique can be applied most easily using heat as a proxy tracer and infrared imaging. This technique works both with the given net heat flux across the ocean (passive thermography, [6]) or infrared radiation applied, e. g., by a CO_2 laser (active thermography, [7, 8]). The resulting imaging technique gives measurements with high temporal and spatial resolution. These techniques have been used recently by a number of research groups both in the field and in lab facilities [1, 3, 5, 15].

Obviously, the four basic measuring techniques investigate air-sea gas transfer at different temporal and spatial scales. Therefore the question arises whether these techniques really do measure the same thing. So far this has been taken for granted. However, discrepancies between results from the active controlled flux technique and mass balance techniques [1, 3] casted doubt on the simple equivalence of all these techniques. This paper addresses this question in a general way. Given the fact that air-sea gas transfer is an intermittent process, i. e., underlies statistical fluctuations, and that modern measurement techniques resolve these intermittencies, the general question arises as how to take correct averages for the different techniques. This question has not been addressed previously and will be analyzed theoretically in this paper.

The plan of this paper is as follows. First the intermittency of air-sea gas transfer is discussed in Section 18.2. This includes experimental evidence (Section 18.2.1), a discussion of the characteristic scales that deter-

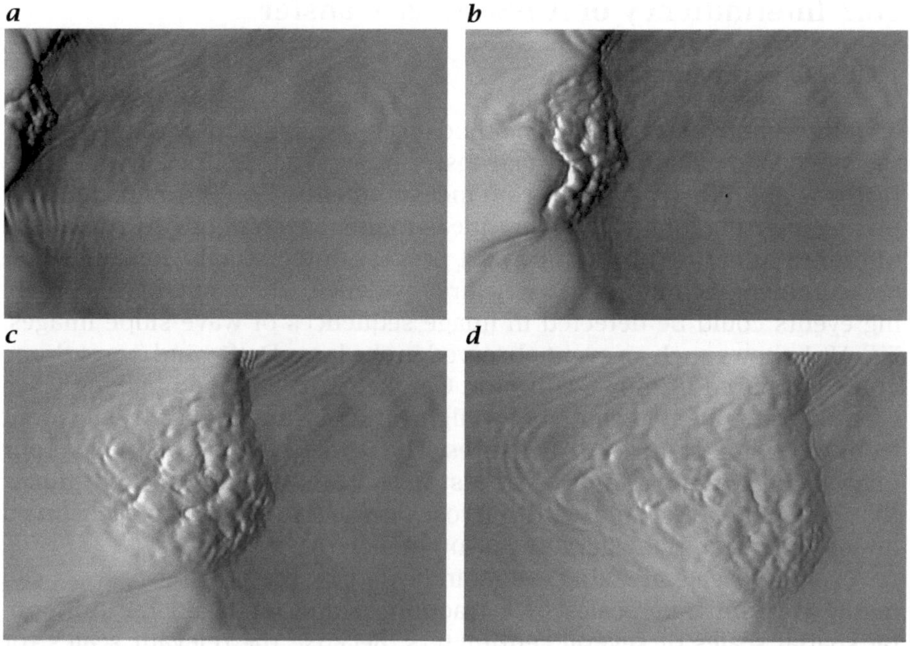

Figure 18.1. Microscale wave breaking observed in the large Delft wind/wave flume at 3.6 m/s wind speed, 100 m fetch on October 5, 1990. Shown is the along-wind wave slope in an image sector of $33 \times 25\,\mathrm{cm}^2$, the time span between each of the images is 0.1 s.

mine the gas exchange process (Section 18.2.2), a discussion of the question as how to average transfer velocities correctly for an intermittent process (Section 18.2.3), and the introduction of simple bimodal probabilistic model that helps to understand the influence of intermittency on the averaging process (Section 18.2.4). In Section 18.3, it is discussed how averaging over an intermittent process influences the measured mean transfer velocities for four basic techniques which are used to investigate gas transfer experimentally. This includes classical mass balance techniques (Section 18.3.1), eddy correlation measurements (Section 18.3.2), measurements of the mean boundary layer thickness (Section 18.3.3), and thermographic techniques (Section 18.3.4). Finally, in Section 18.4 some first direct experimental evidence for the strong intermittent nature of the mass transfer process across the water interface from active thermographic measurements is presented.

18.2 Intermittency of Air-Sea Gas Transfer

18.2.1 Experimental Evidence

As long as only mass balance methods were available for measurements, the stochastic nature of the gas transfer process across the air-water surface was not directly evident from measurements. Therefore it is not surprising that it required imaging measurement techniques to reveal the intermittent nature of the transport process. Interestingly, imaging wave measurement techniques gave the first evidence, where microscale breaking events could be detected in image sequences of wave slope images. Fig. 18.1 shows such an event observed in the large Delft wind/wave flume at a wind speed of only 3.6 m/s on October 5, 1990. The breaking wave leaves a patch behind that shows high turbulent mixing, as it is evident from the irregular surface structures. The whole event shown in the four images covers a time interval of just 0.3 s. Later, patches of high fluxes that could be associated with microscale wave breaking were observed directly in infrared imagery by Jessup et al. [10].

The microscale breaking events indicate that the transfer process can highly vary on time scales of a fraction of a second and the horizontal spatial scales of several centimeters. Because the relevant scales for mesoscale to global modeling are much larger, an adequate integration has to be performed.

18.2.2 Characteristic Scales of the Transfer Process

The flux of chemical species across the air-water interface is characterized by the *transfer velocity*:

$$k = \frac{j}{\Delta c}. \tag{18.1}$$

Because of the division by the concentration difference, this quantity is independent from the concentration difference and has the units of a velocity. It can be thought as the velocity by which an imaginary piston pushes the species through the aqueous mass boundary layer. Therefore it is also known as the "piston velocity".

The characteristic vertical length scale is the thickness of the mass boundary layer, i. e., the layer that is dominated by molecular diffusion. The corresponding time constant indicates the time necessary for the tracer to pass this layer. Independent of any assumptions about the transport mechanism, these scales can be defined as follows. The boundary layer thickness z_* ("film thickness") is given by

$$z_* = \frac{D}{k}, \tag{18.2}$$

where D is the constant of diffusivity.

Geometrically, z_* is given as the interception of the tangent to the concentration profile at the air-water interface with the bulk (or some other reference) concentration level:

$$j = D \left. \frac{\partial c}{\partial z} \right|_{z=0} = D \frac{\Delta c}{z_*} \quad \succ \quad k = \frac{j}{\Delta c} = \frac{D}{z_*}. \tag{18.3}$$

With the help of z_*, the time constant t_* can then be defined as

$$t_* = \frac{z_*}{k} = \frac{D}{k^2}. \tag{18.4}$$

Finally, a horizontal spatial scale can be defined by assuming that the mass boundary layer is advected by a surface velocity u_s

$$x_* = t_* u_s. \tag{18.5}$$

With typical transfer velocities between 5 and 50 cm/h, a molecular diffusion coefficient D of $2 \cdot 10^{-5}\,\mathrm{cm^2/s}$, the time constant t_* is of the order of only 0.1 to 10 s, and the boundary layer thickness z_* is between 14 to 140 μm. Assuming a surface velocity of 0.1 m/s, the horizontal scale x_* is between 0.01 and 1 m.

The above values clearly indicate that air-sea gas transfer is controlled at small temporal and spatial scales. In the context of larger water bodies, i. e., the depth of the mixed surface layer or the depth of rivers and laboratory facilities, however, air-water gas exchange is a slow process. Transfer velocities in the order of 5 and 50 cm/h results in time constants between 2 h and 3 months for water depthes between 1 and 100 m. For biogeochemical mass balances, not the instantaneous transfer velocities at scales natural to the process itself are of importance, but values averaged over much larger spatial and temporal scales. Moreover, the different measuring techniques discussed in the introduction also measure the gas transfer rates inherently at different temporal and spatial scales. In order to compare the results of the different measuring techniques in a correct way, it is imperative to know over which scales and in which way averages are taken.

18.2.3 How to Determine Average Transfer Velocities Correctly?

The averaging of the transfer velocity measurement over an area A and the time span Δt results in

$$\langle k \rangle = \left\langle \frac{j}{\Delta c} \right\rangle = \frac{1}{A \Delta t} \int\limits_A \int\limits_t \frac{j(\boldsymbol{x}, t)}{\Delta c(\boldsymbol{x}, t)} \, \mathrm{d}^2 x \, \mathrm{d}t \, . \tag{18.6}$$

It is important to note that the division of averaged flux densities by an average concentration difference is in general *not* equal to the above correct averaging procedure:

$$\langle k \rangle = \left\langle \frac{j}{\Delta c} \right\rangle \neq \frac{\langle j \rangle}{\langle \Delta c \rangle} . \qquad (18.7)$$

This means that for every measuring technique a careful analysis of fluctuations of both the flux density and the concentration difference is required. The main goal of this investigation is to quantify the extent by which a spatially and/or temporally varying exchange process biases the averaging.

18.2.4 Probabilistic Approach: A Simple Bimodal Model for Intermittency

In order to get a better handle of the variations a simple stochastic model is introduced. We assume the transfer velocity to be a stochastic variable with a probability density function (PDF) $p(k)$. Then the mean transfer velocity is given by

$$\langle k \rangle = \int k\, p(k)\, \mathrm{d}p. \qquad (18.8)$$

The most simple case is to assume a simple bimodal distribution of the transfer velocity:

- A fraction α of the surface and/or time shows enhanced transfer velocity, $f > 1$ times of the mean, thus providing the fraction $\gamma = \alpha f \leq 1$ of the total flux. This corresponds to patches on the water surface with high exchange rates, possibly caused by microscale breaking events.
- The other fraction $1 - \alpha$ shows a reduced transfer velocity $g < 1$ times of the mean. For a correct mean k, g is not a free parameter, but related to α and f by

$$\langle k \rangle = \alpha \cdot f \langle k \rangle + (1 - \alpha) \cdot g \langle k \rangle \quad \rightarrow \quad g = \frac{1 - \alpha f}{1 - \alpha} = \frac{1 - \gamma}{1 - \alpha} .$$

Thus the bimodal model has two free parameters: the enhanced fraction $\gamma \geq \alpha$ of the flux provided by a small fraction $\alpha < 1/2$ of the surface. In the limit of $\alpha \ll 1$, the model is determined only by γ. This simple bimodal model will be used in the following to give a quantitative estimate of the intermittence on the averaging procedures for different measuring techniques.

18.3 Averaging with Different Techniques

18.3.1 Mass Balance Techniques

The general assumptions with mass balance techniques are that the mean flux density is estimated from (slow) temporal changes in the concentration in water and that the concentration in the air is assumed to be constant. For short time periods and under spatially homogeneous conditions, the concentration difference can also assumed to be constant and the averaging

$$\langle k \rangle_j = \left\langle \frac{j}{\Delta c} \right\rangle = \langle j \rangle / \Delta c = \langle k \rangle \tag{18.9}$$

results in an unbiased average of transfer velocity.

However, spatially inhomogeneous conditions might yield biased averages. Even in a laboratory facility, conditions are not homogeneous when the transfer velocity depends strongly on fetch. In areas with high transfer rates, the concentration difference diminishes and if this effect is not compensated for rapidly enough by horizontal mixing, the mean transfer velocity is underestimated. It is also noteworthy to mention that in evasion experiments ($\alpha c_a \ll c_w$), only concentration fluctuations in the water body are of importance, while in invasion experiments ($\alpha c_a \gg c_w$) only the concentration fluctuations in the atmosphere can cause a bias in the estimate of the mean transfer velocity.

18.3.2 Eddy Correlation and Profile Techniques

Eddy correlation techniques determine a mean flux density directly from the correlation of concentration and vertical velocity fluctuations:

$$\langle j \rangle = \langle c' w' \rangle \ . \tag{18.10}$$

When the concentration difference is homogenous and temporally constant, the eddy correlation technique gives, as mass balance techniques, an unbiased mean transfer velocity in Eq. (18.9). The same arguments apply for atmospheric profile techniques where the flux density is determined from the mean steepness of the profile. However, horizontal inhomogeneities in the concentration fields can cause similar biases in the averages as with mass balance techniques.

18.3.3 Boundary Layer Thickness and Concentration Fields

Visualization techniques using fluorescent dyes yield mean vertical concentration profiles. If these profiles show a sufficiently good vertical resolution, it is possible to determine the mean boundary layer thickness $\langle z_* \rangle$ from the profile after Eq. (18.3). According to Eq. (18.2), averaging the mean boundary layer thickness, however, results in an averaging of the inverse transfer velocity:

$$\langle z_* \rangle = \left\langle \frac{D}{k} \right\rangle = D \left\langle \frac{1}{k} \right\rangle . \tag{18.11}$$

Now, the bimodal model, introduced in Section 18.2.4, can be used to estimate to what extend the averaging over the inverse transfer velocity causes a bias. This can be done by relating $1/\langle 1/k\rangle$ to $\langle k\rangle$. The calculations give

$$\frac{1/\langle 1/k\rangle}{\langle k\rangle} = \frac{1-\gamma}{(1-\alpha)^2+\alpha^2(1-\gamma)/\gamma} \approx (1-\gamma) \quad \alpha \ll \gamma < 1\,. \tag{18.12}$$

The bias can be quite significant. We assume, e. g., that 25% of the surface area has two times the mean gas transfer rate and 75% of surface only 2/3 of the mean gas transfer rate. Then both fractions of the surface provide half of the flux each, $\alpha = 1/4$, $\gamma = 1/2$, and averaging of $1/k$ gives only 0.8 times the correct mean transfer velocity. If a small fraction $\alpha \ll 1$ of the surface provides a fraction γ of the flux, the mean is lower by the factor $1-\gamma$ and the mean transfer velocity can be underestimated considerably. From this train of thoughts, we can conclude that a mean mass boundary layer thickness estimated from mean vertical profiles *without* knowing the fluctuations, it is not possible to determine a reliable mean transfer velocity.

The correct way to estimate an unbiased transfer velocity is to average the concentration gradient at the interface, because according to Eq. (18.3) this quantity is directly proportional to the flux density and thus the same averaging procedure is applied as for the mass balance and eddy correlation techniques:

$$\langle k\rangle_g = \left\langle \frac{D\dfrac{\partial c}{\partial z}}{\Delta C} \right\rangle = \frac{D}{\Delta c}\left\langle \frac{\partial c}{\partial z} \right\rangle . \tag{18.13}$$

18.3.4 Thermography

With thermographic techniques, there are essentially two ways to determine the transfer velocity. With the static technique a *constant* flux density j is given (passive technique) or applied (active technique) and the mean concentration difference is determined. The concentration difference across the boundary layer can most simply be obtained from surface measurements if it is possible to switch the constant flux density on and off. When no flux density across the thermal boundary layer occurs, the surface concentration is equal to the bulk concentration. If the flux cannot be switched, it is still possible to estimate the bulk concentration from measurements of the distribution of the surface concentration under the assumption that surface renewal takes place. Then the mean of the probability density function gives the mean surface concentration and the minimum value the bulk concentration. Further details can be found in Schimpf et al. [13].

In any case, the concentration difference across the boundary layer is now fluctuating and must be averaged. This corresponds again to an averaging of the inverse transfer velocity:

$$\langle \Delta c \rangle = \left\langle \frac{j}{k} \right\rangle = j \left\langle \frac{1}{k} \right\rangle \quad (18.14)$$

and the same bias occurs as with averaging of the boundary layer thickness.

The correct averaging procedure would be to average the inverse concentration difference:

$$\left\langle \frac{1}{\Delta c} \right\rangle = \left\langle \frac{k}{j} \right\rangle = \frac{\langle k \rangle}{j}. \quad (18.15)$$

The second way to determine the transfer velocity using thermographic techniques is to probe the temporal response of the mass boundary layer on a temporally changing heat flux density. This could either be a short pulse [7] or a periodic variation [12]. Essentially, this technique applies methods of linear system theory to the investigation of the exchange process through the aqueous heat boundary layer and measures the amplitude damping and phase shift of the periodic temperature variation at the water surface (the output of the "black box") as a function of the frequency of the periodically varying heat flux density (the input).

A theoretical treatment of the transport process through a mass boundary layer with a periodically varying concentration flux density at the interface has been first given by Jähne et al. [9]. Both turbulent diffusion models and extended surface renewal models were investigated. For a nonreactive species and negligible horizontal variations, the non-steady 1-D transport equations for the mean concentration C are given by

$$\begin{array}{ll} \text{(K)} & \dfrac{\partial C}{\partial t} = \dfrac{\partial}{\partial z}\left[(D + K(z))\dfrac{\partial C}{\partial z}\right] \\ \text{(SR)} & \dfrac{\partial C}{\partial t} = D\dfrac{\partial^2 C}{\partial z^2} - \lambda(z) C \end{array} . \quad (18.16)$$

The z-coordinate is defined zero at the interface and positive downwards, $K(z)$ is a depth dependent turbulent diffusion coefficient, and $\lambda(z)$, a depth dependent surface renewal rate. Generally, power laws are assumed for the z dependencies of the turbulent diffusivity and the surface renewal rate:

$$\begin{array}{llll} \text{(K)} & K(z) = \alpha_m z^m & \text{with} & m \geq 2 \\ \text{(SR)} & \lambda(z) = \gamma_p z^p & \text{with} & p \geq 0 \end{array} . \quad (18.17)$$

The classical surface renewal model (depth independent surface renewal rate λ) is given with $p = 0$, while at a surface covered with a surface film, the surface renewal rate at the surface must be zero and thus $p > 0$. With the eddy diffusion models, $m = 2$ at a clean free water surface and $m = 3$ at a film covered surface.

These equations can be made dimensionless with the scale parameters for the transfer across the boundary layer: z_*, t_*, and the concentration difference ΔC. Then the dimensionless quantities $C_+ = C/\Delta C$, $T_+ = t/t_*$, and $Z_+ = z/z_*$ are obtained. In addition, a power law is assumed for the z dependencies of the turbulent diffusivity and the surface renewal rate. Then the equations reduce to

$$\begin{aligned} &\text{(K)} \quad \frac{\partial C_+}{\partial t_+} = \frac{\partial}{\partial z_+}\left[\left(1 + a_{m+} z_+^m\right)\frac{\partial C_+}{\partial z_+}\right] \\ &\text{(SR)} \quad \frac{\partial C_+}{\partial t_+} = \frac{\partial^2 C_+}{\partial z_+^2} - \gamma_{p+} z_+^p C_+ \end{aligned} \tag{18.18}$$

For a solution of these equation with periodic boundary conditions, it is useful to apply a Fourier transformation in time to Eq. (18.18). This reduces the partial equations to ordinary differential equations with complex Fourier amplitudes $\tilde{c}(\omega_+)$

$$\begin{aligned} &\text{(K)} \quad \mathrm{i}\omega_+ \tilde{c} = \frac{\partial}{\partial z_+}\left[\left(1 + a_{m+} z_+^m\right)\frac{\partial \tilde{c}}{\partial z_+}\right] \\ &\text{(SR)} \quad \left(\mathrm{i}\omega_+ + \gamma_{p+} z_+^p\right)\tilde{c} = \frac{\partial^2 \tilde{c}}{\partial z_+^2} \end{aligned} \tag{18.19}$$

where $\omega_+ = \omega t_*$ is the dimensionless frequency normalized by the time constant t_*. The boundary conditions are given by the periodically varying flux density applied at the surface:

$$\begin{aligned} &\left.\frac{\partial \tilde{c}}{\partial z_+}\right|_{z_+=0} = \tilde{j}(\omega_+) = 1 \\ &\lim_{z_+\to\infty} \tilde{c}(z_+) = 0 \end{aligned} . \tag{18.20}$$

A simple analytical solution is possible for the classical surface renewal model (SR, $p = 0, \gamma_{0+} = 1$):

$$\tilde{c}(z_+, \omega_+) = \frac{\exp\left[-z_+\sqrt{(1 + \mathrm{i}\omega_+)}\right]}{\sqrt{(1 + \mathrm{i}\omega_+)}}. \tag{18.21}$$

The penetration depth $z_e = (1 + \mathrm{i}\omega_+)^{-1/2}$ is a complex quantity. For a better understanding of the solution, it is useful to part this expression into an amplitude and a phase function:

$$\sqrt{(1 + \mathrm{i}\omega_+)} = (1 + \omega_+^2)^{1/4} \exp(\mathrm{i}\phi) \quad \text{with} \quad \phi = (\arctan \omega_+)/2. \tag{18.22}$$

Then the amplitude and phase (argument) of the solution are

$$\begin{aligned} |\tilde{c}(z_+, \omega_+)| &= \frac{\exp\left[-z_+\sqrt{\left(\sqrt{1 + \omega_+^2} + 1\right)/2}\right]}{(1 + \omega_+^2)^{1/4}} \\ \phi(z_+, \omega_+) &= -\frac{1}{2}\arctan \omega_+ - z_+\sqrt{(\sqrt{1 + \omega_+^2} - 1)/2}. \end{aligned} \tag{18.23}$$

Table 18.1. Quantities that can be measured from amplitude and phase shift response of the temperature surface on a periodically varying heat flux density with a frequency $\omega_+ = \omega t_*$.

Quantity/Frequency	$\omega_+ << 1$ (low)	$\omega_+ >> 1$ (high)
Amplitude	$\langle k \rangle_c = j \big/ \rho c_p \langle \Delta T \rangle$	$j = \rho c_p \Delta T (D\omega)^{1/2}$
	Transfer velocity k	Flux calibration
Phase shift	$\phi = \kappa \omega t_*$	$\pi/4$
	Time constant t_*	Verification

These equations say that an exponentially decaying temperature wave is penetrating from the interface into the water. The higher the frequency, the lower are the amplitude and the penetration depth. At the surface (z_+), the equations reduce to

$$\begin{aligned} |\tilde{c}(0, \omega_+)| &= (1 + \omega_+^2)^{-1/4} \\ \phi(0, \omega_+) &= -\frac{1}{2} \arctan \omega_+ . \end{aligned} \tag{18.24}$$

All other models give a similar response (Table 18.1, Fig. 18.2). At low frequencies of the applied heat flux density, the amplitude of the temperature response is constant and equal to the response at a constant flux density (Fig. 18.2a). With increasing frequency, the amplitude decreases gradually and the phase shift increases first linearly. At high frequencies $\omega_+ = \omega t_* > 1$, the phase shift reaches a maximum value of $\pi/4 (45^o)$ and the amplitude continues to decrease with the square root of the frequency. At high frequencies, all models give identical results. Because of the low penetration depth, the penetration of the heat is limited to the vicinity of the surface that is dominated by molecular diffusion. The phase shift reaches a maximum value of $\pi/4$. This can be used to verify the phase measurements. The amplitude decreases with the square root of the frequency and depends only on the diffusion coefficient for heat in water. Therefore the measured temperature amplitude can be used to calibrate the applied flux density (Table 18.1).

The phase shift at low frequency is directly proportional to the frequency

$$\phi = -\kappa \omega t_* . \tag{18.25}$$

The constant κ depends on the model and lies in a narrow range between 0.33 and 0.55 [9]. The measurement of the phase shift at low frequencies thus offers a second possibility to determine the transfer velocity k via the time constant t_*. This technique results in an averaging of the inverse square transfer velocity. From Eqs. (18.25) and (18.4), we obtain

$$\langle \phi \rangle = -\kappa \omega \langle t_* \rangle = -\kappa \omega D \left\langle \frac{1}{k^2} \right\rangle . \tag{18.26}$$

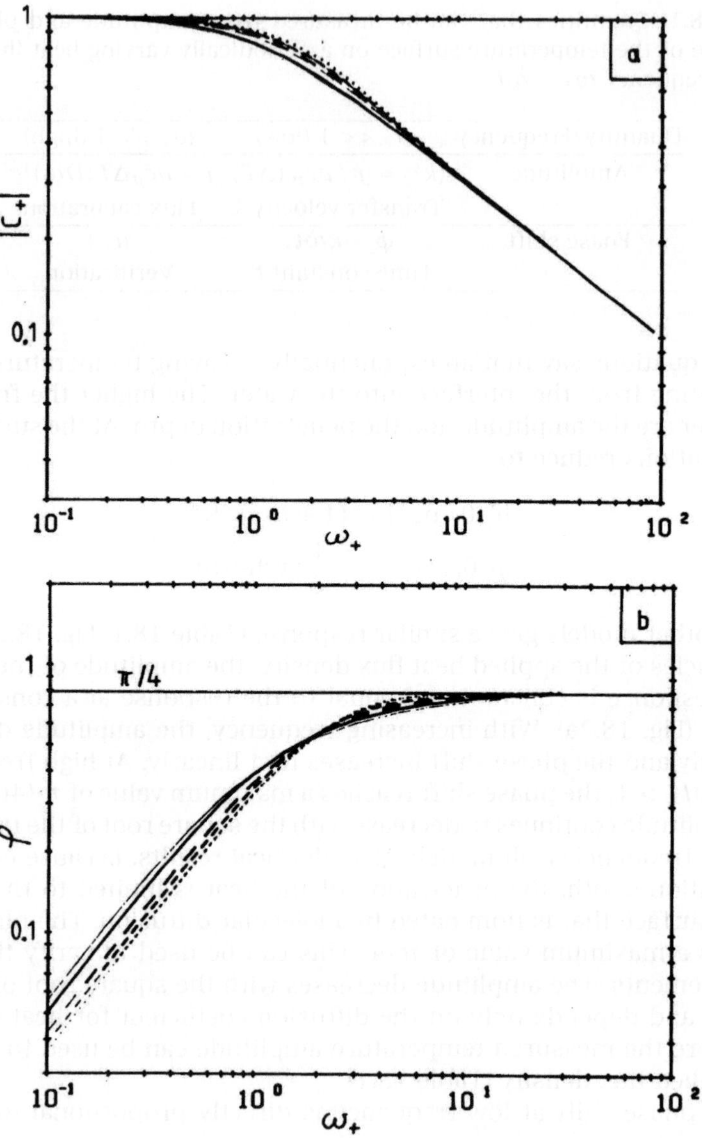

Figure 18.2. Amplitude attenuation (a) and phase shift between the flux density at the water surface and surface concentration periodic (b) with the boundary condition of a forced sinusoidal flux density at the surface. The results of model calculations are shown in double logarithmic plots as a function of the dimensionless frequency $\omega_+ = \omega t_*$. Line key: thick lines: surface renewal models; thin lines: diffusion models; line type indicates the Schmidt number exponent n: solid line: 1/2; long-dashed line 3/5; short dashed: 2/3; dotted: 3/4 (after Jähne et al. [9]).

This averaging results in an even more serious bias for mean transfer velocity than averaging the inverse transfer velocity:

$$\frac{1/\langle 1/k^2\rangle}{\langle k\rangle^2} = \frac{(1-\gamma)^2}{(1-\alpha)^3 + \alpha^3(1-\gamma)^2/\gamma^2} \approx (1-\gamma)^2 \quad \alpha \ll \gamma < 1 \quad (18.27)$$

For the same conditions as discussed in Section 18.3.3 ($\alpha = 0.25, \gamma = 0.5$), the phase shift is $7/4 = 1.75$ too high, and the mean transfer velocity is only $\sqrt{4/7} \approx 0.76$ of the true value, which is only about 5% lower than 0.8 for averaging of the inverse transfer velocity in Eq. (18.12). This means that measurements of the transfer velocity from either the amplitude response or the phase shift are biased by about the same factor towards lower values and there is no way to infer the intermittency from biases of mean values directly.

The discussion in this section concentrated on active thermography using periodically varying heat fluxes. However, for measurements with short laser pulses from which time constant can be derived as well [1, 7], should show the same bias as the measurements with periodically varying heat fluxes.

18.4 Experimental Results from Active Thermography

The first measurements with active thermography using a periodically varying heat flux density were reported by Jähne et al. [9]. At that time, the instrument accuracy was sufficient to capture mean transfer velocities but the data quality was not high enough to allow a detailed analysis of the amplitude damping and the phase response.

Recently, some preliminary measurements were conducted in the Heidelberg Aeolotron. The setup is shown in Fig. 18.3. A 100 W CO_2 laser with a beam expanded to a line in vertical vertical direction and rapidly scanned in horizontal direction was used to illuminate the marked patch of the image in the lower left section on the water surface and further to the left. The wind is blowing from the right to the left.

Infrared images were taken with an Amber Radiance infrared camera, which takes 60 frames per second with a resolution of 256×256 pixel and a NEDT of less than 20 mK. The switching frequencies for the heating of the water surface by the laser ranged from 0.025 Hz to more than 10 Hz. Further details of the setup and measurements can be found in Popp [12].

Some original infrared images are shown in Figs. 18.4 and 18.5. At 2 m/s wind speed the homogeneously heated patch is still very homogeneous after 0.5 s (Fig. 18.4a). Obviously the penetration of the heat is still confined to the part of the mass boundary layer that is still dominated by molecular diffusion. Half a second later (Fig. 18.4b), the first structures become visible and they are fully developed 3 s after start of the heating (Fig. 18.4f). The characteristic time constant t_* was determined to be

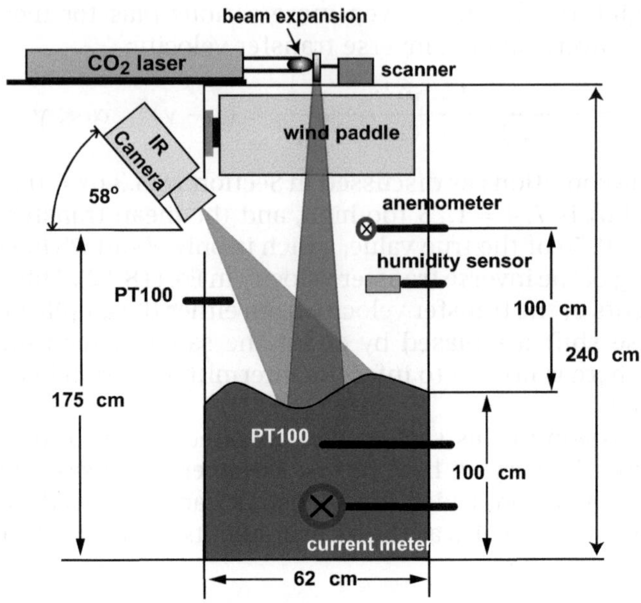

Figure 18.3. Experimental setup for active thermographic measurements in the Heidelberg Aeolotron.

about 3 s. Thus this infrared image sequence nicely demonstrates how the turbulence increases with increasing distance from the water interface. It can also be observed that the shear current in the boundary layer causes the heated boundary layer to be convected to the left beyond the heated patch.

At 7 m/s, already 0.5 s after the start of the heating, fine scale turbulent structures become visible. This clearly indicates that the turbulence penetrates much closer to the water surface and that the boundary layer is much thinner. The characteristic time constant t_* was determined to be about 0.14 s. Therefore the boundary layer is already in equilibrium with the applied heat flux after 0.5 s and all infrared images in Fig. 18.5 look the same. Because of the shorter characteristic time constant, the heated boundary layer is convected only a shorter distance. The surface velocity u_s is higher, but it increases only with the friction velocity ($u_s \propto u_*$), while the time constant t_* is proportional to the inverse squared friction velocity ($t_* \propto u_*^{-2}$). According to Eq. (18.5) this leads to a horizontal length scale $x_* \propto u_*^{-1}$.

The measurements of the amplitude damping as a function of the switching frequency at wind speeds ranging from 2.9 to 6.2 m/s are shown in Fig. 18.6. The measurements show little scatter and agree well with the model predictions. The constant amplitude response gradually changes to an amplitude damping that increases with the square root of the fre-

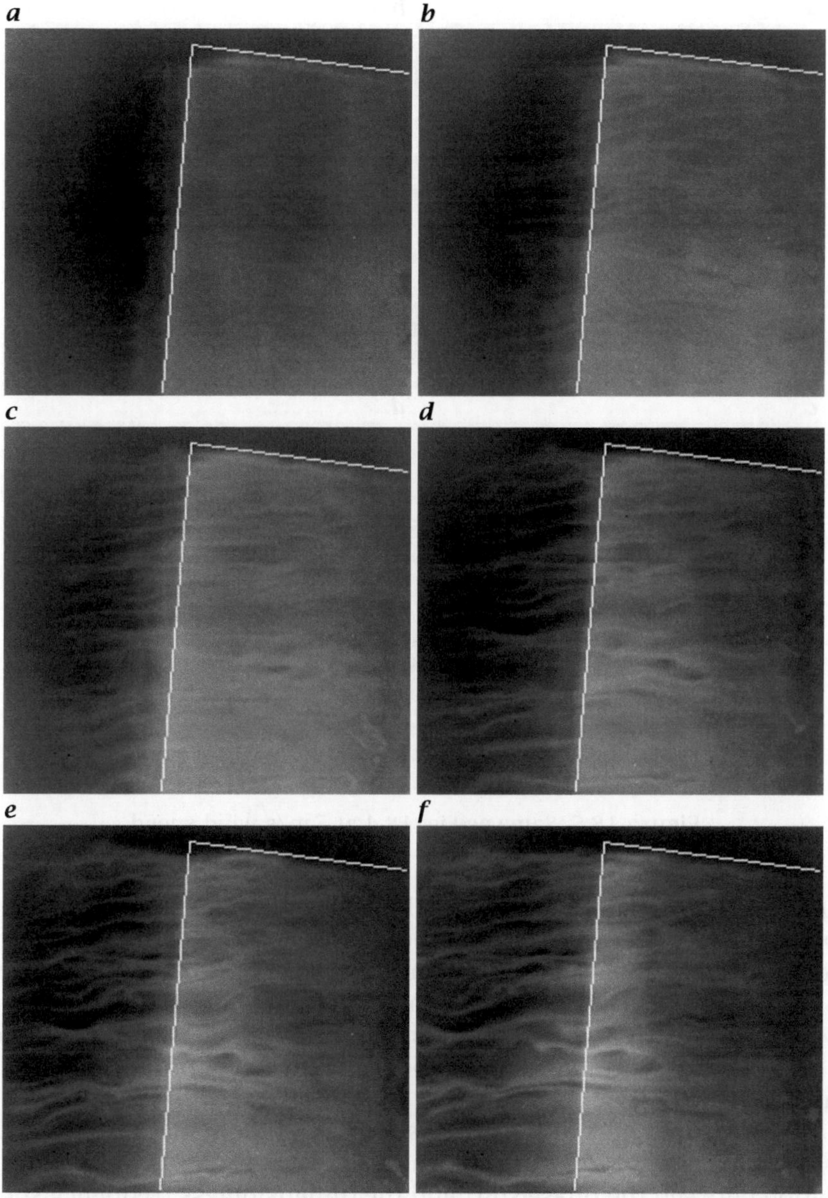

Figure 18.4. Infrared images of the water surface taken after with heating of the marked area of the image by a scanning an expanded 100 W CO_2 laser beam. The images show a sector of about $25 \times 25\,cm^2$ and are taken at a wind speed of 2 m/s 0.5, 1.0, 1.5, 2.0, 2.5, and 3.0 s after the laser heating has been started. The wind is blowing from the left to the right.

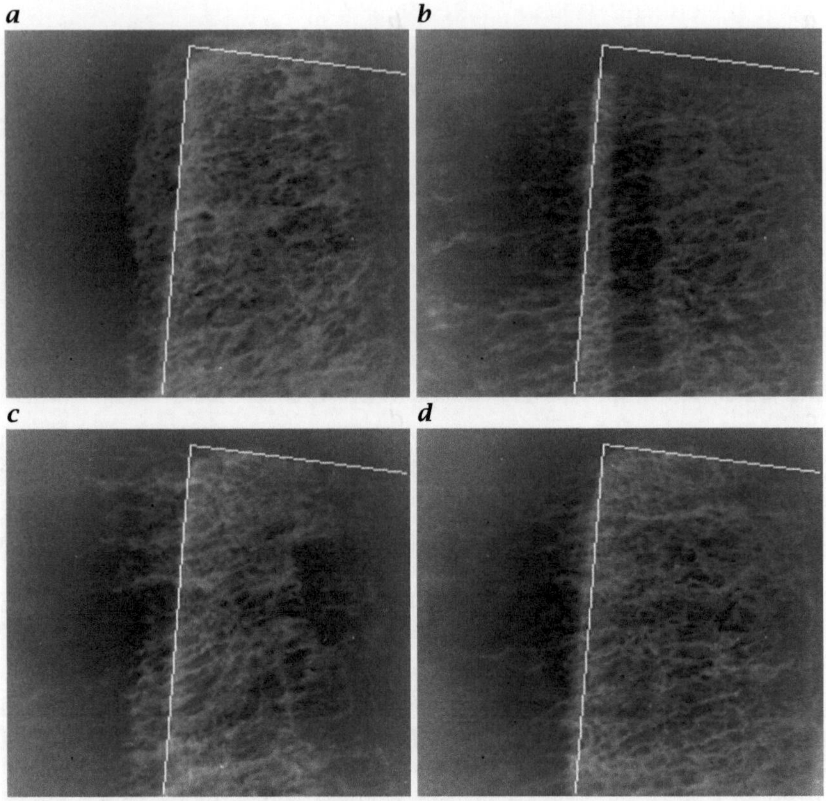

Figure 18.5. Same as Fig. 18.4 at 7 m/s wind speed.

quency as predicted. In this range the measurements at all wind speeds coincide. This directly proves that the transport very close to the interface is dominated by molecular diffusion.

At low frequencies in the constant range, the amplitude response is lower at higher wind speeds. Because the heat flux density is the same at all wind speeds, the amplitude response is inversely proportional to the transfer velocity.

The measurements of the phase speed as a function of the dimensionless frequency $\omega_+ = \omega t_*$ are shown in Fig. 18.7. In this presentation, the relation between the phase shift and the dimensionless frequency should be the same for all wind speeds, except for some variation at low frequencies because of different values of κ in Eq. (18.25). The experimental values generally agree with the model predictions. Especially at high frequencies, the scatter is low and the maximum values agree very well with the predicted value of $\pi/4$. At low frequencies, the scatter is quite large, so that different models still cannot be distinguished. Further investiga-

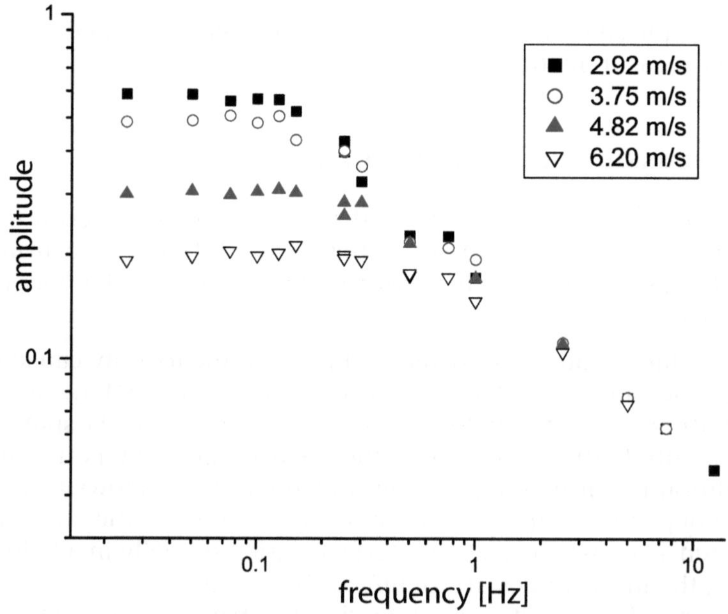

Figure 18.6. Measured amplitude damping as a function of the frequency $\nu = \omega/2\pi$ for wind speeds as indicated.

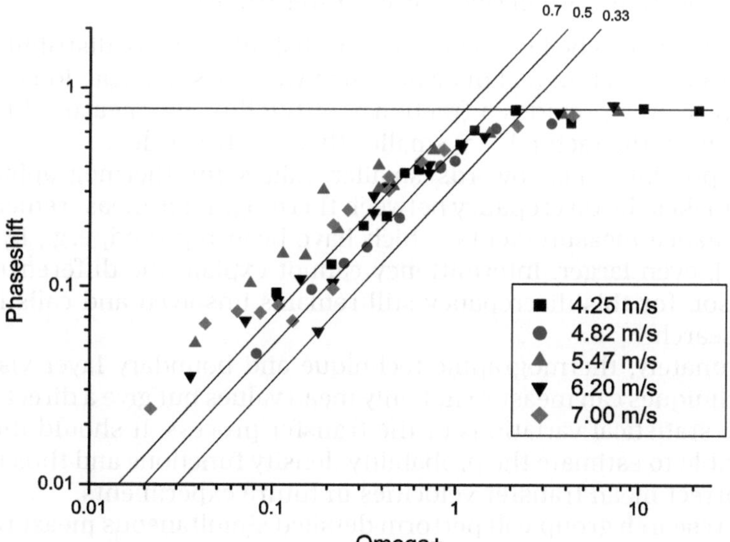

Figure 18.7. Measured phase shift as a function of the dimensionless frequency ω_+ for wind speeds as indicated. The solid lines mark the theoretical relation given by Eq. (18.25) for values of κ of 0.7, 0.5, and 0.33.

tions are required to determine the cause for this scatter and to reduce it in further measurements.

18.5 Discussion and Conclusions

This paper theoretically investigated the influence of intermittency on determining average transfer velocities using different measuring techniques. It was shown that all measuring techniques can be biased by intermittency:

- Mass balance and eddy correlation measurements only cause no bias in the measurement of the mean transfer velocity as long as the concentration difference between the air and the water is spatially and temporally homogeneous over the measurement interval. When this condition is not met, e. g., if the transfer velocity shows a significant fetch dependency in a linear wind wave facility or the concentration difference across the air-sea interface varies spatially in a field experiment, the mean values are significantly biased.
- Mean transfer velocities calculated either from mean boundary layer thicknesses or from thermographic techniques, which compute the mean transfer velocity either from concentration differences of from time constants, are biased toward lower values.

The effects can be large and if the probability density distribution can be described by a bimodal distribution in which a small fraction $\alpha \ll 1$ of the surface area provides a fraction γ of the flux, the measured transfer velocity is by the factor $1 - \gamma$ smaller than the true value.

The possible bias towards smaller values for thermographic techniques makes the discrepancy between thermographic measurements and mass balance measurements, which have been reported, e. g., by Asher et al. [1], even larger. Intermittency cannot explain the difference. Thus the reason for this discrepancy still remains unsolved and calls for further research.

Fortunately, thermographic technique and boundary layer visualization techniques can measure not only mean values but give a direct insight into the statistical variations of the transfer process. It should therefore be possible to estimate the probability density functions and thus to compute correct mean transfer velocities in future experiments.

Our research group will perform detailed simultaneous measurements of gas exchange and heat exchange in different wind/wave facilities in conjunction with a better stochastic modeling of the transfer mechanisms in the near future.

Acknowledgement. This work has partly been funded by DFG Research Unit Program FOR240 Image Sequence Analysis to Investigate Dynamic Processes

References

[1] Asher, W. E., A. T. Jessup, and M. A. Atmane. Oceanic application of the active controlled flux technique for measuring air-sea transfer velocities of heat and gases. *J. Geophys. Res.*, 109:C08S12, doi:10.1029/2003JC001862, 2004.

[2] Asher, W. E. and J. Pankow. The interaction of mechanically generated turbulence and interfacial films with a liquid phase controlled gas/liquid process. *Tellus*, 38B:305–318, 1986.

[3] Atmane, M. A., W. E. Asher, and A. T. Jessup. On the use of the active infrared technique to infer heat and gas transfer velocities at the air-water free surface. *J. Geophys. Res.*, 109:C08S14, doi:10.1029/2003JC001805, 2004.

[4] Broecker, W. J., R. Ledwell, T. Takahashi, R. Weiss, L. Merlivat, L. Memery, T.-H. Peng, B. Jähne, and K. O. Münnich. Isotopic versus micrometeorologic ocean CO_2 fluxes: a serious conflict. *J. Geophys. Res.*, 91:10,517–10,527, 1986.

[5] Garbe, C. S., U. Schimpf, and B. Jähne. A surface renewal model to analyze infrared image sequences of the ocean surface for the study of air-sea heat and gas exchange. *J. Geophys. Res.*, 109:C08S15, doi:10.1029/2003JC001802, 2004.

[6] Garbe, C. S., H. Spies, and B. Jähne. Estimation of Surface Flow and Net Heat Flux from Infrared Image Sequences. *J. Mathematical Imaging and Vision*, 19:159–174, 2003.

[7] Haußecker, H., S. Reinelt, and B. Jähne. Heat as a proxy tracer for gas exchange measurements in the field: principles and technical realization. In Jähne, B. and E. Monahan, eds., *Air-Water Gas Transfer, Selected Papers, 3rd Intern. Symp. on Air-Water Gas Transfer*, pp. 405–413. AEON, Hanau, 1995.

[8] Haußecker, H., U. Schimpf, C. S. Garbe, and B. Jähne. Physics from IR image sequences: quantitative analysis of transport models and parameters of air-sea gas transfer. In Donelan, M. A., W. M. Drennan, E. S. Saltzman, and R. Wanninkhof, eds., *Gas Transfer at Water Surfaces*, vol. 127 of *Geophysical Monograph*, pp. 103–108. American Geophysical Union, Washington, DC, 2001. invited paper.

[9] Jähne, B., P. Libner, R. Fischer, T. Billen, and E. Plate. Investigating the transfer processes across the free aqueous viscous boundary layer by the controlled flux method. *Tellus*, 41B:177–195, 1989.

[10] Jessup, A. T., C. J. Zappa, and H. Yeh. Defining and quantifying microscale wave breaking with infrared imagery. *J. Geophys. Res.*, 102: 23,145–54, 1997.

[11] Münsterer, T. and B. Jähne. LIF measurements of concentration profiles in the aqueous mass boundary layer. *Exp. Fluids*, 25:190–196, 1998.

[12] Popp, C. *Untersuchungen zum windinduzierten Wärmeaustausch durch die wasserseitige Grenzschicht an der Wasseroberfläche mittels aktiver Thermographie.* Diss., Univ. Heidelberg, 2006.

[13] Schimpf, U., C. Garbe, and B. Jähne. Investigation of transport processes across the sea surface microlayer by infrared imagery. *J. Geophys. Res.*, 109:C08S13, doi:10.1029/2003JC00180, 2004.

[14] Wolff, L. M. and T. J. Hanratty. Instantaneous concentration profiles of oxygen accompanying absorption in stratified flow. *Exp. Fluids*, 16:385–392, 1994.

[15] Zappa, C. J., W. E. Asher, A. T. Jessup, J. Klinke, and S. R. Long. Microbreaking and the enhancement of air-water transfer velocity. *J. Geophys. Res.*, 109:C08S16, doi:10.1029/2003JC001897, 2004.

[16] Zemmelink, H. J., J. W. H. Dacey, E. J. Hintsa, W. R. McGillis, W. W. C. Gieskes, W. Klaassen, H. W. de Groot, and H. J. W. de Baar. Fluxes and gas transfer rates of the biogenic trace gas DMS derived from atmospheric gradients. *J. Geophys. Res.*, 109:C08S10, doi:10.1029/2003JC001795, 2004.

19

Micrometeorological Approaches to Measure Air-Water CO_2 Fluxes

Wade R. McGillis[1,2] and James B. Edson[3]

[1] Lamont-Doherty Earth Observatory
Columbia University
Palisades, NY 10964, USA
[2] Earth and Environmental Engineering
Columbia Universit
NY, NY 10027, USA
[3] Department of Marine Sciences
University of Connecticut
Groton, CT 06340, USA

Abstract The processes at the ocean surface play an important role in the exchange of carbon dioxide (CO_2) to and from the atmosphere. Despite this, some fundamental mechanisms that control the rate of interfacial transfer have not been well resolved. Globally, the ocean absorbs a significant amount of atmospheric carbon, however, the ocean has strong local sinks and sources. The carbon that is exchanged between the ocean and atmosphere is primarily as a dissolved gas that is not highly soluble in water. This means that molecular diffusion and turbulent mixing at the aqueous boundary layer are the significant transport processes controlling CO_2 exchange rates across the surface. The aqueous boundary layer and the corresponding CO_2 exchange rate are affected by the local climate and environmental conditions in both the atmosphere and ocean. For the first time, the turbulent transport through the marine atmospheric boundary layer has been measured in the North Atlantic and the Equatorial Pacific using micrometeorological techniques to quantify the CO_2 flux. Measurements also provided an understanding of the environmental conditions controlling the exchange rate. The North Atlantic was a large sink of atmospheric CO_2 with a high variability in flux ranging from 1.2-4.2 mol m^2 yr^{-1} that was forced primarily by the wind. The Equatorial Pacific was a strong source of atmospheric CO_2. Here, the CO_2 flux ranged from 3.0-4.2 mol m^2 yr^{-1} and was forced primarily by diurnal cycles.

19.1 Introduction

Currently, it is estimated that the ocean, as a whole, acts as a sink for CO_2, taking up approximately 2-billion tons per year of the approximately 6.5-billion tons of carbon produced by industrial and agricultural activity.

C.S. Garbe, R.A. Handler, B. Jähne (eds.): Transport at the Air Sea Interface
pp. 275-287, 2007,

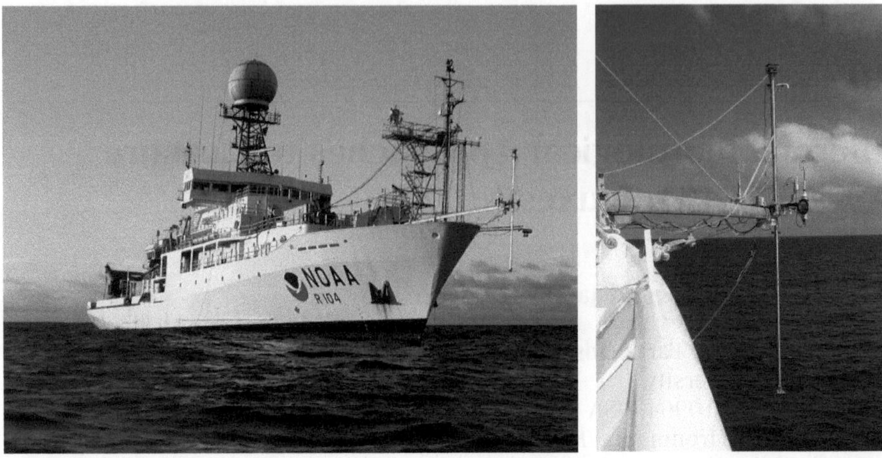

Figure 19.1. The boom system on the bow of the Ronald H. Brown research vessel (left) during the GasEx-2001 open ocean CO2 exchange experiment in the equatorial Pacific (3°S 125°W). The boom was extended 10 m from the side of the vessel. Direct covariance and gradient CO2 flux micrometeorological methods where performed when the bow was manoeuvred into the predominant wind direction. Vertical profiles of atmospheric CO_2 near the ocean surface where measured from the mast located at the end of the boom (right).

However, there is uncertainty in the magnitude and variability of this estimate because the kinetics of ocean-air CO_2 transfer is not well understood [11]. The surface CO_2 flux, F_{CO_2}, can be modeled as

$$F_{CO_2} = k_{600} \Delta C_{CO_2} \left(\frac{Sc_{CO_2}}{660} \right)^{-n}, \tag{19.1}$$

where k_{600} is the gas transfer velocity referenced to Sc=660, ΔC_{CO_2} is the difference between dissolved CO_2 in the bulk seawater and at the ocean surface, Sc_{CO_2} is the CO_2 Schmidt number, and n is a hydrodynamic variable adjusted for flow condition. For turbulent conditions at the ocean surface, n is measured and modeled to be a value of 0.5. The difference of aqueous concentration, ΔC_{CO_2}, is related to the partial pressure difference, ΔpC_{CO_2}, by

$$\Delta C_{CO_2} = s \cdot \Delta pC_{CO_2}, \tag{19.2}$$

where s is the solubility of CO_2, which is a function of temperature and salinity. The surface pCO_2 can vary with temperature, salinity, bubble entrainment, and by biogeochemical changes. In addition, the gas transfer velocity rate, k, can vary with localized environmental conditions, including wind speed [5, 6, 7, 13, 15, 16], atmospheric stability [2], sea state, and aqueous surfactant concentrations [4].

19.2 Ocean-Atmosphere CO_2 Flux Study Sites

GasEx-1998 and GasEx-2001 were open-ocean gas exchange studies focused on shipboard micrometeorological approaches to measure ocean-atmosphere CO_2 fluxes. Before these efforts, few experiments have attempted shipboard micrometeorological techniques to measure gas exchange. The study regions for GasEx-1998 and GasEx-2001 were chosen on the basis of the strength and constancy of the local ΔpCO_2 signal. Large signals were desired to measure CO_2 flux as accurately as possible.

Direct covariance and profile fluxes of CO_2, air and water pCO_2, and wind speed measurements where performed on the NOAA SHIP RONALD H. BROWN (Figure 19.1) during GasEx-1998 in the N. Atlantic at 46° 6' N and 20° 55' W in June and during GasEx-2001 in the Equatorial Pacific at 3° S and 125° W in February. During GasEx-1998, a large, warm core eddy was stable during throughout the experiment. A spring phytoplankton bloom resulted in a large ocean CO_2 undersaturation, which, in turn, provided a large homogeneous sink to perform the experiment. GasEx-2001 took place in a strong atmospheric CO_2 source region with persistent, low windspeeds, and strong diurnal cycles in the South Equatorial Current.

The Takahashi [12] map shows the climatological mean distribution of the net annual sea-air CO_2 flux over the global ocean for the reference year 1995. The net flux has been computed by multiplying the mean monthly sea-air pCO_2 difference and the monthly CO_2 gas transfer rate in each pixel. The sea-air gas transfer rate is generally computed with either the Wanninkhof [13] or Wanninkhof and McGillis [15] model and 10-meter wind speed data. Intense CO_2 source areas are red to yellow and in the equatorial Pacific, Arabian Sea, tropical eastern Atlantic, and northwest subartic Pacific. These are attributed to warming of the upwelled subsurface waters containing high concentrations of respired CO_2. The major sink areas are in the subpolar regions of both hemispheres and are attributed to cold temperatures and intense photosynthetic drawdown of CO_2 during spring-summer months.

19.3 Measurements

Measurements of ΔpCO_2 and windspeed made during the experiments are shown in Figure 19.2. The mean ΔpCO_2 during GasEx-1998 was -85.8±16.0 ppm, while windspeed ranged from 0.85 to 16 ms^{-1}. Wind speeds were measured with a 3-axis ultrasonic anemometer and corrected for ship motion [7]. Note, that observations show variability in pCO_2 over time scales of less than one day, and variability in wind speed over time scales of less than one hour. The mean 2001 ΔpCO_2 during GasEx-2001 was +118 ± 3.5 ppm, with a mean, low variability windspeed of 6.1 m s^{-1}.

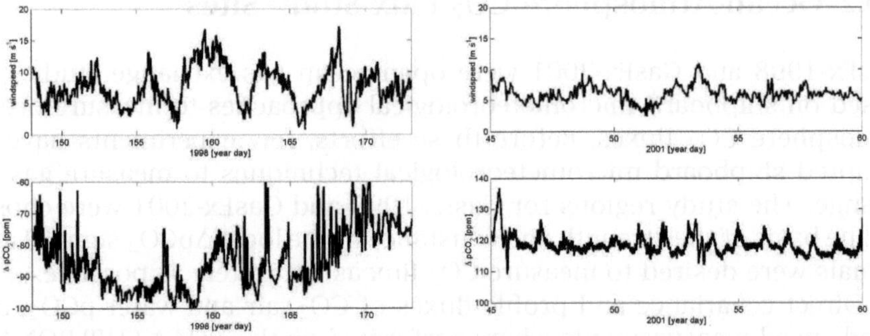

Figure 19.2. Time series of wind speed and air-sea pCO2 difference for (left) GasEx-1998 in the North Atlantic and (right) GasEx-2001 in the Equatorial Pacific. Data are shown in the same vertical spans for comparison. GasEx-1998 conditions included high winds, a CO2 sink, and storm cycles. GasEx-2001 conditions included low winds, a CO2 source, and diurnal cycles.

During GasEx-1998, the temporary decrease in ΔpCO_2 on yearday 156 and 159 were caused by high wind events entraining water masses from below the mixed layer. Overall, the ΔpCO_2 was large for ocean surface waters. After yearday 166 the ΔpCO_2 decreased and greater variability was observed. These features were present because the peak of the algal bloom was over and the warm core eddy was eroding. Note, the ΔpCO_2 varies on times scales less than a day. Wind speeds varied on hour scales and over a wide range with gusts up to 17.7 ms^{-1}. During GasEx-2001, low windspeeds and large solar heating caused cyclical diurnal ΔpCO_2 that was persistent throughout the experiment. Data show significant small-scale variability in the environmental factors forcing oceanic CO_2 transfer.

19.3.1 Direct Covariance

The direct covariance technique provides a means to estimate the ensemble average CO_2 flux across an air-water interface. This technique relates the atmospheric CO_2 concentration fluctuations, c', and the vertical wind fluctuations, w', to the ocean-atmosphere gas flux, F_c, through the expression

$$F_c = \overline{w' \cdot c'} \tag{19.3}$$

By using CO_2 concentrations in that of dry air, this term accounts for mesoscale transport caused by buoyant plumes where non-zero mean vertical flows transport gas [17], and where the mean vertical velocity is dependent on temperature and humidity [3].

In order to estimate CO_2 flux using the direct covariance method, fluctuations in vertical wind velocity and atmospheric CO_2 concentration

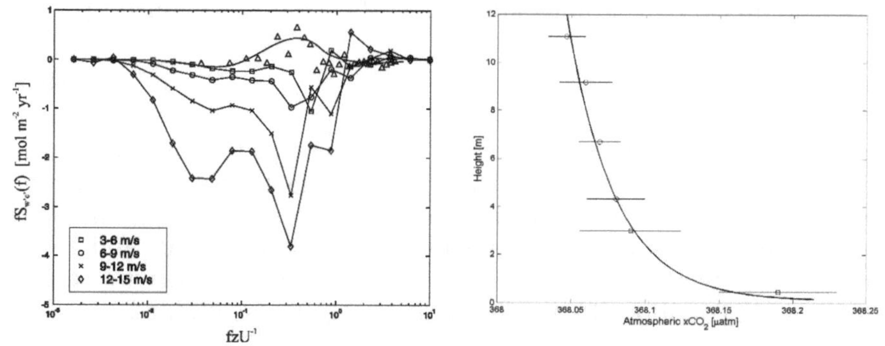

Figure 19.3. Micrometeorological approaches for measuring air-sea CO_2 fluxes. (left) Cospectra of vertical wind fluctuations and atmospheric CO_2 fluctuations measured during GasEx-98 with the direct covariance CO_2 flux system. Ensemble spectral densities for the wind speed range 3-6 m s^{-1} (□), 6-9 m s^{-1} (∘), 9-12 m s^{-1} (×), and 12-15 m s^{-1} (◇). Cospectra of vertical wind fluctuations and zero reference gas measured during calibration intervals (△) are shown for comparison. (right) Atmospheric profile of xCO_2 near the ocean surface. Averaged atmospheric CO_2 profile, showing 1 standard deviation, collected during the cruise from masts on the ship (∘) and the surface processes instrument platform (SPIP) (□). The mean aqueous xCO_2 was 473 µatm.

must be determined. When recording the vertical wind speed from a moving ship, part of the measurement velocity is due to the ship motion itself. Thus, the effects of vertical ship motion must be fully accounted for in windspeed data, before the wind velocity can be computed [1].

The directly measured wc cospectra, averaged in wind speed bins from the entire GasEx-1998 deployment, are shown in Figure 19.3. The cospectral levels increase with increasing wind speed, but the overall shape of each cospectrum is similar. The figure also shows the average wc cospectrum computed during the calibration periods.

19.3.2 Profile Method

A second micrometeorological approach to quantifying surface layer fluxes is the flux-profile method. The flux-profile method uses Monin-Obukov similarity theory and the vertical gradient of the mean concentration for CO_2, $\overline{c}$. The normalized gradient should be a universal function of ξ such that

$$\frac{\kappa z}{c_\star}\frac{\partial \overline{c}}{\partial z} = \varphi_c(\xi) \tag{19.4}$$

where $c_\star$ is the scaling parameter for the CO_2 flux, φ_c is an empirically determined dimensionless gradient function for c [10] and $\xi = z/L$ is the stability function where z is the height above the mean water surface and

L is the Monin-Obukov scaling length. The M-O length is defined by

$$L = \frac{\overline{T_v}}{\kappa g} \frac{u_\star^3}{F_{T_v}} \tag{19.5}$$

where g is the acceleration of gravity, T_v is the virtual temperature, F_{T_v} is the buoyancy flux, $u_\star$ is the velocity scaling parameter also known as the friction velocity, and κ is the von Káráman constant. Note, that the M-O length represents the ratio of mechanical to thermal forcing, and $z = L$ represents the height at which these two forcing mechanisms are equal.

Integrating Equation (19.4), a logarithmic profile is obtained

$$\overline{c}(z) - \overline{c}(z_r) = \frac{c_\star}{\kappa}\left[\ln\left(\frac{z}{z_r}\right) - \psi_c(\xi) + \psi_c(\xi_r)\right] \tag{19.6}$$

where ψ_c is the integral form of φ_c, and $\overline{c}(z_r)$ represents the reference value of CO_2 measured at the fixed height of $z = z_r$. If the value of the M-O length is known from either the direct covariance or bulk fluxes, the scaling parameter, $c_\star$, can be determined as the slope [8]. The flux is then calculated by combining this scaling parameter with the friction velocity

$$F = c_\star \cdot u_\star \tag{19.7}$$

The gas transfer velocity, k, for the profile technique is calculated using the CO_2 flux from equations (19.7) and (19.1).

The profiles are determined using one set of traveling sensors that measure concentrations at various heights within the atmospheric boundary layer. A single set of instruments is used to augment the precision of the measurement since this will eliminate inter-sensor biases. A second set of sensors is held fixed in order to account for the atmospheric variability that is likely to occur over the course of the measurement time. Open-ocean measurements of water vapor gradients show that the atmospheric variability is a factor of 3 greater than the vertical gradient [7]. By using a difference between the atmospheric mean and profiled properties, the atmospheric variability was removed.

GasEx-2001 average profiles of atmospheric CO_2 are also shown in Figure 19.3. Since the profile method relies on mean gradients in the atmospheric surface layer, it has several advantages. Many gases cannot be measured at rates that are fast enough for direct covariance because many gas measurement techniques rely on slow procedures that can also dilute the high frequency response (eg. water vapor removal, mass spectrometry procedures, gas chromatography). Other methods, such as deliberate gas addition rely on measurement procedures that result in large temporal and spatial footprints. The profile method averages over small spatial scales of about 1000 m and temporal scales of 30 minutes, both of which are necessary for sampling processes controlled by atmospheric and ocean surface forcing.

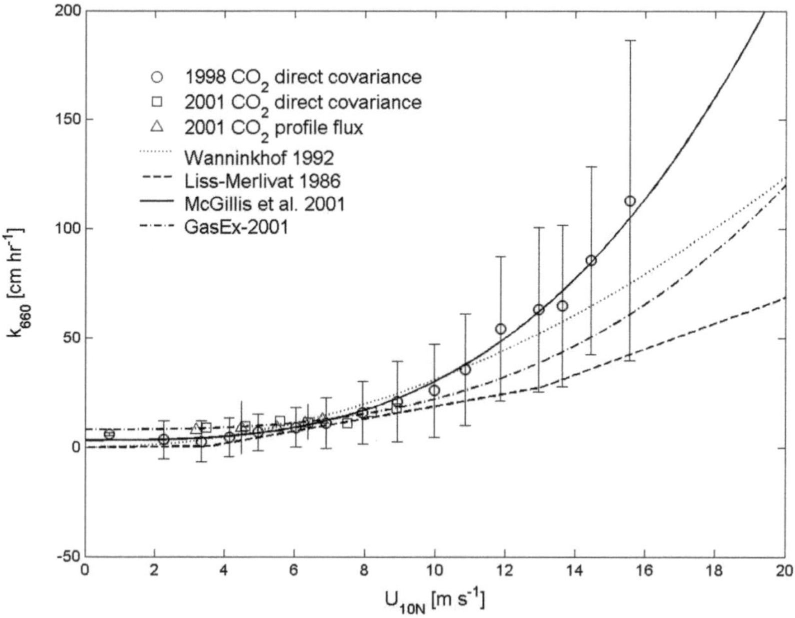

Figure 19.4. Measurements from GasEx-1998 and GasEx-2001. CO_2 gas transfer velocity versus windspeed. A cubic windspeed relationship, with a finite zero-windspeed offset, is fit to this data and shown for comparison. Transfer velocity versus wind speed relationships based on ^{14}C budgets[2] (solid line) and based on lake and wind tunnel experiments[4] (dotted line) are shown for comparison. Enhanced CO_2 fluxes are observed during periods of high windspeeds.

Figure 19.4 presents ensemble results of the calculated CO_2 gas transfer velocity versus the 10-meter atmospheric wind speed. Also presented on this figure are estimates from isotopic and indirect methods previously reported in the literature. Figure 19.4 shows that there is some reconciliation between direct measurements and the models based on wind speeds. For the first time, direct air-sea CO_2 flux measurements are in general agreement with geochemical isotope constraints. At higher wind speeds, the direct measurements exceed the values reported by Liss et al. [6] and Wanninkhof [13]. This may be due, in part, to the presence of breaking waves, which are not accounted for by the Liss et al. [6] and Wanninkhof [13] models.

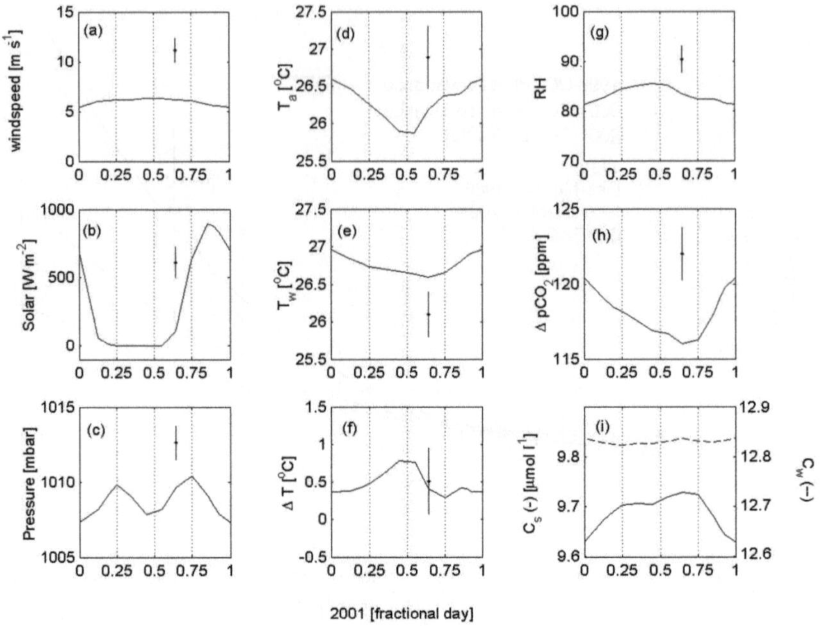

Figure 19.5. Diurnal averages of: (a) windspeed [m s^{-1}]; (b) incident solar radiation [W m^{-2}]; (c) barometric pressure [mbar]; (d) air temperature [°C]; (e) water temperature [°C]; (f) air-sea temperature difference [°C]; (g) relative humidity [%]; (h) pCO2 [ppm]; and (i) Bulk and skin aqueous CO_2 concentrations [mol l^{-1}]. All properties show a coherent trend as a result of the diurnal forcing. The standard deviations of the data (·) are shown for comparison.

19.4 Discussion

Because the ocean surface can still be disrupted by wind-gusts having zero-mean velocity, differential heating, or by surface currents that might enhance mixing by interaction with older, decaying waves, we suggest that the transfer of CO_2 across the aqueous mass boundary layer may still occur at zero mean wind speed. This hypothesis is supported by our data, which demonstrate that the transfer velocity asymptotes to a finite, non-zero value at low wind speeds. If we account for this possibility, a cubic wind speed relationship [9] that fits both GasEx-1998 and GasEx-2001 data becomes $k = 5.1 + 0.024u^3$, where the constant term is a measure of the CO_2 exchange at very low wind speeds.

The physical and biogeochemical climatology in GasEx-2001 was unique because the diurnal cycling was very consistent. This provides an environment to explore diurnal physical, chemical, and biological processes by analyzing daily averages of the data. Figures 19.5 and 19.6 show the

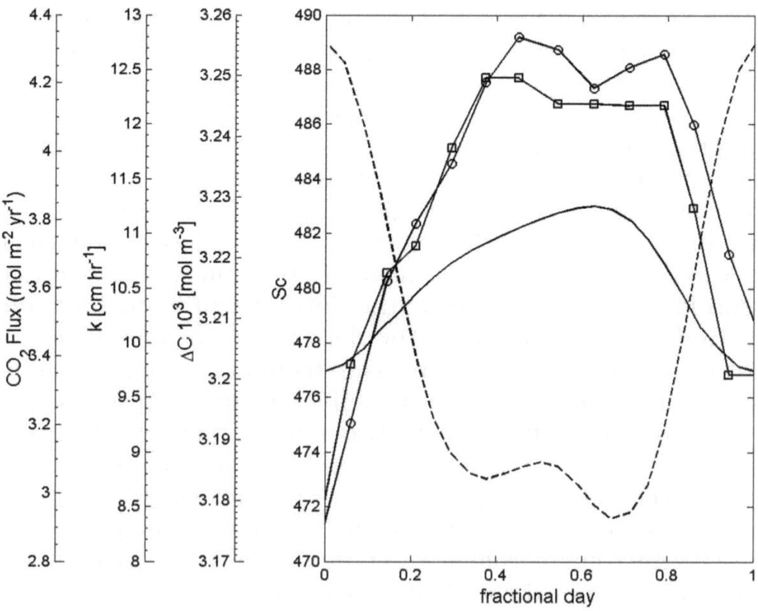

Figure 19.6. Ensemble-averaged diurnal Sc (solid line), bulk-surface aqueous CO_2 difference (dashed-line), gas transfer velocity (○), and air-sea CO_2 flux (□).The diurnal variability of the bulk-skin aqueous CO_2 difference is caused by temperature, pressure, and biological effects. The diurnal effect on gas diffusivity is caused by changes in the surface temperature. The marked increase in nighttime fluxes is > 1.2 mol m^{-2} yr^{-1} representing a 40% increase.

effect of diurnal cycles on the properties controlling CO_2 flux. Average diurnal trends show a small variation in windspeed with a strong cycle in solar incidence.

Figure 19.5 shows the barometric pressure had a diurnal variation caused by atmospheric tides. Nighttime cooling resulted in a decrease in both air temperature and ocean temperature. The mean diurnal ocean temperature was greater than the air temperature with a small increase in the difference at night. The relative humidity also increased at night. These coherent processes resulted in a nighttime decrease in ΔpCO_2. An examination of the CO_2 driving the flux is also shown in Figure 19.5. The diurnal variability of dissolved CO_2 in the bulk water is low. However, the surface dissolved aqueous CO_2 increased at night predominantly because of an increase is gas solubility as a result of cooler surface water. The result on the air-sea difference of dissolved aqueous CO_2 is shown in Figure 19.6.

These diurnal properties had a clear effect on the air-sea CO_2 flux. Figure 19.6 shows that Sc increased and ΔC_{CO_2} decreased at night. The combined effect had only a 2.7% nighttime decrease from the daytime values. However, there was a large increase in the gas transfer velocity, k, which was measured as a 40% increase in nighttime CO_2 fluxes from the Equatorial Pacific. The increase in gas transfer rate is attributed to an increase in nighttime buoyancy fluxes and winds.

At present, the micrometeorological approach is the most capable of resolving surface processes driving the CO_2 exchange between the ocean and the atmosphere over a wide range of environmental conditions. Micrometeorological CO_2 fluxes from the North Atlantic provided information on windspeed dependence. The dependence of CO_2 fluxes on the diurnal cycle was provided from measurements in the Equatorial Pacific. Other successful methods to infer the effect of surface processes on CO_2 fluxes are the deliberate tracer and infrared techniques. However, infrared techniques are limited from high wind conditions and daytime conditions. Deliberate tracer techniques average over several days and cannot resolve wind speed or surface process variability within diurnal cycles.

The GasEx micrometeorological CO_2 flux studies are providing a physics-based quantification of the exchange processes. GasEx results show several possible feedbacks with atmospheric CO_2. If atmospheric CO_2 levels increase global temperatures, there may be increases in atmospheric storms and clouds. In the high latitude sink regions, as in GasEx-1998, the ocean will be a larger sink for atmospheric CO_2 due to enhanced gas exchange. In the tropical source regions, as in GasEx-2001, the ocean will be a reduced source for atmospheric CO_2 due to a decrease in surface heating and a decrease in gas exchange.

Previous oceanic measurements of air-sea gas exchange have relied on techniques averaged over days, weeks, or even months, and have generally used gases with physiochemical properties different than CO_2. Micrometeorological methods allow small-scale quantification of CO_2 behavior, that may provide improvement over these indirect techniques. To date, large uncertainties in the extent of the CO_2 flux between the atmosphere and ocean have prevented us from accurately quantifying how the increasing atmospheric CO_2 burden partitions between the ocean and the terrestrial biosphere. This limits our ability to predict future atmospheric CO_2 levels. The direct covariance technique promises to provide extremely fine scale resolution of changes in air-sea gas flux behavior over time, which will aid in the determination of the small-scale physical and biological processes responsible for mediating changes in the ocean uptake of CO_2.

19.5 Summary

Knowledge of the flux of gas between the atmosphere and the ocean is essential in understanding the sources and sinks of CO_2 is ever important to climate change. Until now, estimates of local air-sea CO_2 flux in the open ocean have been inferred from indirect methods of gas exchange. While indirect methods are necessary to understand many air-water exchange processes, most are unable to directly reproduce all the environmental conditions associated with air-water CO_2 transfer. Results are reported here from a study of a North Atlantic CO_2 sink and an Equatorial Pacific CO_2 source using micrometeorological techniques, and which enable, direct measurement of the CO_2 flux between the ocean and atmosphere over hour time scales. The direct measurements presented are in general agreement with previous estimates based on parameterizations for CO_2 behavior, as well as geochemical isotope constraints.

The air-sea CO_2 flux is predominantly controlled by the CO_2 concentration difference across the ocean surface and the turbulent mixing at the surface of the ocean. The surface turbulence changes on hourly time scale or less. If the physical and chemical processes controlling the flux of CO_2 into the ocean vary on hourly time scales, the CO_2 flux itself must be measured on commensurate time scales. The GasEx micrometeorological CO_2 flux studies provide unique information on oceanic control under different environmental conditions. Micrometeorological flux measurements had not been performed over the open ocean. Results indicate an enhancement in windspeed and diurnal forcing. If the gas exchange enhancement is applied globally, the ocean sources and sinks will be larger than presently estimated. Since sink regions have higher windspeeds, the global ocean will be a greater net sink [14].

Acknowledgement. We thank J. Ware (WHOI) and the crew of the ship Ronald H. Brown, and NOAA for making the field experiment possible. We also thank Jeffrey Hare (NOAA ESRL) Christopher Fairall (NOAA ESRL), and Rik Wanninkhof (NOAA AOML) for supporting measurements. This research was performed and the manuscript prepared with support from: the National Science Foundation grant numbers OCE 03-27256 and OCE 05-26677, and the National Ocean and Atmospheric Administration grant number NA03OAR4320179. The Lamont-Doherty Earth Observatory contribution number is 7015.

References

[1] J. B. Edson, A. A. Hinton, K. E. Prada, J. E. Hare, and C. W. Fairall. Direct covariance flux estimates from moving platforms at sea. *Journal of Atmospheric and Oceanic Technology*, 15:547–562, 1998.

[2] III Erickson, D. J. A stability dependent theory for air-sea gas exchange. *Journal of Geophysical Research*, 98(C5):8471–8488, 1993. doi: 10.1029/93JC00039.

[3] C. W. Fairall, E. F. Bradley, J. S. Godfrey, G. A. Wick, and J. B. Edson. Cool-skin and warm-layer effects on sea surface temperature. *Journal of Geophysical Research*, 101(C1):1295–1308, 1996.

[4] N. M. Frew, P. S. Liss, and R. A. Duce. The role of organic films in air-sea gas exchange. In *The Sea Surface and Global Change*, pages 121–171. Cambridge University Press, Cambridge,UK, 1997.

[5] Jacobs, C.M.J., W. Kohsiek, and W.A. Oost (1999) Air-sea fluxes and transfer velocity of CO_2 over the North Sea: results from ASGAMAGE. Tellus, 51B, 629–641

[6] P. S. Liss, L. Merlivat, and P. Buat-Menard. Air-sea gas exchange rates: Introduction and synthesis. In *The role of air-sea exchange in geochemical cycling*, pages 113–129. Reidel, Boston,MA, 1986.

[7] W. R. McGillis, J. B. Edson, J. E. Hare, and C. W. Fairall. Direct covariance air-sea CO_2 fluxes. *Journal of Geophysical Research*, 106(C8): 16729–16745, 2001.

[8] W. R. McGillis, J. B. Edson, J. D. Ware, J. W. H. Dacey, J. E. Hare, C. W. Fairall, and R. Wanninkhof. Carbon dioxide flux techniques performed during gasex-98. *Marine Chemistry*, 75(4):267–280, 2001.

[9] E. C. Monahan. Occurrence and evolution of acoustically relevant subsurface bubble plumes and their associated, remotely monitorable, surface whitecaps. In B. R. Kerman, editor, *Natural Physical Sources of Underwater Sound*, pages 503–517. Kluwer Academic, 1993.

[10] C. A. Paulson. The mathematical representation of wind speed and temperature profiles in the unstable atmospheric surface layer. *Journal of Applied Meteorology*, 9(6):857–861, 1970. doi: 10.1175/1520-0450(1970)009<0857:TMROWS>2.0.CO;2.

[11] U. Siegenthaler and J. L. Sarmiento. Atmospheric carbon dioxide and the ocean. *Nature*, 365:119–125, 1993.

[12] Takahashi T., Sutherland S.C., Sweeney C., Poisson A., Metzl N., Tilbrook B., Wanninkhof R., Feely R.A., Sabine C., Olafsson J., Nojiri Y. (2002) Global sea-air CO_2 flux based on climatological surface ocean pCO_2 and seasonal biological and temperature effects. *Deep-Sea Res II* 49:1601-1622

[13] R. Wanninkhof. Relationship between gas exchange and wind speed over the ocean. *Journal of Geophysical Research*, 97(C5):7373–7382, 1992.

[14] Wanninkhof, R., S. Doney, T. Takahashi, and W. McGillis (2001) The effect of using averaged winds on global air-sea CO_2 fluxes. In: Gas Transfer at Water Surfaces, M.A. Donelan, W.M. Drennan, E.S. Saltzman and R. Wanninkhof (eds.), Vol. 127, AGU Press, 351–356

[15] R. Wanninkhof and W. R. McGillis. A cubic relationship between gas transfer and wind speed. *Geophysical Research Letters*, 26(13):1889–1892, 1999. doi: 10.1029/1999GL900363.

[16] A. J. Watson, R. C. Upstill-Goddard, and P. S. Liss. Air-sea exchange in rough and stormy seas measured by a dual tracer technique. *Nature*, 349(6305):145–147, 1991.

[17] E. K. Webb, G. I. Pearman, and R. Leuning. Correction of flux measurements for density effects due to heat and water vapour transfer. *Quarterly Journal of the Royal Meteorological Society*, 106:85–100, 1980.

20

Bias of CO_2 Surface Fluxes Estimated by Eddy Covariance due to "Adjustment Fluxes".

Gerhard Peters

Meteorological Institute
University Hamburg
Bundesstrasse 55, 20146 Hamburg, Germany
gerhard.peters@zmaw.de

Abstract The advection of horizontal inhomogeneous CO_2 concentrations by a wind field with vertical gradient causes a height dependent vertical CO_2 flux. It is shown that the corresponding bias between the flux at few meters above the surface and the flux through the surface is in the order of the natural CO_2 flux variability through the air–sea interface, if horizontal CO_2 gradients are assumed that are typical for land conditions or epicontinental seas. Although this bias is a zero mean effect on the long term, it is suspected that it may contribute significantly to the estimation uncertainty of air–sea CO_2 transfer velocities based on eddy correlation flux measurements. A simple model is suggested to retrieve the surface flux from extrapolation of flux measurements at two height levels.

20.1 Introduction

A standard procedure to estimate fluxes of a passive air ingredient through the air-surface interface is the eddy covariance method (ECM) which calculates $F = \overline{w'c'}$ in the air close to the surface. w' and c' are turbulent fluctuations of the vertical wind component w and of the concentration c of the air ingredient respectively. The overline indicates temporal averaging over typically 10 to 60 min. After application of several standard corrections the flux F measured at few meters above the surface is identified with the flux through the surface. Fluxes of some air ingredients including CO_2 through the air/sea interface are limited by the surface transfer resistance and show relatively small flux velocities $k = F/c$, where F is the flux density and c is the mass density of the substance. Although Jones and Smith [2] demonstrated already 1977 that ECM copes also with the small flux amplitudes of CO_2, and although relatively good agreement with geochemical methods was found recently by McGillis et al. [6], there is still controversy about the general validity of air-sea CO_2 flux estimates based on ECM.

C.S. Garbe, R.A. Handler, B. Jähne (eds.): Transport at the Air Sea Interface
pp. 289-296, 2007,

Here only the most important sensor siting conditions and environmental conditions required for ECM are recalled, which are well known for being precarious.

- Measuring "close" to the surface means that the measuring height z_1 is small compared to the height of the atmospheric reservoir z_i , i.e. of the atmospheric boundary layer. At the top z_i of the boundary layer the flux assumes the entrainment flux, which may be of opposite sign as the surface flux.
- If there are horizontal gradients of the surface flux, the method does not represent the local flux density underneath the measuring point rather than a weighted average over some upwind-area, which is usually called the "footprint" [5]. As the footprint increases with increasing measuring height the weighting function changes with height and thus the measured flux as well. Transfer velocities based on locally measured CO_2 concentrations on one hand and area averaged fluxes on the other hand can be seriously biased, if the surface conditions are inhomogeneous on the footprint scale. A first approximation for the bias at measuring height z was provided by Wesely [11] as a function of the corresponding horizontal concentration gradient $\delta c/\delta x$:

$$\Delta F = z\frac{\delta c}{\delta x}u \tag{20.1}$$

where u is the horizontal velocity.

In addition, standard ECM signal processing includes various first-order corrections which are applied to the "raw"-fluxes:

- Although there should be no mean vertical wind component $\overline{w}$ over flat terrain, there is usually some component observed in the measured data due to artificial reasons. To minimize its impact on the estimated surface-flux the coordinate system is rotated until $\overline{w} = 0$ [12].
- Only concentration fluctuations c', which are solely due to fluctuations of the mixing ratio of the considered ingredient, may be attributed to the surface flux. Dilution effects contributing to c' may give rise to differences between the measured flux and the surface flux. Such mechanisms include fluctuations of air temperature or of air constituents as for example water vapour. The correction of such effects is known as Webb-correction [9].
- Spectral corrections are related to sensor characteristics as finite frequency response, path averaging and the separation of sensors for w and c respectively [7].

While there is general consensus that the parameterization of the transfer velocity with the wind speed u_{10} is too simplistic (see for example Jacobs et al. [1] for a comprehensive discussion of further processes controlling the transfer velocity) the large scatter of transfer velocities, which is typically found in ECM field campaigns including a recent long term measurement Weiss et al. [10] calls for the search of further potential sources of error affecting the standard ECM applicability.

In this note a mechanism is discussed, which - to the author's knowledge - was not considered in previous studies, but which may give rise to differences between the measured eddy flux and the surface flux . Their magnitude is comparable to Equation (20.1), and they are particularly significant for ingredients like CO_2 with high transfer resistance at the air-surface interface.

For simplicity homogeneous surface conditions are assumed on scales far beyond the footprint size. Even in this case the atmospheric mean concentration $\overline{c}$ of CO_2 may show a "cloud like" i.e. a horizontally heterogeneous distribution. If the wind field would be uniform, its only effect would be the translation of these structures without changing their form. But due to the non-slip condition at the surface the wind field close to the ground exhibits a vertical gradient, and the differential advection related to the height-dependent wind-speed together with a horizontal gradient of $\overline{c}$ causes the evolution of a vertical gradient of $\overline{c}$. Subsequent vertical turbulent mixing will then reduce this gradient. As explained in the following, it is this mixing (called "adjustment flux" henceforth), which causes a potentially significant difference ΔF between the ECM flux at measuring height z and the surface flux.

20.2 Vertical Flux Gradient due to Differential Advection

A simple conceptual model is considered where the local temporal change $\overline{\dot{c}(z)}$ at the measuring height z is controlled only by two processes, namely horizontal advection by the mean wind and the vertical turbulent flux gradient:

$$\overline{\dot{c}(z)} = \overline{\dot{c}(z)_h} + \overline{\dot{c}(z)_v} \tag{20.2}$$

with

$$\overline{\dot{c}(z)_h} = -\overline{u(z)}\frac{\partial \overline{c(x,z)}}{\partial x} \quad \text{(horizontal advection)} \tag{20.3}$$

and

$$\overline{\dot{c}(z)_v} = -\frac{\partial \overline{w'c'}}{\partial z} \quad \text{(vertical flux-gradient),} \tag{20.4}$$

where x and z are positive in downwind and upward direction respectively. The overlines indicating mean values will be omitted for shortness in the following. In order to obtain the bias of the flux measured at the height z versus the surface flux, Equation (20.2) is rewritten putting the vertical flux-gradient on the left side:

$$\frac{\partial w'c'}{\partial z} = -u(z)\frac{\partial c(x,z)}{\partial x} - \dot{c}(z) \tag{20.5}$$

The flux-bias ΔF at some height z is obtained by integrating Equation (20.5) from the surface to z:

$$\Delta F(z) = -\int_0^z u(\zeta)\frac{\partial c(x,\zeta)}{\partial x}d\zeta - \int_0^z \dot{c}(\zeta)d\zeta \qquad (20.6)$$

It is assumed that the vertical turbulent mixing is sufficiently efficient to maintain well mixed conditions with respect to c against the height dependent advection term $\dot{c}(z)_h$. This implies height-independence of $\dot{c}$ and $\partial c(x)/\partial x$ respectively, which facilitates the integration over z:

$$\Delta F(z) = -\frac{\partial c(x)}{\partial x}\int_0^z u(\zeta)d\zeta - z\dot{c} \qquad (20.7)$$

That means, that stable stratification with suppressed mixing and corresponding errors - which can be severe - are not considered here. Likewise any buoyancy effects on the wind profile are neglected here, as this would probably overstrain this simple conceptual model.

A logarithmic height dependence of the wind speed, applicable in neutral conditions, yields

$$\Delta F(z) = z\left(\frac{u_*}{k}\left(1 - \ln\left(\frac{z}{z_0}\right)\right)\frac{\partial c}{\partial x} - \dot{c}\right) \qquad (20.8)$$

with u_*, k and z_0 are the friction velocity, von Kármán constant and z_0 roughness length respectively.

An alternative convenient form of Equation (20.8) is

$$\Delta F(z) = z\left(\left(\frac{u_*}{k} - u(z)\right)\frac{\partial c}{\partial x} - \dot{c}\right) \qquad (20.9)$$

which contains the wind speed $u(z)$ instead of z_0.

The variables $\dot{c}$, $u(z)$ and u_* can be derived from local measurements. The only unknown on the right hand side of Equation (20.9) is the horizontal concentration gradient $\partial c/\partial x$. Basically two approaches are imaginable to account for $\partial c/\partial x$:

1. Direct measurement by a horizontally distributed array of sensors.
2. Indirect determination by measuring the flux at two or more levels.

Here only the second option is considered. The total vertical turbulent flux, measured at two heights z_1 and z_2, is

$$F(z_{1,2}) = F_s + \Delta F(z_{1,2}) \qquad (20.10)$$

Solving these equations for F_s and elimination of $\partial c/\partial x$ yields:

$$F_s = \frac{F(z_2)U(z_1) - F(z_1)U(z_2) - (\partial\bar{c}/\partial t)\,(z_2U(z_1) - z_1U(z_2))}{U(z_1) - U(z_2)} \qquad (20.11)$$

with $U(z) = z\,(u(z) - u_*/k)$

20.3 Significance of ΔF

As a general rule surface fluxes should be deduced from measurements as close to the surface as possible. In reality trade-offs between various constraints have to be made leading sometimes to measuring heights up to 10 m over ground - or even higher. Here typical conditions over sea are considered where the lower limit should be around 5 meter to protect the instruments from swell.

In order to estimate realistic figures for the flux bias as function of height, Equation (20.8) is rewritten in a slightly modified form:

$$\Delta F(z)/c = z\frac{\partial c}{\partial x}\frac{1}{c}\left(\frac{u_*}{k}\left(1 - \ln\left(\frac{z}{z_0}\right)\right) + u(z_c)\right) \qquad (20.12)$$

with $u(z_c) = -(\partial c/\partial t)(\partial c/\partial x)^{-1}$, which is obviously the velocity, that is required to explain the observed temporal variation of c solely by advection. Likewise z_c can be interpreted as that height, where $u(z)$ assumes the "mean" drift velocity. It should be located somewhere in the midst of the boundary layer.

As an example $z_c = 100$ m, $z_0 = 2 \cdot 10^{-4}$ m, $u_* = 0.2$ ms^{-1} is chosen, and $u(z_c) \approx 6.5$ m/s is assumed, which would fit to the logarithmic wind profile extending up to z_c. For the choice of the horizontal gradient $\partial c/\partial x$ spatial scales must be considered, which fit to the aforementioned averaging times of 10 to 60 min and to the drift velocity $u(z_c)$.

Simulations of atmospheric CO_2 concentrations, provided by Karstens et al. [4], show that the spatial variability is area-dependent: Continental conditions are characterized by strong heterogeneous and time-dependent sources and sinks, which lead to enhanced small- and meso-scale concentration variability, while more homogeneous CO_2 distributions prevail over the open oceans. The variability of CO_2 fields over epicontinental seas like the Baltic Sea resemble more the inhomogeneous continental conditions rather than ocean conditions. This becomes apparent from the inspection of modeled CO_2 fields as shown in Figure 20.1. It represents a snapshot of the horizontal field of CO_2 concentration at 300 m height provided by the regional model REMO with 0.5° and 1 hour resolution [3]. One recognizes strong meso-scale textures over the continents, which cover also the Mediterranean and Baltic Sea, but which are absent over the North Atlantic. CO_2 surface flux estimates by ECM over coastal waters are therefore faced with large, land dominated inhomogeneities on one hand and with marine surface fluxes on the other hand, which are typically small compared to fluxes over vegetation.

Based on these model results a typical gradient of $|(\partial c/\partial x)/c| = 10^{-6}$ m^{-1} is assumed for continental conditions. The bias corresponding to these exemplary parameters is shown in Fig. 20.2 as a function of z. Its significance becomes obvious by comparison with the transfer velocity k of CO_2, which is an upper bound for actually occurring

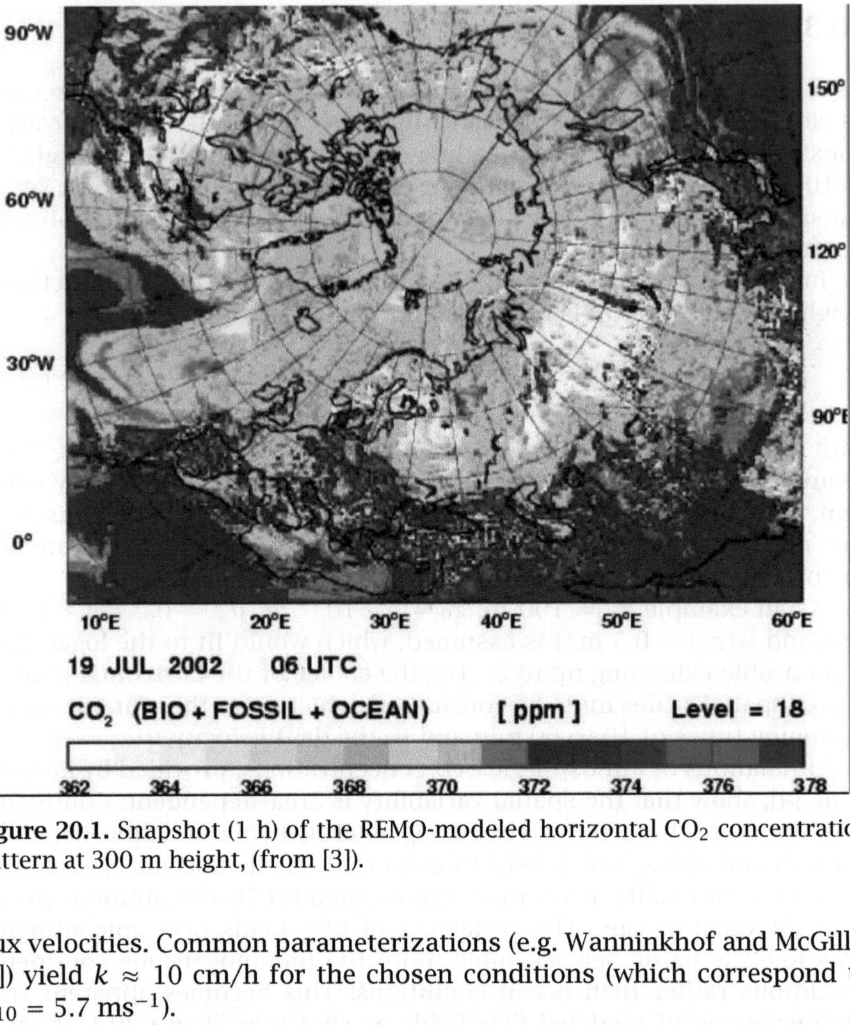

Figure 20.1. Snapshot (1 h) of the REMO-modeled horizontal CO_2 concentration pattern at 300 m height, (from [3]).

flux velocities. Common parameterizations (e.g. Wanninkhof and McGillis [8]) yield $k \approx 10$ cm/h for the chosen conditions (which correspond to $u_{10} = 5.7\ \mathrm{ms}^{-1}$).

The dominating error of ECM measurements is the sampling error, which limits the relative accuracy of flux estimates to about ±20 % (depending on wind speed, stability, measuring height and sampling period). For any meaningful comparison between fluxes measured at two levels, the relative difference must by larger than the sampling error. Therefore the upper measuring height should be as high as possible. On the other hand it should not exceed the 10 m level too much, as the error of extrapolation according to Equation (20.11) becomes increasingly sensitive to the actual wind profile, which may deviate from the logarithmic model - particularly in shallow marine boundary layers. In Fig. 20.2 measuring heights at $z_1 = 5$ m and $z_2 = 10$ m are indicated. One recognizes that $|\Delta F(z_1) - \Delta F(z_2)|$ is sufficiently large compared with $|\Delta F(z_1)|$, which

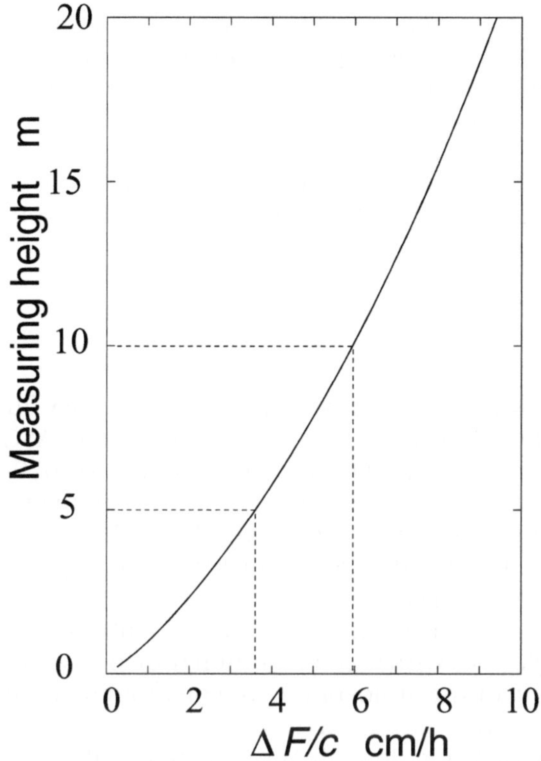

Figure 20.2. Normalized flux bias as a function of measuring height according to Equation (20.12) with $z_c = 100$ m, $z_0 = 2 \cdot 10^{-4}$ m, $u_* = 0.2$ ms^{-1}, $\partial c/\partial x = 10^{-6}$ m^{-1}.

means that this set of measuring heights would be a feasible choice for reducing the estimation-error of F_s.

20.4 Summary

It has been shown that even for horizontal homogeneous conditions with respect to the surface fluxes the ECM flux, measured at few meters above the surface, does not necessarily represent the surface flux. Flux samples may be biased by height dependent adjustment fluxes induced by differential advection. Although this is a zero-mean error in the long term, it may contribute significantly to the uncertainty of measurement results. This effect is expected to be particularly significant for CO_2-fluxes in a coastal marine environment, where the horizontal atmospheric distribution of CO_2 is determined by heterogeneous and instationary sources and sinks over the land, but where the local surface fluxes - being controlled by the air-sea transfer resistance - are small compared to land conditions. The bias considered here is reaching many hundred kilometers into the open sea until the continentally generated CO_2 cloud structures are diluted (see Fig. 20.1). It should not be mistaken for the well–known footprint effect related to internal boundary layers close to surface inhomogeneities as for example coastal lines. Measurement of the turbulent fluxes at more than one height is suggested in order to mitigate the effects on the estimation of the surface flux.

References

[1] Jacobs C., Kohsiek W., Oost W.A. (2002), Air-sea flux and transfer velocity of CO2 over the North Sea: results from ASGA/AGE, *Tellus.* Series B, Chemical and Physical Meteorology 51: 629-641.
[2] Jones E.P., Smith S.D. (1977) A fgirst measurement of sea-air CO_2 flux by eddy correlation, *J. Geophys. Res.*, 82: 5990-5992.
[3] Karstens U. (2006), Max-Planck-Institute for Biogeochemistry, Jena, Personal Communication.
[4] Karstens U., Gloor M., Heimann M., Rödenbeck C. (2006) Insights from simulations with high-resolution transport and process models on sampling of the atmosphere for constraining midlatitude carbon sinks. *J. Geophys. Res.* 111:D12301
[5] Kljun N., Calanca P., Rotach M.W., Schmid H.P. (2004) A Simple Parameterisation for Flux Footprint Predictions. *Boundary-Layer Meteorology* 112:503-523
[6] McGillis, W.R., Edson J.B., Hare J.E., Fairall C.W. (2001) Direct covariance air-sea CO2 fluxes, *J. Geophys. Res.*, 106, 16: 729-16,745
[7] Moore C.J. (1986) Frequency response corrections for eddy correlation systems. *Boundary-Layer Meteorol.* 37: 17 - 35.
[8] Wanninkhof R., McGillis W. (2001) A cubic relationship between air-sea CO_2 exchange and wind speed, *Geophys. Res. Lett.*, 26(13):1889-1892
[9] Webb E, Pearman G, Leuning R (1980) Correction of the flux measurements for density effects due to heat and water vapor transfer, *Quart. J. Roy. Meteorol. Soc.* 106:85-100.
[10] Weiss A., Kuss J., Peters G., Schneider B. (2006) Evaluating transfer velocity-wind speed relationship using a long-term series of direct eddy correlation CO2 flux measurements, *Journal of Marine Systems*, In Press, Corrected Proof, Available online 20 September 2006,
[11] Wesely, M. (1986), Response to "Isotopic Versus Micrometeorologic Ocean CO_2 Fluxes: A Serious Conflict" by W. Broecker et al., *J. Geophys. Res.*, 91, C9: 10,533 - 10,535.
[12] Wilczak J.M., Oncley S.P., Stage S.A. (2004) Sonic Anemometer Tilt Correction Algorithms. *Boundary-Layer Meteorology* 99: 127-150

21

Eddy Correlation Measurements of Sea Spray Aerosol Fluxes

Gerrit de Leeuw[1]*, Marcel Moerman[1], Christopher J. Zappa[2], Wade R. McGillis[2], Sarah Norris[3], and Mike Smith[3]

[1] TNO Defense, Security and Safety, The Hague, The Netherlands
[2] Lamont-Doherty Earth Observatory, Colombia University, Palisades, New York, USA
[3] University of Leeds, School of the Environment , Leeds, UK

Abstract Fluxes of primary marine aerosol in the sub-micron fraction were measured using a flux package consisting of a sonic anemometer, a Condensation Particle Counter (CPC) and an optical particle counter (OPC) equipped with a heated inlet. Whereas the CPC provides the total particle number flux of particles larger than 10 nm, the OPC measures size segregated fluxes for particles between 0.5 and 5 μm radius. By varying the temperature of the OPC inlet, particles of certain composition can be selected. Results are presented with the inlet temperature at 100°C (dry particles retaining most of the aerosol material, representative of sea spray aerosol) or at 300°C which volatilizes all material except sea salt and elementary carbon. Temperature scans confirm the choice of these temperatures to discriminate between sea spray and sea salt. This flux system was deployed at the end of the 560 m long pier of the US Army Corps of Engineers Field Research Facility in Duck (NC, USA). Initial results show the increase of the sea spray fluxes with wind speed u, roughly varying as u^3 for u up to 16 ms^{-1}.

21.1 Introduction

The interest in sea spray aerosol has traditionally been focussed on large particles (> 1 μm) because of their influence, at high wind speeds, on sea-air transfer of water vapour and latent heat (e.g. Andreas, [2]). The overview of sea spray source fluxes presented by Andreas [3], which includes most of the common source functions presented until 1998, shows a variation of about 6 orders of magnitude. It is noted that the formulations of Monahan et al. [15], Fairall et al. [9], Smith et al. [29], Smith and Harrison [27] and Andreas [2] are all within about one order of magnitude. It is further noted that the Andreas [2] source function is a composite of those formulated by Smith et al. [29], Fairall et al. [9] and Wu et al.[32].

* Current address: Finnish Meteorological Institute, Helsinki, Finland & University of Helsinki, Department of Physical Sciences, Helsinki, Finland

C.S. Garbe, R.A. Handler, B. Jähne (eds.): Transport at the Air Sea Interface
pp. 297-311, 2007,

These source functions were determined by different methods. Monahan et al. [15] used the whitecap method that is based on combining the parameterisation of the whitecap cover using field experimental data and the production of sea spray aerosol per unit whitecap area determined from laboratory experiments. This parameterisation applies to particle diameters in the range 0.3-10 µm that is the range of the particle counter used in their laboratory experiments. Fairall et al. [9] and Smith et al. [29] used a balance method, i.e., during steady conditions the production and removal are assumed to be in balance and the production is set equal to the deposition. This method applies only to large particles because the removal for submicron particles is indeterminate since the submicron particles are mainly removed by wet deposition (washed out by rain) and, in the absence of precipitation, it takes a very long time to reach steady state (Hoppel et al. [12]). Wu et al. [32] measured droplets in a laser beam on a float very close to the water surface. These droplets are considered spume droplets that are directly torn off the waves.

More recently, the interest in sea spray has shifted because of its role in chemical reactions [22, 23, 30] and, in particular, its role in climate [13]. Sea salt is the dominant submicrometer scatterer in most ocean regions and dominates the marine boundary layer particulate mass concentration in remote oceanic regions, with a significant fraction occurring in the submicrometer size range [13]. Sea salt contributes 44% to the global aerosol optical depth. Estimates for top-of-atmosphere, global-annual radiative forcing due to sea salt are -1.51 and -5.03 Wm^{-2} for low and high emission values, respectively [13]. Sea spray not only affects climate by scattering of solar radiation, they are also very hygroscopic and act as cloud condensation nuclei, thus contributing to the indirect aerosol effect. The indirect aerosol effect is the most uncertain forcing mechanism in the prediction of climate change. Hence, there is a large demand for improved sea salt source functions in the submicron size range, extending down to ca. 10 nm.

The gap in the submicron size range could be filled by extrapolation of the Monahan et al. [15] source function to smaller sizes. For instance, Guelle et al. [11] used this source function in a global model to represent particles between 0.03 and 8 µm dry radius. Figure 21.1 is a compilation of sea spray source functions published since 1998, focused on sub-micron particles. The data are plotted for their range of validity as indicated in the original articles. For comparison we also show the Monahan et al. [15] source function and its extrapolation to smaller sizes. Fig. 21.1 shows that for particles smaller than 0.1 µm extrapolation of the Monahan source function is a significant overestimate. Most of the source functions shown in Fig. 21.1 were discussed in Schulz et al. [26] and a comparison was made based on the mass fluxes determined from these formulations: De Leeuw et al. [5] (based on surf zone experiments), Vignati et al. [31] (based on inverse modelling), Reid et al. [25] (based on a con-

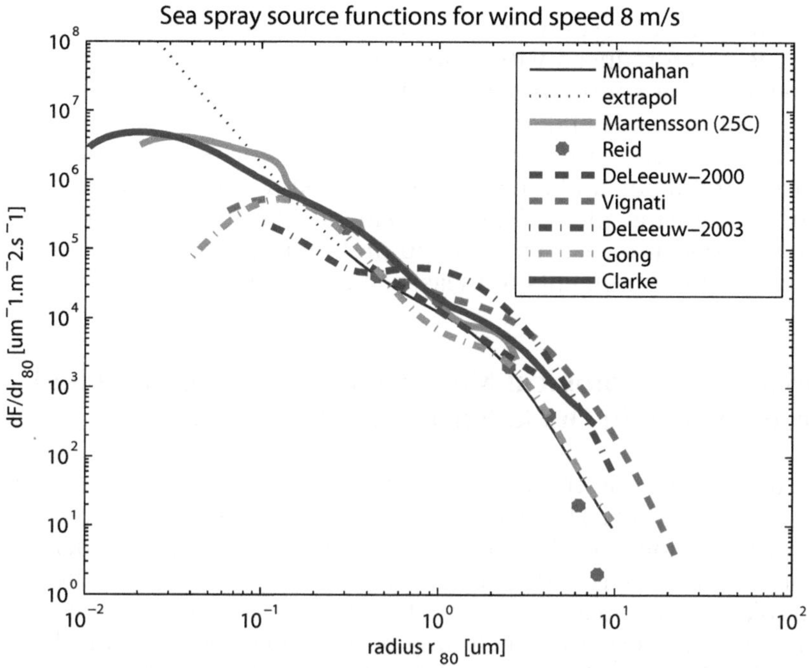

Figure 21.1. Compilation of sea spray source functions published since 1998, see text for explanation of the various curves and the references. The fluxes are represented as function of radius r80, where r80 is the particle radius normalized at a relative humidity of 80%.

centration build up method using air craft measurement in an evolving internal boundary layer) and Nilsson et al. [17] (based on eddy covariance measurements). More recent formulations were provided by De Leeuw et al. [6], by application of the method of Vignati et al. [31] to data collected in the Pacific near Hawaii, by Clarke et al. [4], who determined the production over the surf zone and related that to open ocean production using whitecap cover, similar to De Leeuw et al. [5] but extending the valid interval to the range 0.01-8 μm dry radius. Gong [10] modified the Monahan et al. [15] source function to account for the observed decrease in the particle concentrations for particles smaller than ca. 0.1 μm radius [21] (the same data as used by Vignati et al. [31]). By modifying the Monahan formulation, Gong retained the cubic wind speed dependence. The O'Dowd et al. [21] data apply to dry radius larger than 0.04 μm and the results shown by Gong for smaller sizes may be questioned. Furthermore, Gong introduces a scaling factor Θ which has a large influence on the flux magnitude. In Fig. 21.1 we used $\Theta = 30$; the use of other values would lead to results that may be an order of magnitude different.

The data in Fig. 21.1 show that the discrepancy between different formulations is much reduced with respect to the situation when Andreas did his review. The source functions shown in Fig. 21.1 were obtained using different methods and different physical principles.

Lewis and Schwartz [14] provided a very comprehensive review of sea salt aerosol production and the reader is referred to this book for further detail on mechanisms, methods, measurements and models. In this contribution we present measurements and preliminary results from micrometeorological measurements to measure fluxes of sea salt aerosol.

21.2 Micro-Meteorological Measurements at the FRF Field Research Facility in Duck, North Carolina (USA)

Of particular interest is the eddy covariance method that was first attempted by Nilsson et al. [17] using an ultrasonic anemometer (Sonic) and a CPC (Condensation Particle Counter) to measure the total number flux of sea spray particles larger than 10 nm. The advantage of this method, as opposed to the whitecap method, is that all particles within the detectable size range may be measured, including both bubble-mediated produced sea spray droplets and spume droplets which are torn off the waves at wind speeds exceeding $\approx 9\ \mathrm{ms}^{-1}$. With respect to the balance method the advantage of micrometeorological techniques is that fluxes are directly measured, with the additional advantage that sea salt particles can be separated from sea spray (containing all components) using the volatility technique as described below. The volatility technique can be used with sampling frequencies that are fast enough for application of eddy correlation (≈ 1 Hz or slightly faster) as opposed to other techniques to determine chemical composition which require sampling times of the order of minutes to hours.

The Nilsson et al. [17] flux package was complemented with an optical particle counter equipped with an inlet heated to 300°C, to drive off volatile species and retain the non-volatile refractory fraction, which over the ocean is generally sea salt aerosol. Thus a direct measure is obtained of the production of sea salt, unperturbed by other species common to clean marine regions (Clarke et al., [4]), as explained in section 21.3. This was first tested on FLIP (FLoating Instrument Plaform) during an experiment in the Pacific Ocean near Hawaii in August/September 2001 (Anderson et al. [17]) and the results were reported in De Leeuw et al. [6] and Nilsson et al.[18].

Based on this first experience this flux package was further improved and deployed from the 560 m long pier at the US Army Corps of Engineers Field Research Facility (FRF) in Duck (NC, USA, `http://www.frf.usace.army.mil/`). This site was selected because of the long pier and supporting meteorological and oceanographic measurements. The experiments took place in November 2004 (UNISOURCE; toward a universal sea spray source function [7]) and in October 2005 (WASFAB; Waves, Air-Sea Fluxes, Aerosols, and Bubbles [8, 33]), with the goal to further test the methods and to obtain a comprehensive data set to further constrain the sea salt source function. In particular, the analysis is intended to test which parameters, other than wind speed and friction velocity, influence the sea spray source function, such as wave height, age and steepness, sea surface temperature, salinity, and atmospheric parameters such as wind speed, fetch, micrometeorological fluxes, thermal stability [28]. In this paper the experiments at the FRF pier are described and preliminary results are presented as function of wind speed.

The flux package was mounted on a meteorological mast at the far end of the pier at a height of 16.2 m above mean sea level. A second aerosol flux package was added, where the aerosol size spectra were measured with a frequency of 10 Hz using a CLASP (Compact Lightweight Aerosol Spectrometer Probe) developed by the University of Leeds ([19, 20]). CLASP is based on fast response MetOne particle counters (0.1- 3.5 μm radius). This suite of aerosol flux instruments was further complemented by a Licor LI-7500 open path sampler for water vapor and CO_2 fluxes. Supporting instruments included a sea spray package consisting of PMS OAP, FSSP and PCASP, a TSI APS, a PMS OPC based volatility system that continuously cycled between 50°C and 700°C to determine the contributions of volatile components and sea salt to the total particle size distribution, an aethalometer to determine the absorbing aerosol fraction and a meteorological station. During WASFAB a boom extending ≥ 8 m from the pier end was added with another set of instruments, including a Riegl altimeter to measure wave height, a Licor LI-7500 open path sampler, and a sonic anemometer [33]. The height up and distance out from the pier were selected to minimize air-flow perturbation due to obstruction by the pier and buildings at the end of the pier. Other wave measurements are available from buoys and meteorological and oceanographic parameters are available from the standard suite of instrumentation maintained at FRF.

21.3 Sea Salt Flux Measurements

Aerosol in the marine atmosphere is a mixture of sea spray and other aerosol types such as sulphates produced from DMS, advected anthropogenic aerosol, and mixtures of all of these due to atmospheric processes. Sea spray may contain a significant fraction of organic aerosol [24]. A particle counter with a heated inlet provides a means to determine the relative contributions of some of these aerosol types. Chemical species such as sulphate and sea salt volatilize at specific temperatures.

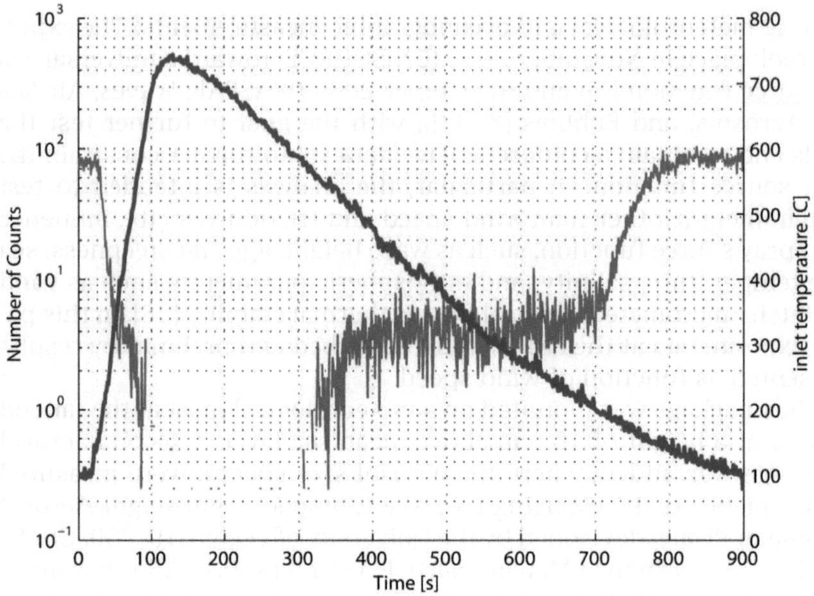

Figure 21.2. Volatility scan, see text for explanation.

Fig. 21.2 illustrates the effect of the temperature on the number of particles counted by the OPC. Fig. 21.2 shows an average of 4 temperature scans, of 15 minutes each, on September 10, 2001, 04:45 GMT, during the FLIP experiment referred to above. The horizontal axis shows the time, at $t = 0$ the temperature (dark line, scale on secondary y axis) is 100°C. The temperature of the inlet tube is increased fast to 750°C, after which controlled cooling takes place. The temperature of the heated inlet tube is measured and has been checked at the inside of the tube to be accurate within 1°C. When heating starts, the particle counts (gray line, scale on primary y axis) drop to zero (some counts are observed in this average which are due to noise: the OPC was regularly checked to count zero particles when a null filter is applied at the inlet, i.e. there are no leaks). When the temperature decreases to below about 600°C the particle counts rise and remain relatively constant until the temperature drops below about 160°C, when the counts rise fast. The temperatures of 160°C and 600°C characterize the volatilization of sulphate and sea salt aerosol. Hence, when measurements are made with the inlet temperature fixed at 300°C, only sea salt is measured (scans to 750°C confirmed that no other species are present and counts drop to zero). When the inlet temperature is fixed to 100°C, the sulphate fraction is measured also and the aerosol is representative for the total dry fraction.

To measure the sea salt fluxes, the aerosol was aspirated trough a ¼" tube with its inlet mounted just below the sonic anemometer sample

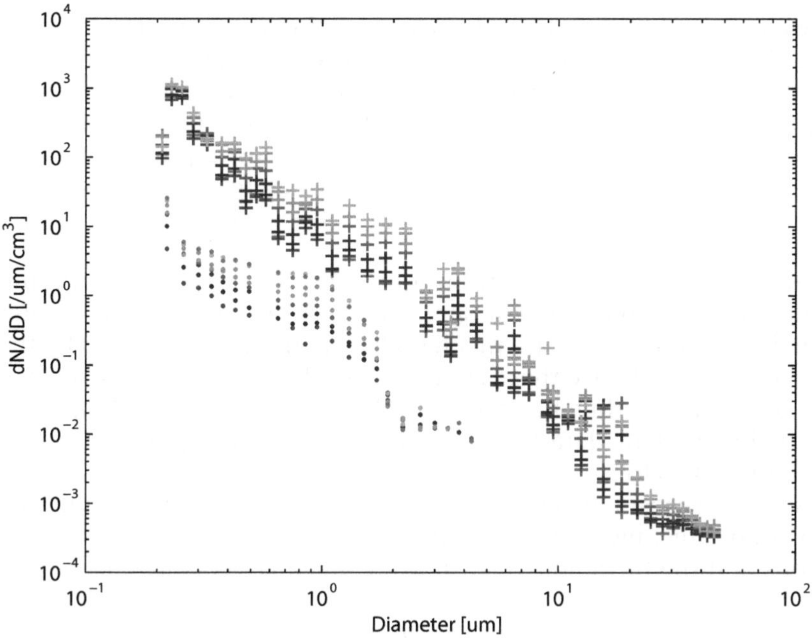

Figure 21.3. Aerosol number size distributions binned in 1 ms^{-1} wind speeds intervals ranging from 5 to 11 ms^{-1} (bottom to top): the lower set of curves shows sea spray aerosol measured with the volatility system with inlet heated to 300°C (refractory particles), the upper set of curves shows concentrations of ambient particles measured with the CSASP.

head. The ca. 10 m long inlet tube connected to the heated inlet of the PCASP that measures particles in the size range between 0.05 and 5 μm radius. This set up was chosen because in practice it was not feasible to mount the volatility system on the mast near the sonic. The inlet tube was mounted vertical with no sharp bends to avoid particle losses. The particles are dried in the volatility system, reducing their sizes by roughly a factor of 2-4 (for ambient relative humidity > 80%, which is common over the ocean). Fig. 21.3 shows particle size distributions measured with the PCASP with the heated inlet and with a CSASP that measures in the radius range 0.1-10 μm. The latter has a very short inlet tube in which losses are negligible and relative humidity has been checked to stay close to ambient. The data are shown for size distributions averaged in 1 ms^{-1} wind speed bins. The concentrations of the refractory fraction are roughly 2 orders of magnitude smaller than for the ambient data, or rather, the spectra are shifted to smaller particles due to volatilization. As the particles become larger the refractory and ambient concentrations are closer. As expected, most of the volatile fraction (i.e. the material that volatilizes at the temperature set at the inlet, including most of the organic material, sulphates and water vapor) is in the accumulation mode while for larger particles the sea spray fraction becomes more dominant. Furthermore, the refractory aerosol concentration drops suddenly for particles greater than 2 μm diameter, indicating that large particles are not reaching the sample volume, whereas for particles between 1 and 2 μm the concentrations decrease. This is suspected to be due to increasing losses as the particles are larger. Hence, for this set-up, only particles smaller than 1 μm diameter were reliably sampled.

The ambient particle size distributions show a mode in the supermicron range, i.e. the common sea spray mode at a few μm. For smaller particles the concentrations are much higher than those measured by the volatility system and hence these are non-sea-salt particles. The sea spray aerosol spectrum can thus be constructed from the volatility data up to 1 μm and the ambient data larger than 2 μm, and interpolation of the intermediate values [6].

The sea salt aerosol spectral fluxes were directly measured using the eddy covariance technique in which the turbulent fluctuating aerosol concentrations and the turbulent vertical wind speed are correlated:

$$\mathrm{cov}(w, c) = \langle w' \cdot c' \rangle, \tag{21.1}$$

Where $w' = w - \langle w \rangle$ is the turbulent fluctuating vertical wind speed, w is the instantaneous value of the vertical wind speed, and $\langle w \rangle$ is the mean vertical wind speed ($\langle w \rangle = 0$ by definition). Likewise, $c' = c - \langle c \rangle$ is the turbulent fluctuating particle concentration for a certain size interval. Using particle counters, the size segregated flux, or flux spectrum, can be obtained. Because ambient measurements could not be made with the inlet close to the sonic sample volume, in the analysis of the current data we only use the data collected with the volatility system. During the UNISOURCE experiments in November 2004 all measurements were made with the inlet heated at a constant temperature of 300°C, complemented with scans between 100 and 750°C. During the WASFAB experiments, measurements additionally were made with the inlet at 100°C.

Equation (21.1) was applied to the data after flow lag correction to account for the time it takes for the aerosol to be transported from the inlet near the sonic to the particle counter. The eddy covariance technique applies only for steady conditions, i.e. no low frequency trends are present in the data. In the analysis half-hour data series were used. The raw data were de-trended, the mean wind components (u, v, w), wind speed and wind direction were computed. Rotation and tilt angle corrections were applied to compute the true stream wise, lateral and vertical wind components and the corrected data were used to calculate variances and co-variances.

21.4 Results

Initial results from micrometeorological measurements show that in on-shore winds the conditions are representative for open ocean conditions for the range of wind speeds encountered (up to 16 ms^{-1}). For instance, drag coefficients compare favorably with the TOGA-COARE 2.6 parameterization for drag coefficient over the open ocean [33]. The relevant spectra of micrometeorological parameters (sensible and latent heat fluxes and the respective scaling parameters) as well as aerosol concentrations show the expected increase with wind speed and the power spectra show the expected frequency dependence.

For aerosol fluxes this behavior is less clear. To a large extent this is due to the very small number concentrations retained after volatilization. The statistics were improved by taking several size channels together in a larger bin. Nevertheless, the results are quite variable and no clear trends were visible at first sight. In large part this could be due to the influence of various sources, in particular breaking waves in open water, at shallow sand banks and in the surf zone. Waves breaking in the surf zone produce large amounts of sea spray aerosol [4, 5] and a careful selection was made to ensure that only data for on-shore wind directions were considered for analysis. The measurements were made on a mast that is mounted at the SE corner of the pier, which extends into the ocean at an angle of 80°from North. Hence wind directions between 10°and 150°can be considered coming from the open ocean. Also, the influence of waves breaking on sand banks was excluded.

During the UNISOURCE experiments in November 2004 two storms were experienced with measurements in sustained wind speeds of up to 16 ms^{-1} and then gradually decreasing. During part of the second storm the instruments were removed from the mast because of lightning warnings. Therefore, the analysis presented here focuses on the first storm, in the period 8-11 November. Wind speed increased from 5 ms^{-1} to about 15 ms^{-1} in 2-3 hours, around noon (time reference is UTC), while the wind direction veered from SW to NNE. The wind was steady from NNE from 8 November 14:00 until early in the morning of 11 November, while the wind speed gradually decreased to lower than 4 ms^{-1}. This period was analyzed for aerosol fluxes. The results in Fig. 21.4 show the covariances $\langle w'c' \rangle$ (in 10^6 $m^{-2}s^{-1}$) for four size bins with different grouping of refractory particle sizes, plotted vs. wind speed. Because the production of sea spray aerosol is expected to vary with wind speed as u^3, the shape of this curve is shown for comparison. Fig. 21.4 shows that the data roughly follow this behavior for wind speed > 7 ms^{-1}, i.e. when significant wave breaking occurs and hence sea spray aerosol is expected to be produced. The increase is significant for wind speed larger than 10-12 ms^{-1}. Least squares fits of a power law function to the data show that the fluxes increase as u^b, where b varies between 2.9 and 3.4 for different size bins.

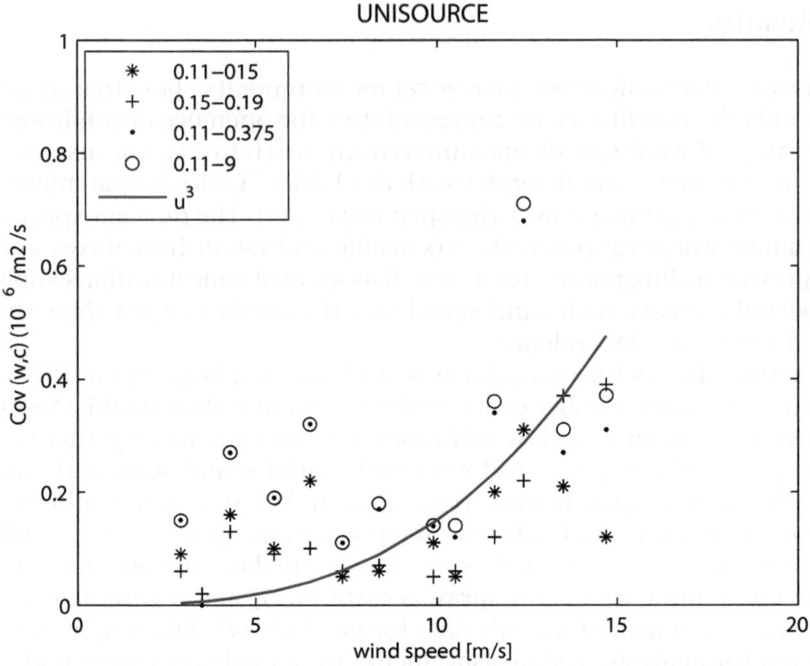

Figure 21.4. Covariances $\langle w'c' \rangle$ for the particles binned in size fractions indicated in the legend (particle sizes in diameter). The u^3 curve is shown for guidance. UNISOURCE measurements 8-11 November 2004.

For the largest size bin (0.11-9 µm diameter), which have the best statistics with standard deviations of about 50%, b has a value of 3.0.

For wind speed lower than 7 ms^{-1}, the sea spray fluxes show no clear trend with wind speed. At such low wind speed whitecapping becomes smaller as does the associated aerosol production. It is noted that a threshold value of 7 ms^{-1} for whitecapping is somewhat arbitrary and values as low as 4 ms^{-1} for the onset of whitecapping are mentioned in the literature. However, the aerosol counts are already low, even at the highest wind speeds, and hence the current technique can only be used with some degree of reliability when the production is significant.

During WASFAB some experiments were conducted where the statistics were improved by using the OPC inlet temperature at 100°C. This was occasionally done when particle concentrations were very low in situations with onshore wind and long trajectories over the ocean, complementary to measurements with the inlet heated to 300°C. Results for the 0.11-9 µm size bin are shown in Fig. 21.5. A fit to these data also indicates a u^3 dependence, but the number of data points and the wind speed range are insufficient to be conclusive. With the inlet at 300°C, we further observe a kink around 8 ms^{-1}, for which we have no explanation.

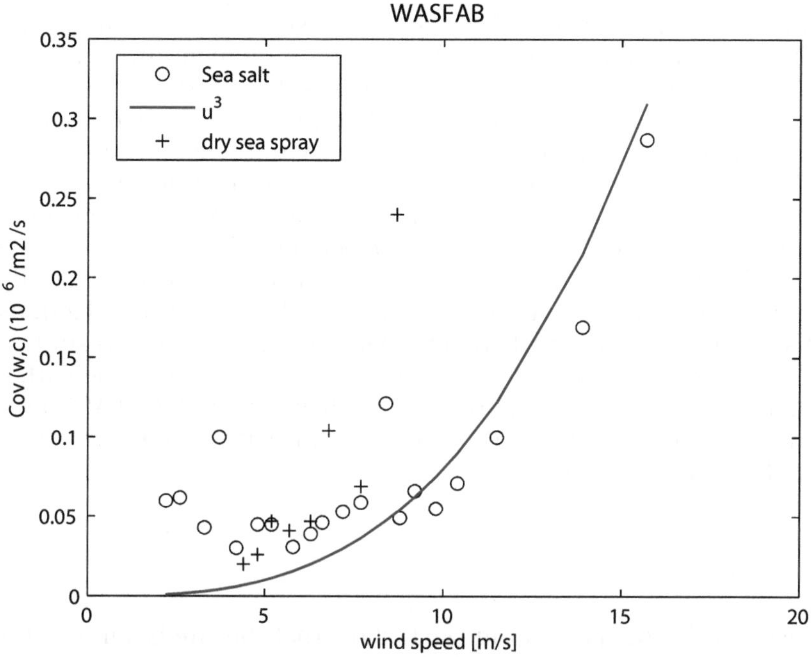

Figure 21.5. Covariances $\langle w'c' \rangle$ for sea salt particles in the 0.1-9 µm size fraction, measured during WASFAB in October 2005, with the inlet heated to 300°C (red dots). The blue symbols show fluxes measured with the inlet heated to 100°C. u^3 is plotted for guidance.

21.5 Discussion

The results presented in this paper are preliminary. The flux determined from UNISOURCE and WASFAB show large differences at the highest wind speeds. Ambient conditions other than wind speed were not taken into account and we cannot yet offer an explanation for observed differences. Only few cases have been analyzed, covering two periods spanning together about 5 days out of almost 2 months of measurements. The analysis focused on conditions of on-shore winds that have been shown to be representative of open ocean conditions at this site.

The results are of similar magnitude as those presented in Fig. 21.1, when accounting for the refractory fraction, i.e. dry particles from which the volatile substances have been removed. Whereas the drying effect would account for roughly a factor of 2 when they are grown to a relative humidity of 80%, the volatile fraction may account for significant additional mass loss. This concerns not only sulphates, but also organic species that are part of the sea spray aerosol and may take up 70% of the aerosol mass [24].

The measurement of fluxes of sea spray aerosol is more complex than for other air-sea interaction parameters such as momentum, heat, water vapour and gases such as CO_2, for which eddy covariance measurements have been developed over the last two decades. These quantities can be measured with quite reasonable accuracy to the extent that unattended flux systems can be deployed. For aerosols the complexity is in the large number of degrees of freedom, such as size, chemical composition, swelling due to humidity effects, etc. This paper shows an attempt to overcome these by using a particle counter with a heated inlet to physically select particles of certain composition and size. Due to these selections the statistics are very low and the standard deviation in the results is large (50-100% in the data presented here). This can be overcome with a fast particle counter with a high flow rate, such as the CLASP. CLASP measures particles in the size range 0.1-3.5 μm radius with a frequency of 10 Hz and a flow rate of 50 cm^3/s.

21.6 Conclusion

The experiments at the FRF Pier in Duck show that the site is suitable for air-sea interaction experiments, including the measurement of sea spray fluxes. Data obtained show that the site is representative of the open ocean. Sea spray aerosol fluxes have the expected u^3 dependence, but careful data selection is needed because of the insensitivity of the aerosol measurements. Careful data selection is needed to avoid spurious effects, e.g., due to wave breaking on sand banks or in the surf zone. Data averaging is needed to achieve better statistics. The techniques employed here work for sub-micron particles. For larger particles, their fluxes may be measured using *in situ* sampling. A problem may be to sample the particles and the vertical wind speed in the same eddy, without distortion of the flow by the particle counter.

Acknowledgement. The work of TNO was supported by the US Office of Naval Research, Grant N00014-96-1-0581. The work of LDEO was financially supported by the US Office of Naval Research Young Investigator Program (Grant N00014-04-1-0621) and by the National Science Foundation (grant number OCE-05-26677). Carl Miller and FRF colleagues are gratefully acknowledged for their continuous support. The WASFAB experiments were undertaken with participation of the University of New South Wales (Australia: Mike Banner and Morison).

References

[1] Anderson, K., P. Caffrey, A. Clarke, K. Crahan, K. Davidson, A. De Jong, G. de Leeuw, D. Dion, S. Doss-Hammel, P. Frederickson, C.

Friehe, D. Hegg, T. Hristov, D. Khelif, J. Reid, S.Reising, E. Terrill, and D. Tsinikidis, 2004. The Rough Evaporation Duct (RED) Experiment; An Assessment of Boundary Layer Effects in a Trade Winds Regime on Microwave and Infrared Propagation over the Sea. *Bulletin of the American meteorological Socieity (BAMS)*, September 2004.

[2] Andreas, E. L., 1998. A new sea spray generation function for wind speeds up to 32 m s^{-1}, *J. Phys. Oceanogr.*, 28, 2175-2184.

[3] Andreas, E.L., 2002. A review of the sea spray generation function for the open ocean. *Atmosphere-Ocean Interactions*, Volume 1, W.A. Perrie, Ed., WIT Press, Southampton, UK, pp. 1-46.

[4] Clarke, A.D., S.R. Owens and J. Zhou, 2006. An ultrafine sea-salt flux from breaking waves: implications for cloud condensation nuclei in the remote marine atmosphere. *J. Geophys. Res.*, 111, D06202, doi:10.1029/2005JD006565, 14 pp.

[5] De Leeuw, G., F.P. Neele, M. Hill, M.H. Smith and E. Vignati, 2000. Sea spray aerosol production by waves breaking in the surf zone. *J. Geophys. Res.*, 105 (D2), 29397-29409.

[6] De Leeuw, G., M. Moerman, L. Cohen, B. Brooks, M. Smith and E. Vignati, 2003. Aerosols, bubbles and sea spray production studies during the RED experiments, *Proceedings AMS conference*, Long Beach, CA, 9-13 Feb.

[7] De Leeuw, G., M.M. Moerman,. M.H. Smith, S. Norris, J. Lingard, J. Gunby and C. Zappa, 2005. Primary marine aerosol production studies fro Duck (NC). *Abstracts of the European Aerosol Conference 2005*, ISBN 9080915939, abstract nr 247.

[8] De Leeuw, G., M.M. Moerman, M.H. Smith, S.J. Norris, C.J. Zappa, M.L. Banner, R.P. Morison, and W.R. McGillis, 2006. Sea spray aerosol production from breaking waves at open ocean and in the surf zone. *Ocean Sciences Meeting of the American Geophysical Union*, Honolulu, HI, February, 2006.

[9] Fairall, C.W., J.D. Kepert and G.J. Holland, 1994. The effect of sea spray on surface energy transports over the ocean, *Global Atmos. Ocean Sys.*, 2, 121-142.

[10] Gong, S.L., 2003. A paramterization of sea-salt aerosol source function for sub- and super-micron particles. *J. Geophys. Res.*, 17, doi:10.1029/2003GB002079, 7 pp.

[11] Guelle, W., M. Schulz, Y. Balkanski and F. Dentener, Influence of the source formulation on modeling the atmospheric global distribution of sea salt aerosol, *J. Geophys. Res.*, 106, 27509-27524, 2001.

[12] Hoppel, W. A., G. M. Frick, and J. W. Fitzgerald, 2002. Surface source function for sea-salt aerosol and aerosol dry deposition to the ocean surface, *J. Geophys. Res.*, 107(D19), 4382, doi:10.1029/2001JD002014.

[13] Intergovernmental Panel on Climate Change (IPCC), *Climate Change 2001: The scientific Basis,* (Edited by Houghton et al.), Cambridge Univ. Press, New York, 2001.

[14] Lewis, E. R.; Schwartz, S. E. Sea salt aerosol production: mechanisms, methods, measurements and models- a critical review; American Geophysical Union: Washington DC, 2004; Vol. Geophysical monograph 152.

[15] Monahan, E.C., D.E. Spiel, and K.L. Davidson, 1986. A model of marine aerosol generation via whitecaps and wave disruption, In: *Oceanic whitecaps and their role in air-sea exchange processes,* Edited by E.C. Monahan and G. MacNiocaill, Reidel, Dordrecht, The Netherlands, 167-174.

[16] Mårtensson, M., E. D. Nilsson, G. de Leeuw, L.H. Cohen, and H-C Hansson, 2003. Laboratory simulations of the primary marine aerosol generated by bubble bursting, *JGR-Atmospheres* 108 (D9), 10.1029/2002JD002263.

[17] Nilsson, E. D., U. Rannik, E. Swietlicki, C., Leck, P.P., Aalto, J., Zhou, J., and M. Norman, 2001. *J. Geophys. Res.*, 106, D23, 32139-32154.

[18] Nilsson, E.D., Mårtensson, M., Van Ekeren, S., De Leeuw, G., Moerman, M., O'Dowd, C., Flanagan, R., and Geever, M., 2003. *Proc. AMS conference*, Long Beach, CA.

[19] Norris, S., 2006. Sea Spray. PhD Thesis, University of Leeds, School of Earth and Environment, 2006

[20] Norris, S., M. Smith, G. de Leeuw and M. Moerman (2006). Near-surface Air-sea Aerosol Concentrations and Fluxes Using a Novel Fast-response Particle Spectrometer. *Ocean Sciences Meeting of the American Geophysical Union*, Honolulu, HI, February 2006.

[21] O'Dowd, C.D., M.H. Smith, I.E. Consterdine, and J.A. Lowe, Marine aerosol, sea salt, and the marine sulphur cycle: a short review, *Atmospheric Environment* 31, 73-80, 1997.

[22] O'Dowd, C.D., J.A. Lowe and M.H. Smith, Coupling of sea-salt and sulphate interactions and its impact on cloud droplet concentration predictions, *Geophy. Res. Lett.*, 26, (9), 1311-1314, 1999a.

[23] O'Dowd, C.D., J.A. Lowe and M.H. Smith, Observations and modelling of aerosol growth in marine stratocumulus, *Atmos. Environ.* 33, (18), 3053-3062, 1999b.

[24] O'Dowd, C.D., J.L. Jimenez, R. Bahreini, R.C. Flagan, J.H. Seinfeld, K. Hämeri, L. Pirjola, M. Kulmala, S.G. Jennings and T. Hoffman, 2002. Marine aerosol formation from biogenic iodine emissions. *Nature* 417, 632-636.

[25] Reid, J.S., H.H. Jonsson, M.H. Smith and A. Smirnov, 2001. Evolution of the vertical profile and flux of large sea-salt particles in the coastal zone, *J. Geophys. Res.* 106, 12,039-12,053.

[26] Schulz, M., G. de Leeuw, and Y. Balkanski, 2004. Sea-salt aerosol source functions and emissions, in *Emissions of Atmospheric Trace*

Compounds, eds. C. Granier, P. Artaxo and C. Reeves, Kluwer, 2004. pp. 333-359.
[27] Smith, M. H., and N. M. Harrison, 1998. The sea spray generation function, *J. Aerosol Sci.*, 29, Suppl. 1, S189-S190.
[28] Smith and De Leeuw, 2004. Workshop *Toward a Universal Sea Spray Source Function*, Skipton, UK (11-13 May, 2004).
[29] Smith, M. H., P. M. Park, and I. E. Consterdine, 1993. Marine aerosol concentrations and estimated fluxes over the sea, *Q. J. R. Meteorol. Soc.*, 119, 809-824.
[30] Sørensen, L.L., S. Pryor, G. de Leeuw, and M. Schulz (2005). Flux divergence for nitric acid in the marine atmospheric surface layer. *J. Geophys. Res.* 110, D15306, doi:10.1029/2004JD005403, 12 pp.
[31] Vignati, E., G. de Leeuw and R. Berkowicz, 2001. Modeling coastal aerosol transport and effects of surf-produced aerosols on processes in the marine atmospheric boundary layer, *JGR-Atmospheres*, Vol. 106, D17 (September 16), pages 20225-20238.
[32] Wu, J., J. J. Murray, and R. J. Lai, Production and distributions of sea spray, *J. Geophys. Res.*, 89, 8163-8169, 1984.
[33] Zappa, C. J., F. A. Tubiana, W. R. McGillis, J. Bent, G. de Leeuw, and M. M. Moerman, 2006. Investigating wave processes important to air-sea fluxes using infrared techniques. *Ocean Sciences Meeting of the American Geophysical Union*, Honolulu, HI, February 2006.

Author Index

Subject Index

MIX
Papier aus verantwortungsvollen Quellen
Paper from responsible sources
FSC® C105338

If you have any concerns about our products,
you can contact us on
ProductSafety@springernature.com

In case Publisher is established outside the EU,
the EU authorized representative is:
Springer Nature Customer Service Center GmbH
Europaplatz 3, 69115 Heidelberg, Germany

Printed by Libri Plureos GmbH
in Hamburg, Germany